Beginning
AutoCAD 2006

Beginning AutoCAD 2006

Bob McFarlane
MSc, BSc, ARCST
CEng, FIED, RCADDes
MIMechE, MIEE, MBCS, MCSD, FRSA

Autodesk®
Authorised Author

Routledge
Taylor & Francis Group

LONDON AND NEW YORK

First published by Newnes

First edition 2006

This edition published 2011 by Routledge
2 Park Square, Milton Park, Abingdon, Oxon, OX14 4RN
711 Third Avenue, New York, NY 10017, USA

Routledge is an imprint of the Taylor & Francis Group, an informa business

Notice
No responsibility is assumed by the publisher for any injury and/or damage to persons
or property as a matter of products liability, negligence or otherwise, or from any use
or operation of any methods, products, instructions or ideas contained in the material
herein. Because of rapid advances in the medical sciences, in particular, independent
verification of diagnoses and drug dosages should be made

British Library Cataloguing in Publication Data
A catalogue record for this book is available from the British Library

Library of Congress Cataloging in Publication Data
A catalog record for this book is available from the Library of Congress

ISBN-13: 978-0-75-066957-3
ISBN-10: 0-75-066957-8

Contents

Preface

AutoCAD 2006 incorporates several new features which will increase user draughting skills and improve productivity. Some of these new features are:

Create

1   Command enhancement of several common commands including:
    a) join segments
    b) create rectangles by area and angle of rotation
    c) copy option with rotate and scale.

2   Dimension enhancement:
    a) arc length dimension
    b) jogged radius dimension
    c) fixed length extension lines
    d) different linetypes for dimension and extension lines
    e) ability to flip the dimension arrow.

3   Hatching enhancement:
    a) editing hatch boundaries
    b) calculation of hatch area
    c) creation of several hatch areas as separate objects
    d) ability to specify the hatch origin
    e) recreate a hatch boundary.

4   Multiline text enhancements:
    a) in-place editor
    b) bullets and numbering.

Manage

1   Customise the interface:
    a) create and save user-defined workspaces
    b) manage customised user-interface elements
    c) toggle drawing aids with override keys.

2   General enhancements:
    a) easier process of creating schedules and bill-of-materials with attribute extraction
    b) scale list manager for viewports, page layouts and plotting.

3   Interface improvements:
    a) Locking of toolbar and palette positions.

Produce

1   Dynamic blocks:
    a) blocks defined with custom properties
    b) part of blocks can be moved, rotated and stretched
    c) visibility parameter for blocks
    d) block look-up table.

2 Dynamic input:
   a) commands entered via a tooltip
   b) co-ordinate data displayed as cursor moved
   c) co-ordinate entry 'tied' to dimensions
   d) selection preview to highlight objects.

3 General enhancements:
   a) Simple formulae can be inserted into tables for calculations.

4 New tools:
   a) mathematical and trigonometric calculations possible
   b) retrieval of back-up files.

5 Undo/Redo operations
   a) Possible to undo/redo zoom and pan operations in a single action.

Share

1 Plot and publish tools:
   a) Publish a 3D DWF.

Many of these new features will be discussed in this book.

Note the following:

1 This book is intended for:
   a) new users to AutoCAD who have access to AutoCAD 2006
   b) experienced AutoCAD users wanting to upgrade their skills from previous releases
   c) readers who are studying for a formal CAD qualification at City and Guilds, BTEC or SQA level
   d) training centres offering CAD topics
   e) undergraduates and post-graduate students at higher institutions who require AutoCAD draughting skills
   f) industrial CAD users who require both a text book and a reference source.

2 The objective of this book is to introduce the reader to the essential basic 2D draughting skills required by every AutoCAD user, whether at the introductory, intermediate or advanced level. Once these basic skills have been 'mastered', the user can progress to the more 'demanding' topics such as 3D modelling, customisation and AutoLISP programming.

3 As with all my AutoCAD books, the reader will learn by completing worked examples, and further draughting experience will be obtained by completing the additional activities which complement many of the chapters. All drawing material has been completed using Release 2006 and all work has been checked to ensure there are no errors.

4 Your comments and suggestions for work to be included in any future publications would be greatly appreciated.

Bob McFarlane

Acknowledgements

It would not have been possible for me to complete the various exercises and activities in this book without the inspiration from all other AutoCAD authors. It is very difficult to conceive new ideas with CAD and I am very grateful for the ideas from these other authors. A special mention must be given to Dennis Maguire and his book 'Engineering Drawing from First Principles using AutoCAD' published by Arnold.

Chapter 1

Using the book

The aim of this book is to assist the reader to use AutoCAD 2006 with a series of interactive exercises. These exercises will be backed up with activities, thus allowing the reader to 'practice the new skills' being demonstrated. While no previous CAD knowledge is required, it would be useful if the reader knew how to use:

a) the mouse to select items from the screen
b) Windows concepts, e.g. maximise/minimise screens.

Concepts for using this book

There are several simple concepts with which the reader should become familiar, and these are:

1   Menu selection will be in bold type, e.g. **Draw**.

2   A menu sequence will be in bold type, e.g. **Draw-Circle-3 Points**.

3   User keyboard entry will also be highlighted in bold type, e.g.:
    a) co-ordinate entry: **125,36**; **@100,50**; **@200<45**
    b) command entry: **LINE**; **MOVE**; **ERASE**
    c) response to a prompt: **15**.

4   Icon selection will be in bold type, e.g. select the **LINE** icon from the Draw toolbar.

5   The AutoCAD 2006 prompt text will be displayed in typewriter face, e.g.:
    a) *prompt*   Specify first point
    b) *prompt*   Specify second point of displacement.

6   The symbol <R> or <RETURN> will be used to signify pressing the RETURN or ENTER key. Pressing the mouse right button will also give the <RETURN> effect – called right click.

7   The term pick is continually used with AutoCAD, and refers to the selection of a line, circle, text item, dimension, etc. The mouse left button is used to pick an object – called left click.

8   Keyboard entry can be **LINE** or **line**. Both are acceptable.

Saving drawings

All work should be saved for recall at some later time, and drawings can be saved:
a) to a formatted medium (zip disc, CD, memory stick, etc.)
b) in a named folder on the hard drive.

It is the user's preference as to which method is used, but for convenience purposes only I will assume that a named folder is being used. This folder is named **MYCAD** and when a drawing is being saved or opened, the terminology used will be:
a) save drawing as **MYCAD\WORKDRG**
b) open drawing **MYCAD\EXER1**.

Chapter 2

The AutoCAD 2006 graphics screen

In this chapter, we will investigate the graphics screen and the user-interface. We will also discuss some of the basic AutoCAD terminology.

Starting AutoCAD 2006

1 AutoCAD 2006 is started:
   a) from the Windows 'Start screen' with a double left-click on the AutoCAD 2006 icon
   b) by selecting the windows taskbar sequence:
      Start-Programs-AutoCAD 2006-AutoCAD 2006 (or similar).

2 Both methods will briefly display the AutoCAD 2006 logo and then:
   either   *a*) the actual graphics screen
   or   *b*) the Startup dialogue box.

3 If the Startup dialogue box is displayed, then select Cancel at present. This will allow the user access to the graphics screen. We will discuss the Startup dialogue box later in this chapter.

The graphics screen

Figure 2.1 displays the basic AutoCAD 2006 graphics screen. Your screen may differ slightly, but the general layout will be the same. The numbered items are:

1 The AutoCAD title bar
2 The graphic screen menu bar
3 The Standard toolbar
4 The Layer information toolbar
5 The Properties toolbar
6 The Text Styles toolbar
7 The standard 'windows buttons'
8 The Windows taskbar
9 The Status bar
10 The Layout tabs
11 The Command prompt window area
12 The Co-ordinate system icon
13 The on-screen cursor
14 The Drawing area
15 Scroll bars at right and bottom of drawing area
16 The Object Snap toolbar (floating)
17 The Sheet Set Manager palette
18 The Tool palette
19 The Workspaces toolbar (floating)
20 The Draw toolbar (docked)
21 The Modify toolbar (docked)
22 The Grips/Pickfirst boxes

Title bar

The title bar is positioned at the top of the screen and displays the AutoCAD 2006 icon, the AutoCAD Release version and the current drawing name.

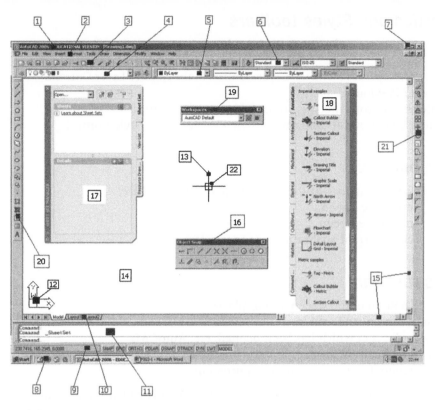

Figure 2.1 The AutoCAD 2006 graphics screen.

Menu bar

1 The screen menu bar displays the default AutoCAD menu headings. By moving the mouse into the menu bar area, the cursor cross-hairs change to a **pick arrow** and with a left-click on any heading, the relevant '**pull-down**' menu will be displayed. The full menu bar headings are:
File Edit View Insert Format Tools Draw Dimension Modify Window Help

2 Menu bar notes:
   a) Pull-down menu items with '...' after their name result in a dialogue box being displayed when the item is selected, i.e. left-clicked.
   b) Pull-down menu items with ▶ after their name result in a further menu being displayed when the item is selected. This is termed a cascade menu effect.
   c) Menu items in BOLD type are available for selection.
   d) Menu items in GREY type are not available for selection.
   e) Menu bar and pull-down menu items are selected (picked) with a mouse left-click.
   f) Pull-down menus are often called 'drop-down' menus.

The Standard toolbar

The Standard toolbar is normally positioned below the screen menu bar and allows the user access to several button icon selections including New, Open, Save, Print, etc. By moving the cursor pick arrow onto an icon and 'leaving it for about a second', the icon name will be displayed in yellow (default). The Standard toolbar can be positioned anywhere on the screen or 'turned off' if required by the user. It is recommended that the Standard toolbar be displayed at all times and positioned below the screen menu bar (as the default).

Layer, Properties and Styles toolbars

These are normally positioned below and to the side of the Standard toolbar. Icon selections are:

a) Layers: Layer Properties Manager, Layer Control, Make Object's Layer Current, Layer Previous.
b) Properties: Color Control, Linetype Control, Lineweight Control.
c) Styles: Text Style Control, Dimension Style Control, Table Style Control.

The Windows buttons

The Windows buttons are positioned at the right of the title bar, and are:

a) left button: minimise screen
b) centre button: maximise screen
c) right button: close current application.

The Windows taskbar

1   This is situated at the bottom of the screen and displays:
    a) the Windows 'Start button' and icon
    b) the name of any application which has been opened, e.g. AutoCAD
    c) the time and the sound control icons
    d) other icons/information dependant on user requirements.

2   By left-clicking on 'Start', the user has access to the other Programs which can be run 'on top of AutoCAD', i.e. multi-tasking.

The Status bar

Positioned above the Windows taskbar, the status bar gives the user:

a) on-screen cursor X, Y and Z co-ordinates information at the left
b) access to the drawing aid buttons, e.g. SNAP, GRID, ORTHO, POLAR, OSNAP, OTRACK, DYN, LWT
c) access to the MODEL/PAPER space toggle.

Layout tabs

Allows the user to 'toggle' between model and paper space for drawing layouts. The layout tabs will be discussed in a later chapter.

Command prompt window area

1   The command prompt area is where the user 'communicates' with AutoCAD 2006 to enter:
    a) a command, e.g. LINE, COPY, ARRAY
    b) co-ordinate data, e.g. 120,150; @15<30
    c) a specific value, e.g. a radius of 25.

2   The command prompt area is also used by AutoCAD to supply the user with information, which could be:
    a) a prompt, e.g. **Specify first point**
    b) a message, e.g. **object does not intersect an edge**.

3   The command area can be increased in size by 'dragging' the bottom edge of the drawing area upwards. I recommend a 2 or 3 line command area display.

4   The command prompt area can be toggled on/off with a **CTRL and 9** key press.

The co-ordinate system icon

This is the X–Y icon at the lower left corner of the drawing area. This icon gives information about the co-ordinate system in use. The default setting is the traditional Cartesian system with the origin (0,0) at the lower left corner of the drawing area. The co-ordinate icon display can be altered by the user.

The cursor cross-hairs

Used to indicate the on-screen position, and movement of the pointing device will result in the co-ordinates in the status bar changing. The 'size' of the on-screen cursor can be increased or decreased to suit user preference and will be discussed later.

The drawing area

This is the user's drawing sheet and can be any size required. In general we will use A3-sized paper, but will also investigate very large and very small drawing paper sizes.

Scroll bars

Positioned at the right and bottom of the drawing area and are used to scroll the drawing area. They are very useful for larger sized drawings and can be 'turned-off' if they are not required.

Toolbars

By default, AutoCAD 2006 displays the Draw and Modify toolbars although users may have them positioned differently from that shown in Figure 2.1. Other toolbars may also be displayed, and Figure 2.1 displays the Object Snap and Workspaces toolbars. Toolbars will be discussed later in this chapter.

Tool palettes

AutoCAD 2006 displays the Sheet Set Manager and all palettes by default. The user's screen may not display any tool palettes. Palettes can be cancelled, minimised or repositioned by the user at any time.

The Grips/Pickfirst box

The user may have a small square box 'attached' to the cursor cross-hairs. This box may be the Grips and/or the Pickfirst box, both aids to the user.

Terminology

AutoCAD 2006 terminology is basically the same as previous releases, and the following gives a brief description of the items commonly encountered by new users to AutoCAD.

Menu

1 A menu is a list of options from which the user selects (picks) the one required for a particular task.

2 Picking a menu item is achieved by moving the mouse over the required item and left-clicking.

3 There are different types of menus, e.g. pull-down, cascade, screen, toolbar button icon.

Command

1   A command is an AutoCAD function used to perform some task. This may be to draw
   a line, rotate a shape or modify an item of text. Commands can be activated by:
   a) selection from a menu
   b) selecting the appropriate icon from a toolbar button
   c) entering the command from the keyboard at the command line
   d) entering the command abbreviation
   e) using the Alt key as previously described.

2   Only the first three options will be used in this book.

Objects

Everything drawn in AutoCAD 2006 is termed an **object** (or **entity**) e.g. lines, circles,
text, dimensions, hatching, etc. are all objects. The user 'picks' the appropriate
entity/object with a mouse left-click when prompted.

Default setting

All AutoCAD releases have certain values and settings which have been 'preset',
these being essential for certain operations. Default values are displayed with <>
brackets, but the actual value can be altered by the user as and when required. For
example:

1   From the menu bar select **Draw-Polygon** and:
   prompt   _polygon Enter number of sides<4>
   respond   **press the ESC key** to cancel the command.

2   *Notes*
   a) <4> is the default value for the number of sides
   b) _polygon is the active command.

3   At the command line enter **LTSCALE <R>** and:
   prompt   Enter new linetype scale factor<1.0000> (or other value)
   enter   **0.5 <R>**.

4   *Notes*
   a) <1.0000> is the LTSCALE default value on my system
   b) we have altered the LTSCALE value to 0.5.

The escape (Esc) key

This is used to cancel any command at any time. It is very useful, especially when the
user is 'lost in a command'. Pressing the Esc key will cancel any command and return
the command prompt line.

Icon

An icon is a menu item in the form of a picture contained on a button within a named
toolbar. Icons will be used extensively referred to throughout the book.

Cascade menu

A cascade menu is obtained when an item in a pull-down menu with ▶ after it's name
is selected.

1   From the menu bar select the sequence **Draw-Circle** and the cascade effect as Figure
   2.2(a) will be displayed.

Figure 2.2   (a) Cascade menu and (b) a shortcut menu.

2   Cancel the cascade effect by:
   a) moving the pick arrow to any part of the screen and left-clicking
   b) pressing the Esc key – cancels the 'last' cascade menu, so two escapes are required.

Shortcut menu

1   A shortcut menu allows quick access to commands that are relevant to the current activity.

2   Shortcut menus are displayed with a right-click:
   a) within the drawing area with or without any objects selected
   b) within the drawing area during a command
   c) within the text and command windows
   d) within areas and on icons in Design Center
   e) on a toolbar, tool palette, model or layout tabs
   f) on the status bar or the status bar buttons.

3   Shortcut menus typically include options to:
   a) repeat the last command entered
   b) cancel the current command
   c) display a list of recent user input
   d) cut, copy, and paste from the Clipboard
   e) select a different command option
   f) undo the last command entered.

4   Figure 2.2(b) displays a typical shortcut menu.

Dialogue boxes

A dialogue box is always displayed when an item with '…' after it's name is selected.

1   Select the menu bar sequence **Format-Units** and:
   prompt   `Drawing Units dialogue box` as Figure 2.3
   respond   **select Cancel** to 'remove' the dialogue box from the screen.

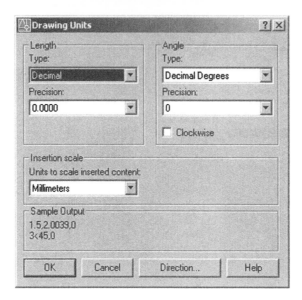

Figure 2.3   The Drawing Units dialogue box.

2   Dialogue boxes allow the user to:
   a) alter parameter values
   b) toggle an aid ON/OFF
   c) select an option from a list.

3   Most dialogue boxes display the options OK, Cancel and Help which are used as follows:
   OK: accept the values in the current dialogue box
   Cancel: cancel the dialogue box without any alterations
   Help: gives further information in Windows format. The Windows effect can be cancelled with **File-Exit** or using the Windows Close button from the title bar (the right-most button).

Toolbars

1   Toolbars are aids for the user. They allow the AutoCAD 2006 commands to be displayed on the screen in button icon form. The required command is activated by picking (left-click) the appropriate button. The icon command is displayed as a **tooltip** in yellow (the default colour) by moving the pick arrow onto an icon and leaving it for a second.

2   There are 30 toolbars available for selection. The toolbars normally displayed by default when AutoCAD 2006 is started are Standard, Layers, Properties, Styles, Modify and Draw.

3   Toolbars can be:
   a) displayed and positioned anywhere in the drawing area
   b) customised to the user preference.

4   To activate a toolbar, right-click any displayed toolbar and:
   prompt   toolbar shortcut menu – Figure 2.4(a)
   with   *a*) list of all available toolbars
                b) active toolbars indicated with a √
   respond   pick any toolbar name
   and   1 toolbar displayed in drawing area
                2 shortcut menu cancelled.

5   When toolbars are positioned in the drawing area as the Object Snap toolbar they are called **FLOATING** toolbars.

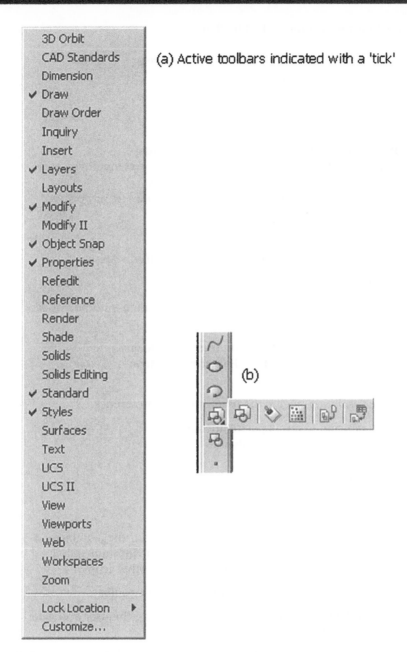

Figure 2.4 (a) The toolbar shortcut menu and (b) the Insert Block Fly-out menu.

6   Toolbars can be:
  a) Moved to a suitable position on the screen by the user. This is achieved by moving the pick arrow into the blue title area of the toolbar and holding down the mouse left button. Move the toolbar to the required position on the screen and release the left button.
  b) Altered in shape by 'dragging' the toolbar edges sideways or downwards.
  c) Cancelled at any time by picking the 'Cancel box' at the right of the toolbar title bar.

7   It is the user's preference as to what toolbars are displayed at any one time. In general I always display the Draw, Modify and Object Snap toolbars and activate others as and when required.

8   Toolbars can be **DOCKED** at the edges of the drawing area by moving them to the required screen edge. The toolbar will be automatically docked when the edge is reached.

9 Figure 2.1 displays a floating and several docked toolbars:
   a) Docked:
      1. Standard, Layers, Properties and Styles at the top of the screen
      2. Draw and Modify at either side of the screen
      3. These toolbars 'were set' by default.
   b) Floating: The Object Snap toolbar.

10 *Notes*
   a) Toolbars **do not have to be used** – they are an aid for the user.
   b) While all commands are available from the menu bar, it is recommended that tool-bars are used, as they greatly increase draughting productivity.
   c) When used, it is the user's preference as to whether they are floating or docked.

Tool Palette

1 A tool palette is an efficient method of organizing and sharing various items.

2 Tool palettes can be customised by the user.

3 By selecting **Auto-hide** from the title bar, the tool palette can be minimised/maximised.

4 Tool palettes can be:
   a) cancelled by selecting the Close (topmost) button from the title bar
   b) activated from the menu bar with **Tools-Tool Palettes Window**
   c) positioned by the user.

5 Like toolbars, the tool palettes do not need to be used. It is user preference.

6 The objects which the user can add to a tool palette include:
   a) geometric objects such as lines, circles, and polylines
   b) dimensions
   c) blocks, hatches, solid fills and gradient fills.

Fly-out menu

When a button icon is selected an AutoCAD command is activated. If the icon has a ◢ at the lower right corner of the icon box, and the left button of the mouse is held down, a **FLY-OUT** menu is obtained, allowing the user access to other icons:

1 Move the cursor pick arrow onto the Insert Block icon of the Draw toolbar.

2 Hold down the left button and a fly-out menu is displayed allowing the user access to another four icons as Figure 2.4(b).

3 Move the cursor to a clear area of the graphics screen and release the left button.

Wizards

The Wizards give access to various parameters necessary allowing the user to:

1 Start a drawing session, e.g. units, paper size, etc.

2 Create layouts, a new sheet sets

3 Publish to the web and add plotters.

Template

A template allows the user access to different drawing standards with different sized paper, each template having a border and title box. AutoCAD 2006 supports several drawing standards including ANSI, DIN, Gb, ISO, JIS and Metric. Templates will be used for our drawing activities.

Toggle

This is the term used when a drawing aid is turned ON/OFF and usually refers to:
a) pressing a key
b) activating a parameter in a dialogue box, i.e. a tick/cross signifying ON, no tick/cross signifying OFF.

Function keys

Several of the keyboard function keys can be used as aids while drawing, these keys being:

F1   accesses the AutoCAD 2006 Help menu
F2   flips between the graphics screen and the AutoCAD Text window
F3   toggles the object snap on/off
F4   toggles the tablet on/off (if attached)
F5   toggles the isoplane top/right/left – for isometric drawings
F6   co-ordinates on/off toggle
F7   grid on/off toggle
F8   ortho on/off toggle
F9   snap on/off toggle
F10   polar tracking on/off toggle
F11   toggles object snap tracking off
F12   the dynamic input toggle.

Help menu

AutoCAD 2006 has an 'on-line' help menu which can be activated at any time by selecting from the menu bar **Help-Help** or pressing the F1 function key. The Help dialogue box will be displayed as two distinct sections:
a) Left: with four tab selections – Contents, Index, Search, Ask Me
b) Right: details about the topic selected.

File types

1   When a drawing has been completed it should be saved for future recall.

2   All drawings are called files.

3   AutoCAD 2006 supports different file formats, including:
    .dwg: AutoCAD drawing
    .dws: AutoCAD Drawing Standard
    .dwt: AutoCAD Template Drawing template file
    .dxf: AutoCAD Data Exchange Format.

Saved drawing names

1   Drawing names should be as simple as possible.

2   While operating systems support file names which contain spaces and full stops(.), I would not recommend this practice.

3   The following are typical drawing file names which I would recommend be used: EX1; EXER-1; EXERC_1; MYEX-1; DRG1, etc.

4   When drawings have to be saved during the exercises in the book, I will give the actual named to be used.

Finally

1 At this time, we have:
   a) started AutoCAD
   b) investigated the graphics screen
   c) discussed some terminology
   d) quit AutoCAD.

2 We are now ready to draw some AutoCAD objects.

Drawing and erasing objects and using the selection set

In this chapter we will investigate how to:

1 draw and erase lines and circles

2 use the selection set – a very powerful user aid when modifying a drawing.

Starting a new drawing with Wizard

1 Start AutoCAD and:

    prompt   Startup dialogue box with four selections:
                  Open a Drawing; Start from Scratch; Use a Template; Use a Wizard
    respond   pick Use a Wizard icon (right-most icon)
    prompt   Startup – Use a Wizard dialogue box
    respond   *a*) pick Quick Setup – Figure 3.1(a)
                  b) pick OK
    prompt   Quick Setup (Units) dialogue box
    respond   *a*) Select Decimal Units – Figure 3.1(b)
                  b) pick Next>
    prompt   Quick Setup (Area) dialogue box
    respond   *a*) enter Width: 420
                  b) enter Length: 297 – Figure 3.1(c)
                  c) pick Finish.

2 A blank AutoCAD 2006 drawing screen should be returned with the Standard, Layers, Properties and Styles toolbars at the top of the screen, and the docked Modify and Draw toolbars.

3 *Notes*:

    a) The toolbars which are displayed will depend on how the last user 'left the system'. If you do not have the Draw and Modify toolbars displayed then:

        1. right-click in any displayed toolbar

        2. activate the Draw and Modify toolbars with a mouse left-pick

        3. position the toolbars to suit, i.e. floating or docked.

    b) After selecting options from the Startup dialogue box, the New Features Workshop screen may be displayed. The user should read the options, decide on which should remain active, then close the dialogue box.

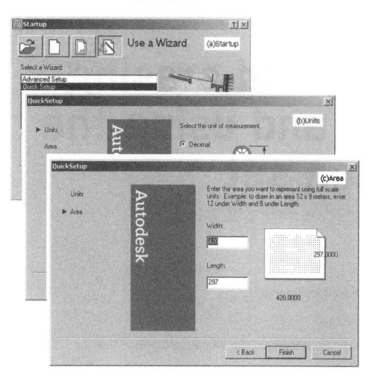

Figure 3.1 The Use a Wizard startup with (a) Quick Setup (b) Units and (c) Area dialogue boxes.

Drawing line objects

1 A line requires a start point and an end point.

2 To draw a line with AutoCAD, activate (pick) the LINE icon from the Draw toolbar and:
   a) **the command prompt window displays: _line Specify first point**
   b) small tooltip boxes 'may be' attached to the cursor cross-hairs
   c) forget about these tooltips at present.

3 You now have to pick a **start** point for the line so:
   a) move the pointing device and pick (left-click) any point within the drawing area
   b) **the command prompt window displays: Specify next point or [Close/Undo]**.

4 You now have to pick the end point of the line to be drawn so:
   a) move the pointing device away from the start point
   b) a line will be dragged from your start point to the on-screen cursor position
   c) this drag effect is termed **RUBBERBAND**
   d) left-click at any other point on the screen and **this is your first AutoCAD 2006 object**.

5 The line command is still active with the rubberband effect and the prompt line is still asking you to specify the next point.

6 Continue moving the mouse about the screen and pick points to give a series of 'joined lines'.

7 Finish the LINE command with a right-click on the mouse and:
   a) shortcut menu will be displayed as Figure 3.2(a)
   b) pick Enter to end the LINE command and a 'blank' command line will be returned.

8 *a*) from the menu bar select **Draw-Line** and the Specify first point prompt will again be displayed at the command prompt
   b) draw some more lines to the end the command by pressing the RETURN/ENTER key
   c) the LINE command will be 'stopped', but no shortcut menu will have appeared.

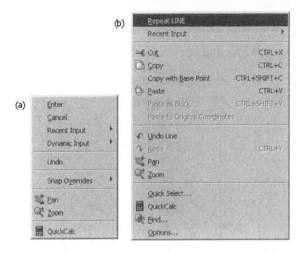

Figure 3.2 The shortcut menus to (a) end the LINE command and (b) repeat the LINE command.

9 *a*) at the command line enter **LINE <R>** and draw a few more lines
   b) end the command with a right-click and pick Enter from the shortcut menu.

10 *a*) right-click the mouse to display a shortcut menu as Figure 3.2(a)
   b) pick **Repeat LINE**
   c) draw some more lines then end the command with a right-click and pick Enter.

11 *Notes*
   a) The different ways of activating the LINE command:
      1. with the LINE icon from the Draw toolbar
      2. from the menu bar with **Draw-Line**
      3. by entering LINE and the command line
      4. with a right-click of the mouse (if LINE was the last command).
   b) The two options to 'exit/stop' a command:
      1. with a right-click of the mouse – shortcut menu displayed
      2. by pressing the RETURN/ENTER key – no shortcut menu.
   c) When a command has been completed, a mouse right-click will display a shortcut menu as Figure 3.2(b) with the **LAST COMMAND** available for selection, e.g. Repeat LINE.

Drawing circle objects

1 All circles have a centre point and a radius/diameter.

2 There are several options for drawing a circle with AutoCAD, but at present, select the CIRCLE icon from the Draw toolbar and:
   prompt   1. _circle Specify centre point for circle or [3P/2P/Ttr (tan tan radius)]
             2. tooltips may be displayed.
   respond **pick any point on the screen as the circle centre**
   prompt   Specify radius of circle or [Diameter]
   respond **drag out the circle and pick any point for radius**.

3 From the menu bar select **Draw-Circle-Center, Radius** and:
   a) pick a centre point
   b) drag out a radius.

4 At the command prompt enter **CIRCLE <R>** and create another circle anywhere on the screen.

5 Using the icons, menu bar or keyboard entry, draw some more lines and circles until you are satisfied that you can activate and end the two commands.

6 Figure 3.3(a) displays some AutoCAD line and circle objects.

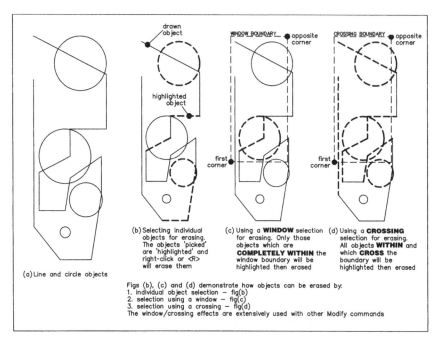

Figure 3.3 Line and circle objects for use with the drawing, erasing and selection set exercise.

Erasing objects

Now that we have drawn some lines and circles, we will investigate how they can be erased, which seems rather silly? The erase command will be used to demonstrate different options available to us when it is required to modify a drawing. The actual erase command can be activated by one of three methods:
a) picking the ERASE icon from the Modify toolbar
b) with the menu bar sequence **Modify-Erase**
c) entering **ERASE <R>** at the command line.

1 Before continuing with the exercise, select from the menu bar the sequence **Tools-Options** and:

prompt   Options dialogue box
respond   pick the Selection tab and ensure:
      1. Selection preview: both not active, i.e. blank boxes
      2. Noun/verb selection: not active
      3. Use shift to add to selection: not active
      4. Press and drag: not active
      5. Implied windowing: active, i.e. tick in box
      6. Object grouping: active
      7. Associative hatching: active
      8. Pickbox size: set to suit or accept default
      9. pick Apply then OK when complete.

2 Now continue with the erase exercise.

3 Ensure you still have several lines and circles on the screen. Figure 3.3(a) is meant as a guide only.

4   From the menu bar select **Modify-Erase** and:

    prompt   `Select objects`

    and   cursor cross-hairs replaced by a 'pickbox' which moves as you move the mouse

    respond   **position the pickbox over any line and left-click**

    and   the following will happen:

                 a) the selected line will 'change appearance', i.e. be 'highlighted'

                 b) the prompt displays `Select objects: 1 found`

                 c) then: Select objects

    respond   continue picking lines and circles to be erased (about six) and each selected object will be highlighted.

5   When you have selected enough objects, right-click the mouse.

6   The selected objects will be erased, and a blank Command prompt window area will be returned blank.

7   Figure 3.3(b) demonstrates the individual object selection erase effect.

OOPS

1   Suppose that you had erased the wrong objects.

2   Before you **DO ANYTHING ELSE**, enter **OOPS <R>** at the command line.

3   The erased objects will be returned to the screen.

4   Consider this in comparison to a traditional draughtsperson who has rubbed out several lines/circles. They would have to redraw each one.

5   *Notes*:

    a) OOPS is used to restore objects erased by the LAST erase command.

    b) It must be used **IMMEDIATELY** after the last erase command.

    c) It **must be entered from the keyboard**.

Erasing with a window/crossing effect

Individual selection of objects is satisfactory if only a few objects have to be modified (remember that we have only used the erase command so far). When a large number of objects require to be modified, the individual selection method is very tedious, and AutoCAD overcomes this by allowing the user to position a 'window' over an area of the screen which will select several objects 'at the one pick'.

To demonstrate the window effect, ensure you have several objects (about 20) on the screen and refer to Figure 3.3(c).

1   Select the ERASE icon from the Modify toolbar and:

    prompt   `Select objects`

    enter   **W <R>** (at the command line) – the window option

    prompt   `Specify first corner`

    respond   **position the cursor at a suitable point and left-click**

    prompt   `Specify opposite corner`

    respond   **move the cursor to drag out a blue coloured window (rectangle) and left-click**

    prompt   `??? found` and certain objects highlighted

    then   `Select objects`, i.e. any more objects to be erased?

    respond   **right-click or <R>**.

2   The highlighted objects will be erased.

3   At the command line enter **OOPS <R>** to restore the erased objects.

4   From the menu bar select **Modify-Erase** and:
    prompt   Select objects
    enter   **C <R>** (at the command line) – the crossing option
    prompt   Specify first corner
    respond   **pick any point on the screen**
    prompt   Specify opposite corner
    respond   **drag out a green coloured rectangular window and pick the other
                 corner**
    prompt   ??? found and highlighted objects
    respond   **right-click**.

5   The objects highlighted will be erased – Figure 3.3(d).

Note on window/crossing

1   The window/crossing concept of selecting a large number of objects will be used exten-
    sively with the modify commands, e.g. erase, copy, move, scale, rotate, etc. The objects
    which are selected when **W or C** is entered at the command line are as follows:
    (W) for window: all objects **completely within** the window boundary are selected
    (C) for crossing: all objects **completely within and also which cross** the
                        window boundary are selected.

2   The window/crossing option **IS ENTERED FROM THE KEYBOARD**, i.e. W or C.

3   Figure 3.3 demonstrates the single object selection method as well as the window and
    crossing methods for erasing objects.

4   The rectangular colour defaults are blue for the window selection and green for the
    crossing.

5   *Automatic window/crossing*
    In the example used to demonstrate the window and crossing effect, we entered a W
    or a C at the command line. AutoCAD allows the user to activate this window/crossing

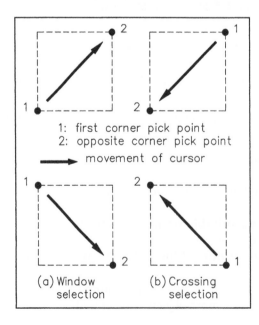

Figure 3.4   Automatic window/crossing selection.

effect automatically by picking the two points of the 'window' in a specific direction. Figure 3.4 demonstrates this with:

a) the window effect by picking the first point anywhere and the second point either upwards or downwards to the right

b) the crossing effect by picking the first point anywhere and the second point either upwards or downwards to the left.

The selection set

Window and crossing are only two options contained within the selection set, the most common selection options being:

Crossing, Crossing Polygon, Fence, Last, Previous, Window and Window Polygon.

During the various exercises in this book, we will use all of these options but will only consider three at present, so:

1   Erase all objects from the screen using a window or crossing selection.

2   Refer to Figure 3.5(a) and draw some new lines and circles – the actual layout is not important, but try and draw some objects 'inside' others.

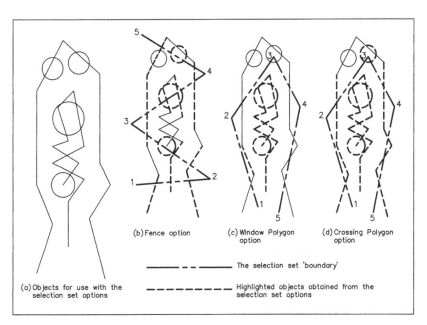

Figure 3.5   Investigating the fence, window polygon and crossing polygon selection set options.

3   Refer to Figure 3.5(b), select the ERASE icon from the Modify toolbar and:

prompt	Select objects
enter	**F <R>** – the fence option
prompt	First fence point
respond	**pick a point 1**
prompt	Specify endpoint of line or [Undo]
respond	**pick a suitable point 2**
prompt	Specify endpoint of line or [Undo]
respond	**pick point 3, then points 4 and 5 then right-click**
prompt	Shortcut menu
respond	**pick Enter**

prompt   ??? found and certain objects highlighted
respond   **right-click or <R>.**

4   The highlighted objects will be erased.

5   Enter **OOPS <R>** to restore these erased objects.

6   Menu bar with **Modify-Erase** and referring to Figure 3.5(c):
   prompt   Select objects
   enter   **WP <R>** – the window-polygon option
   prompt   First polygon point
   respond   **pick a point 1**
   prompt   Specify endpoint of line or [Undo]
   respond   **pick points 2,3,4,5 then right-click and pick Enter**
   prompt   ??? found and objects highlighted
   respond   **right-click** to erase the highlighted objects.

7   **OOPS <R>** to restore the erased objects.

8   *a)* activate the ERASE command
   b) enter **CP <R>** at command line – crossing-polygon option
   c) pick points in order as Figure 3.5(d) then right-click and pick Enter from shortcut menu
   d) right-click to erase the highlighted objects.

9   The fence/window polygon/crossing-polygon options of the selection set are very useful when the 'shape' to be modified does not permit the use of the normal rectangular window. The user can 'make their own shape' for selecting objects to be modified.

10 The complete selection set options can be viewed by entering **SELECT <R>** at the command line and:
   prompt   Select objects
   enter   ? <R> – the query entry
   prompt   list of selection set options, activated by entering the
                 option CAPITAL letters
   respond   press the ESCAPE key to end the command.

Activity

Spend some time using the LINE, CIRCLE and ERASE commands and become proficient with the various selection set options for erasing – this will greatly assist you in later chapters.

Proceed to the next chapter but do not exit AutoCAD if possible.

Chapter 4

The 2D drawing aids

Now that we know how to draw lines and circles, we will investigate the aids which are available to the user. AutoCAD 2006 has several 2D drawing aids which are generally toggled on/off using the status bar buttons and include:

1 Grid
   a) allows the user to place a series of imaginary dots over the drawing area
   b) the grid spacing can be altered by the user at any time while the drawing is being constructed
   c) as the grid is imaginary, it does not appear on the final plot.

2 Snap
   a) allows the user to set the on-screen cursor to a pre-determined point on the screen
   b) the snap spacing can be altered at any time by the user
   c) when the snap and grid are set to the same value, the term **grid lock** is often used.

3 Ortho
   a) an aid which allows only horizontal and vertical movement of the on-screen cursor

4 Polar tracking
   a) allows objects to be drawn at specific angles along an alignment path
   b) the user can alter the 'polar angle' at any time.

5 Object Snap
   a) the user can set a snap relative to a pre-determined geometry
   b) this drawing aid will be discussed in detail in a later chapter.

6 Dynamic Input
   a) provides tooltip display information near the on-screen cursor
   b) this information is updated as the cursor moves
   c) the tooltip allows the user direct keyboard entry.

Note: It would be helpful to the user if this chapter and the following chapter could be completed 'at a single sitting' as we will create several drawings, which will be used to demonstrate saving completed work.

Getting ready

1 Still have some line and circle objects from Chapter 3 on the screen?

2 Menu bar with **File-Close** and:
   prompt   AutoCAD Message dialogue box with Save changes options
   respond   **pick No** – more on this in the next chapter.

3 Begin a new drawing with the menu bar sequence **File-New** and:
   prompt   Create New Drawing dialogue box
   respond   *a*) pick Use a Wizard
              b) pick Quick Setup
              c) pick OK
   prompt   Quick Setup (Units) dialogue box

respond   pick Decimal then Next>
prompt   Quick Setup (Area) dialogue box
respond   *a*) set Width: 420 and Length: 297
          b) pick Finish.

4   A blank drawing screen will be displayed.

5   Menu bar with **Draw-Rectangle** and:
    prompt   Specify first corner point and enter **0,0 <R>**
    prompt   Specify other corner point and enter **420,297 <R>**.

6   Menu bar with **View-Zoom-All** and the rectangle shape will 'fill the screen'. This rectangle will be 'our drawing paper'.

Grid and snap setting

The grid and snap spacing can be set by different methods and we will investigate setting these aids from the command line and from a dialogue box:

1   At the command line enter **GRID <R>** and:
    prompt   Specify grid spacing (X) or [On/OFF/Snap/Aspect]<10.000>
    enter   **20 <R>**.

2   At the command line enter **SNAP <R>** and:
    prompt   Specify grid spacing (X) or [On/OFF/Aspect/Rotate/Style/
               Type]<10.000>
    enter   **20 <R>**.

3   Refer to Figure 4.1 and use the LINE command to draw the letter H using the grid and snap settings of 20.

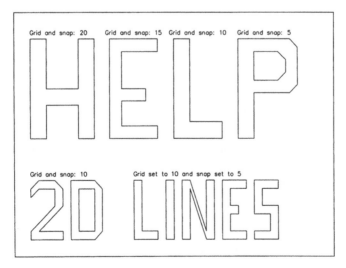

Figure 4.1   Using the GRID and SNAP drafting aids to draw lines.

4   Using keyboard entry, change the grid and snap spacing to 15.

5   Use the LINE command and draw the letter E.

6   From the menu bar select **Tools-Drafting Setting**s and:
    prompt   Drafting Settings dialogue box with four tabs:
             Snap and Grid, Polar Tracking, Object Snap, Dynamic Input

respond   **activate the Snap and Grid tab**

prompt   Snap and Grid tab Drafting Settings display with (from
            step 4):
            a) Snap on with X and Y spacing 15
            b) Grid on with X and Y spacing 15

respond   *a*) alter the Snap X and Y spacing to 10
            b) alter the Grid X and Y spacing to 10
            c) ensure Rectangular grid snap active (Figure 4.2)
            d) pick OK.

7   Use the LINE command to draw the letter L.

8   Now use the Drafting Settings dialogue box to set both the grid and snap spacing to 5
    and draw the letter P.

9   *Task*: Refer to Figure 4.1 and:
    a) with the grid and snap set to 10, draw the 2D line shapes
    b) with the grid set to 10 and the snap set to 5, complete LINES to your own design
       specification
    c) when complete, do not erase any of the objects.

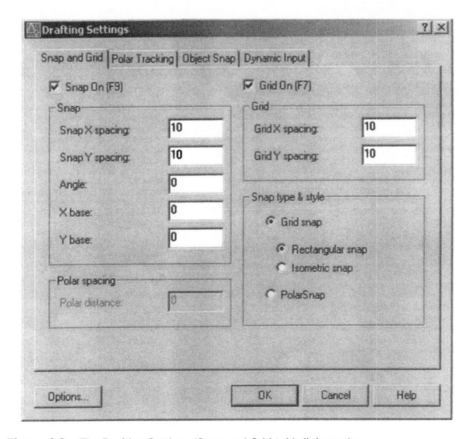

Figure 4.2   The Drafting Settings (Snap and Grid tab) dialogue box.

Drawing lines with the Polar Tracking drafting aid

1   The screen should still display HELP 2D LINES.

2   Menu bar with **File-New** and:

prompt   Create New Drawing dialogue box

respond   a) pick Start from Scratch icon (second left)
           b) pick Metric
           c) pick OK
and   blank drawing screen returned.

3   Set the grid and snap to 20.

4   Right-click on **POLAR in the Status bar**, pick Settings and:
   prompt   `Drafting Settings dialogue box with Polar Tracking tab`
           `active`
   respond   a) Polar Tracking On (F10) active, i.e. tick
           b) Polar Angle Settings: incremental angle – scroll and pick 45
           c) Object Snap Tracking Settings: track using all polar angle settings active
           d) Polar Angle measurement: absolute active
           e) dialogue box as Figure 4.3(a)
           f) pick Options.
   prompt   `Options dialogue box with Drafting tab active`

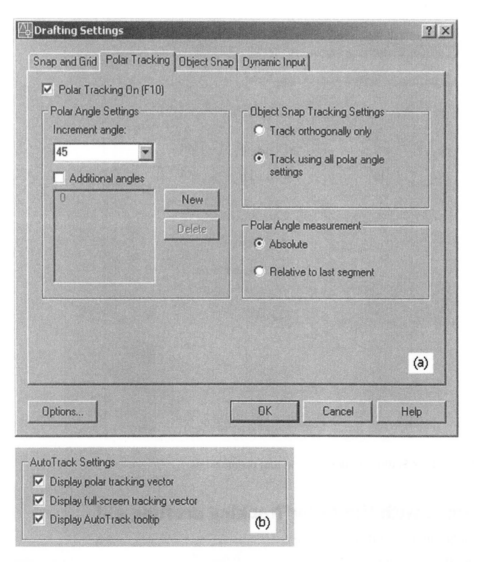

Figure 4.3   The Drafting Settings dialogue box with (a) Polar Tracking tab active and (b) AutoTrack Settings of the Options dialogue box.

respond   Autotrack Settings:
          a) Display polar tracking vectors: active
          b) Display full-screen tacking vectors: active
          c) Display AutoTrack tooltip: active – Figure 4.3(b)
          d) pick OK.
prompt   `Drafting Settings dialogue box with Polar Tracking tab`
          `active`
respond   pick OK.

5 Activate the LINE command and pick any suitable grid/snap start point towards the top of the screen.

6 *a*) move the cursor horizontally to the right and observe the polar tracking tooltip information displayed
  b) move until the tracking data is Polar: 80.0000 < 0° as Figure 4.4(a) then left-click
  c) the drawn line segment has been created with the polar tracking drawing aid.

7 Move the cursor vertically downwards until 40.0000 < 270° is displayed as Figure 4.4(b) then left-click.

8 *a*) move the cursor downwards and to the right until a 225 degree angle is displayed as Figure 4.4(c)
  b) enter 50 from the keyboard
  c) the entered value of 50 is the length of the line segment.

9 *a*) move upwards to left until a 135 degree angle is displayed as Figure 4.4(d)
  b) enter 30 from the keyboard.

10 Complete the polar tracking line segments with an angle of 0 and a keyboard entry of 100.

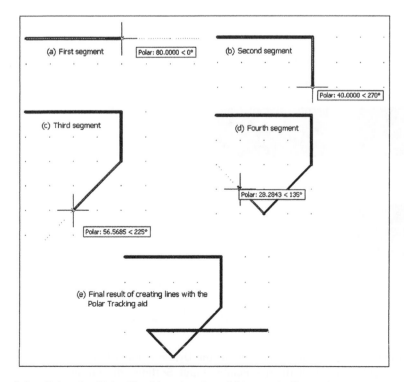

Figure 4.4   Using the Polar Tracking drawing aid to create line segments.

11 The result of drawing the fine line segments with polar tracking is Figure 4.4(e).

12 Note that the polar tracking aid displays information of the format **100.0000<90** and:
  a) 100 is the length of the line segment being drawn
  b) 90 is the angle of the line segment relative to the positive x-axis (more on this later).

13 When this exercise is complete, proceed to the next exercise but try not to exit AutoCAD.

The Dynamic Input drafting aid

1 The screen should still display the line segments created with the polar tracking aid.

2 Menu bar with **File-New** and:
  prompt   Create New Drawing dialogue box
  respond   1. pick Start from Scratch icon (second left)
            2. pick Metric
            3. pick OK
  and   blank drawing screen returned.

3 Right-click on **DYN in the Status bar**, pick Settings and:
  prompt   Drafting Settings dialogue box with Dynamic Input tab active
  with   Four distinct sections, these being:
            1. Enable Pointer Input
            2. Enable Dimension Input where possible
            3. Dynamic Prompts
            4. Drafting Tooltip Appearance – Figure 4.5(a)
  respond   1. Enable Pointer Input active (tick)
            2. Enable Dimension Input not active
            3. Dynamic Prompts not active
            4. pick Enable Pointer Input Settings

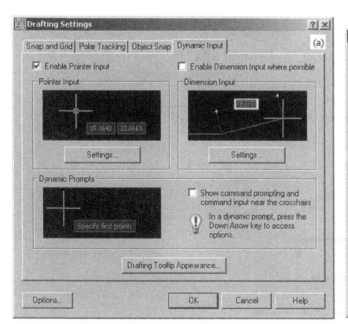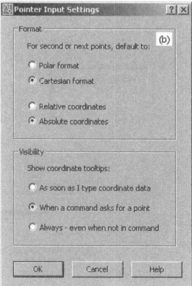

Figure 4.5   The Drafting Settings dialogue box with (a) the Dynamic Input tab displaying Enable Pointer Input active and (b) the Pointer Input Settings displaying Cartesian, Absolute and Asks for point active.

prompt   Pointer Input Settings dialogue box
respond   1. Format:
          a) Cartesian format active
          b) Absolute co-ordinates active
       2. Visibility: Show co-ordinate tooltips:
          When a command asks for a point active – Figure 4.5(b)
       3. pick OK
prompt   Drafting Settings dialogue box (Dynamic Input tab active)
respond   pick OK.

4   With snap and grid on and set to your own values, select the LINE command and:
      a) note the cursor display with the tooltip similar to Figure 4.6(a1)
      b) select a start point
      c) drag out the line and pick a suitable end point – Figure 4.6(a2).

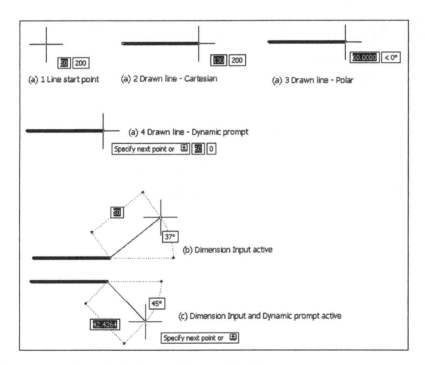

Figure 4.6   Tooltip display with Dynamic Input active.

5   *Task*
    Investigate the other options in the Dynamic Input tab by drawing some more line segments and note the appearance of the tooltip. Figure 4.6 displays the tooltips for:
    a) Pointer Input active and:
       1. line start point – Cartesian format and Absolute co-ordinates active
       2. drawn line – Cartesian format and Absolute co-ordinates active
       3. drawn line – Polar format and Relative co-ordinates active
       4. drawn line – Cartesian, Relative and Dynamic Prompt active
    b) Dimension Input active
    c) Dimension Input and Dynamic Prompt active.

6   This introduction to the Dynamic Input drafting aid is complete as we will use the aid in more detail when we create objects in a later chapter.

Toggling the drawing aids

1 The various drawing aids can be toggled ON/OFF with:
   a) the function keys:
   F7   grid
   F8   ortho
   F9   snap
   F10 polar tracking
   F12 dynamic input
   b) the Drafting Settings dialogue where a tick in the box signifies that the aid is on, and a blank box means the aid is off
   c) the status bar for Grid, Ortho, Snap, Polar, Dyn and:
      1. a left-click toggles the aid On/Off
      2. a right-click displays the shortcut menu and:
         a) the aid can be toggled
         b) Settings will activate the appropriate dialogue box.

2 My preference is to set the grid and snap spacing values from the dialogue box or command line then use the function keys to toggle the aids on/off as required.

3 *a*) take care if the ortho drawing aid is on
   b) ortho only allows horizontal and vertical movement and lines may not appear as expected
   c) I tend to work with ortho off.

4 The Drafting Settings dialogue box can be activated:
   a) from the menu bar with Tools-Drafting Settings
   b) with a right-click on Snap or Grid from the Status bar and then picking Settings.

5 We have not investigated Object Snap Tracking. This will be discussed in a later chapter.

Now proceed to the next chapter without leaving AutoCAD if possible.

Chapter 5

Saving and opening drawings

AutoCAD 2006 allows multiple drawings to be opened during a drawing session. It is thus essential that all users know how to save and open a drawing, and how to exit AutoCAD correctly. In this and all the following chapters, all drawing work will be saved to the named folder **MYCAD**.

Saving a drawing and exiting AutoCAD

1 If the previous chapter work has been followed correctly, the user has three drawings opened:
   a) line segments drawn using the Dynamic Input drawing aid
   b) line segments drawn using the Polar Tracking drawing aid
   c) the HELP 2D LINES.

2 Menu bar with **File-Exit** and:
   prompt   AutoCAD message box – similar to Figure 5.1.

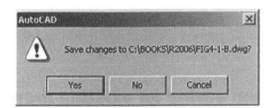

Figure 5.1   The AutoCAD message display.

3 This dialogue box is informing the user that since starting the current drawing session, changes have been made and that these have drawing changes not yet been saved. The user has to respond to one of the three options which are:
   Yes   picking this option will save a drawing with the name displayed, i.e. Drawing1.dwg or similar
   No   selecting this option means that the alterations made will not be saved
   Cancel   returns the user to the drawing screen.

4 At present, pick Cancel as we want to investigate how to save a drawing.

5 Select from the menu bar **File-Save As** and:
   prompt   Save Drawing As dialogue box
   respond   *a*) Scroll at Save in by picking the arrow at right
            b) pick (left-click) the C: drive to display folder names
            c) double left-click on the MYCAD folder (which is empty at present)
            d) File name: alter to DRG2
            e) File type: scroll and select AutoCAD 2004 Drawing (*.dwg)
            f) pick Save.

6 The screen drawing will be saved to the named folder, but will still be displayed on the screen.

7 Menu bar with **File-Close** and:
   a) the line segments drawing will disappear from the screen
   b) the line segments drawn with the Polar Tracking aid will be displayed.

8 Menu bar with **File-Save As** and using the Save Drawing As dialogue box:
   a) ensure the MYCAD folder is current
   b) alter File name to DRG1
   c) ensure file type is AutoCAD 2004 Drawing (*.dwg)
   d) pick Save
   e) menu bar with File-Close to display the HELP 2D LINES drawing.

9 Repeat step 8 with:
   a) folder MYCAD current
   b) MYFIRST as the drawing name and AutoCAD 2004 Drawing (*.dwg) as the file type.

10 Now menu bar with **File-Exit** to exit AutoCAD.

11 *Notes*
   a) when multiple drawings have been opened in AutoCAD 2006, the user is prompted to save changes to each drawing before AutoCAD can be exited
   b) the Save Drawing As dialogue box has:
      1. other typical Windows options:
         History, My Documents, Favorites, etc.
      2. icon selections for:
         a) back to previous folder
         b) up one level
         c) search the web
         d) delete
         e) create new folder
         f) views: list, details, thumbnails, preview (details and preview usually active)
         g) tools.

Opening, modifying and saving an existing drawing

While AutoCAD is used to create drawings, it also extensively used to modify existing drawings. To demonstrate this concept:

1 Start AutoCAD and:
prompt	Startup dialogue box
respond	**pick the Open a Drawing tab**
prompt	Startup: Open a Drawing dialogue box
respond	**pick Browse**
prompt	Select File dialogue box
respond	*a*) scroll at Look in
	b) pick (left-click) the C: drive
	c) double left-click the MYCAD folder
and	all saved drawings displayed
respond	*a*) pick MYFIRST
	b) preview displayed – Figure 5.2
	c) pick Open.

2 The HELP 2D LINES drawing will be displayed.

3 Erase (with a window selection set option) the 2D and LINES effect, leaving HELP.

Figure 5.2 The Select File dialogue box with drawing C:\MYCAD\MYFIRST selected.

4   Menu bar with **File-Save As** and:
   prompt   Save Drawing As dialogue box
   with   MYFIRST.dwg as the File name
   respond   **pick Save**
   prompt   Save Drawing As message dialogue box – Figure 5.3
   with   C:\MYCAD\MYFIRST.dwg already exists (or a similar C: path name)
                 Do you want to replace it?
   respond   **Do nothing at present**.

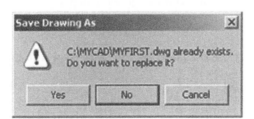

Figure 5.3 The Save Drawing As message.

5   This dialogue box is very common with AutoCAD and it is important that the user understands the three options:
   Cancel does nothing and returns the dialogue box
   No   returns the dialogue box allowing the user to alter the file name which should be highlighted
   Yes   will overwrite the existing file name and replace the original drawing with any modifications.

6   At this stage, respond to the message with:
   a) pick No
   b) alter the file name to **MYFIRST1**
   c) pick Save.

7   What have we achieved?
    a) we opened drawing MYFIRST from the C:\MYCAD folder
    b) we altered the drawing layout
    c) we saved the alterations as MYFIRST1
    d) the original MYFIRST drawing is still available and has not been modified.

8   Menu bar with **File-Open** and:
    prompt   Select File dialogue box
    respond   *a*) pick DRG1 and note the preview
                b) pick Open.

9   The screen will display the line segments drawn with Polar Tracking.

10 We now have two opened drawings:
    a) DRG1
    b) MYFIRST1 (or is it MYFIRST?).

Closing files

We will use the two opened drawings to demonstrate how AutoCAD should be exited when several drawings have been opened in the one drawing session.

1   Erase the line segments with a window selection – easy?

2   Menu bar with **File-Close** and:
    prompt   AutoCAD Message dialogue box
    with   Save changes to C:\MYCAD\DRG1.dwg message
    respond   **pick No** – can you reason out why we pick No?

3   The screen will display the MYFIRST (HELP) modified drawing, i.e. MYFIRST1.

4   Menu bar with **File-Close** and a blank screen will be returned with a short menu bar display:

    File, View, Window, Help

5   Question: why no 'Save changes' AutoCAD message with step 4?

6   Select **File** from the menu bar and:
    prompt   pull-down menu
    with   selections: New, Open and the last 9 (in my case) used AutoCAD drawings
    respond   **pick MYFIRST1**
    and   HELP drawing displayed.

7   *a*) menu bar with File and pick MYFIRST: HELP 2D LINES displayed
    b) menu bar with File and pick DRG1: polar tracking line segments displayed
    c) three drawings have now been opened with DRG1 displayed.

8   Menu bar with:
    a) **File-Close** to close DRG1 and display MYFIRST
    b) **File-Close** to close MYFIRST and display MYFIRST1
    c) **File-Close** to close MYFIRST1 and display a blank screen
    d) **File-Exit** to exit AutoCAD.

9   All drawings having been closed correctly and AutoCAD has been exited properly.

Save and Save As

1   The menu bar selection of **File** allows the user to pick either **Save** or **Save As**.

2   New AutoCAD users should be aware of the difference between these two options.

Save

1 Will save the current drawing with the same name with which the drawing was opened.

2 No dialogue box will be displayed.

3 The original drawing will be automatically overwritten if alterations have been made to it.

Save As

1 Allows the user to enter a drawing name via a dialogue box.

2 If a drawing already exists with the entered name, a message is displayed in a dialogue box.

It is strongly recommended that the SAVE AS selection is used at all times.

Assignment

You are now in the position to try a drawing for yourself, so:

1 Start AutoCAD and select Start from **Scratch-Metric-OK**.

2 Refer to activity drawing 1 (all activity drawings are grouped together at the end of this book).

3 Set a grid and snap spacing to suit, e.g. 10 and/or 5.

4 Menu bar with **Draw-Rectangle** and create a rectangle from 0,0 to 420,297.

5 Using only the LINE and CIRCLE commands (and perhaps ERASE if you make a mistake):
   a) draw the given shapes
   b) the size and position are not really important at this stage, the objective being to give you a chance to practice drawing using the drawing aids. All the shapes should 'fit into' the rectangle
   c) when the drawing is complete, save it as C:\MYCAD\ACT1.

Chapter 6

Standard sheet 1

1   Traditionally one of the first things that a draughtsperson does when starting a new drawing is to get the correct size sheet of drawing paper. This sheet will probably have borders, a company logo and other details already printed on it. The drawing is then completed to 'fit into' the pre-printed layout material.

2   A CAD drawing is no different from this, with the exception that the user does not 'get a sheet of paper'. Companies who use AutoCAD will want their drawings to conform to their standards in terms of the title box, text size, linetypes being used, the style of the dimensions, etc.

3   Parameters which govern these factors can be set every time a drawing is started, but this is tedious and against CAD philosophy. It is desirable to have all standard requirements set automatically, and this is achieved by making a drawing called a standard sheet, prototype drawing or template.

4   Standard sheets can be 'customised' to suit all sizes of paper e.g. A0, A1, etc. as well as any other size required by the customer. These standard sheets will contain the companies settings, and the individual draughtsman can add their own personal settings as required. It is this standard sheet which is the CAD operators 'sheet of paper'.

5   We will create an A3 standard sheet and save it as a template file for all future drawing work. At this stage, the standard sheet will not have many 'settings', but we will continue to refine it and add to it as we progress through the book.

Terminology

Several new concepts and terms will be used in this chapter. These concepts will be discussed in greater detail in later chapters, but it is important for the user to have an understanding of these new concepts.

a) Drawing
   A drawing is a saved file with the extension **.dwg** and all existing work has been with .dwg files
b) Template
   A template is a saved file with the extension **.dwt**. Templates have settings and default values and are used for all **new** drawing work
c) Layers
   Layers are used to enable the user to draw objects with different linetypes and colour
d) Model-Paper Space
   AutoCAD has two drawing environments, Model space and Paper space and:
   1. Model space: used to complete actual drawing work
   2. Paper space: used to 'lay out' the drawing paper.
e) Viewports
   These are areas that display different views of the drawing being created. This may be for a scale effect or for different views of a 3D model.

Creating an A3 template file

We will create our template using an existing AutoCAD template file so:

1 Start AutoCAD and:
   prompt   Startup dialogue box
   respond   1. pick Use a Template
               2. scroll and pick: **Iso a3 – named plot styles.dwt**
               3. pick OK.

2 The screen will display the AutoCAD ISO A3 template file with:
   a) a title box with several XXX items of text
   b) various coloured line effects
   c) a typical drawing grids effect to identify sections of a drawing
   d) the icon at the lower left of the drawing area indicates that the paper space environment is active.

3 Menu bar with **View-Zoom-All** and the drawing will 'fill the screen'.

4 *a*) Paper space is used to layout the drawing sheet while model space is for creating objects.
   b) To 'enter' model space left-click **Paper from the Status bar** and:
      1. the traditional 2D icon will be displayed
      2. the word MODEL is displayed in the status bar
      3. an area of the 'paper' is outlined. This is the model space viewport used for all drawing work.

5 Menu bar with **Tools-Drafting Settings** and from the Drafting Settings dialogue box select:
   a) Snap and Grid tab with:
      1. Snap on with X and Y spacing set to your requirements e.g. 5
      2. Grid on with X and Y spacing set to your requirements e.g. 10
      3. Rectangular snap style active.
   b) Polar Tracking tab with the Polar Tracking aid set to suit e.g. off
   c) Object Snap tab with:
      1. Object Snap off
      2. Object Snap Tracking off
      3. All snap modes off.
   d) Dynamic Input tab with the Dynamic Input aids set to suit e.g. all off
   e) pick OK.

6 Menu bar with **Format-Unit**s and:
   prompt   Drawing Units dialogue box
   respond   1. Length: *a*) Type: Decimal
                            b) Precision: scroll and pick 0.00
              2. Angle: *a*) Type: Decimal Degrees
                            b) Precision: scroll and pick 0.0.
              3. Insertion scale units: Millimeters
              4. pick OK and the status bar will display co-ordinate data in the form: 80.00, 50.00, 0.00.

7 Menu bar with **View-Display-UCS Icon-Properties** and:
   prompt   UCS Icon dialogue box
   respond   1. select the UCS icon style to suit (I recommend 2D)
               2. pick OK.

8 At the command line enter:
   a) **GRIPS <R>** and set to 0
   b) **PICKFIRST <R>** and set to 0.

9 Display toolbars to suit. I would suggest:
   a) Standard and Properties docked at the top as default
   b) Draw and Modify docked to suit
   c) Other toolbars will be displayed as required.

10 At the command line enter **–LAYER <R>** and:

prompt	`Enter an option [?/Make/Set/New/............`
enter	**N <R>** – the new layer option
prompt	`Enter name list for new layer(s)`
enter	**OUT <R>** – for outline
prompt	`Enter and option`
enter	**C <R>** – the colour option
prompt	`New colour [Truecolour/Colourbook]`
enter	**RED <R>**
prompt	`Enter name list of layer(s) for colour 1 (red)`
enter	**OUT <R>**
prompt	`Enter an option`
enter	**S <R>** – the set option
prompt	`Enter layer name to make current`
enter	**OUT <R>**
prompt	`Enter an option`
enter	**<R>** to end the command.

11 *a*) The Layers toolbar at the top of the screen should display a red square and the word OUT
   b) All objects drawn will be in red as the created layer OUT is current (as I have stressed, this will be discussed in greater detail when we investigate layers).

12 Menu bar with **View-Zoom-All.**

13 Menu bar with File-Save As and:

prompt	`Save Drawing As dialogue box`
respond	1. File type: scroll and pick AutoCAD Drawing Template (*.dwt)
	2. Save in: scroll and pick MYCAD as the named folder
	3. File name: enter A3PAPER
	4. pick Save.
prompt	`Template Description dialogue box`
respond	1. Description: My A3 paper template file with various settings and a created layer OUT
	2. Measurement: Metric – Figure 6.1
	3. pick OK.

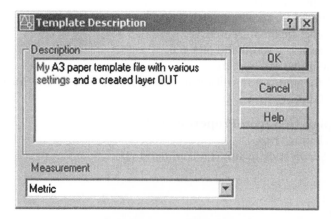

Figure 6.1   The Template Description box for A3PAPER.

14 *Notes*

a) Template files are generally saved to the AutoCAD Template folder.

b) We have saved our A3PAPER template file to our named MYCAD folder for 'ease of access'.

c) The created template file can also be saved as a drawing file with the sequence:
1. Menu bar with **File-Save As** then:
2. File type: AutoCAD 2004 Drawing (*.dwg)
3. Save in: scroll and pick our named folder C:\MYCAD
4. File name: A3PAPER
5. pick Save.

d) This completes the creation of our standard sheet at this stage

e) Although we have activated several toolbars in our standard sheet, the user should be aware that these may not always be displayed when your standard sheet drawing is opened. AutoCAD displays the screen toolbars which were active when the system was 'shut down'. If other CAD operators have used 'your machine', then the toolbar display may not be as you left it. If you are the only user on the machine, then there should not be a problem. Anyway you should know how to display toolbars?

You can now exit AutoCAD or continue to the next chapter.

Chapter 7

Line and circle object creation

1   The line and circle objects so far created were drawn at random on the screen without any attempt being made to specify position or size.

2   To draw objects accurately, co-ordinate input is required and AutoCAD 2006 allows different 'types' of co-ordinate entry including:
    a) **Absolute**, i.e. from an origin point
    b) **Relative** (or incremental), i.e. from the last point referenced.

3   In this chapter we will use our A3PAPER template file to create a working drawing and use it for future work.

Conventions

When using co-ordinate input, the user must know the positive and negative directions for both linear and angular input. The two conventions are as follows:

1   Co-ordinate axes
    The X–Y axes convention used by AutoCAD is shown in Figure 7.1(a) and displays four points with their co-ordinate values. When using the normal X–Y co-ordinate system:
    a) a positive X direction is to the right, and a positive Y direction is upwards
    b) a negative X direction is to the left, and a negative Y direction is downwards.

2   Angles
    When angles are being used:
    a) positive angles are anti-clockwise
    b) negative angles are clockwise
    c) Figure 7.1(b) displays the angle convention of four points with their polar co-ordinate values.

Getting started

1   If your template file from Chapter 6 is displayed, proceed to step 4.

2   *a*) If AutoCAD is active, then close any existing drawing then menu bar with **File-Open** and:
    prompt   Select File dialogue box
    respond   1. file type: scroll and pick Drawing Template (*.dwt)
                  2. look in: scroll and pick C:\MYCAD
                  3. pick A3PAPER and note Preview
                  4. pick Open.

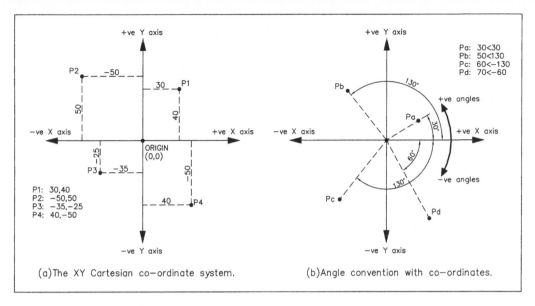

Figure 7.1 Co-ordinate and angle conventions.

b) If AutoCAD is not active, then start AutoCAD and:
    prompt   Startup dialogue box
    respond   1. select Use a Template
             2. pick Browse and:
                 prompt   Select a template file dialogue box
                 respond   *a*) scroll and activate the C:\MYCAD named folder
                         b) pick A3PAPER
                         c) pick Open.

3   The A3PAPER template file will be displayed with:
    a) model space active
    b) layer OUT current.

4   Display the Draw and Modify toolbars and position to suit and decide if you want to use polar tracking. Ensure that the Object Snap modes are off – they should be.

5   *Note*
    Step 2 is how all new drawing work will be started with the phrase 'open your A3PAPER template file'.

6   Refer to Figure 7.2.

LINES

Lines require the user to specify a start and end point for every line segment to be drawn and there are several methods for specifying these points.

Absolute co-ordinate entry

1   This method uses the traditional X–Y Cartesian system, where the origin point is (0,0) at the lower left corner of the drawing area. This origin point can be 'moved' by the user as you will discover later in this chapter. The user specifies the X and Y co-ordinates of every point relative to the (0,0) origin.

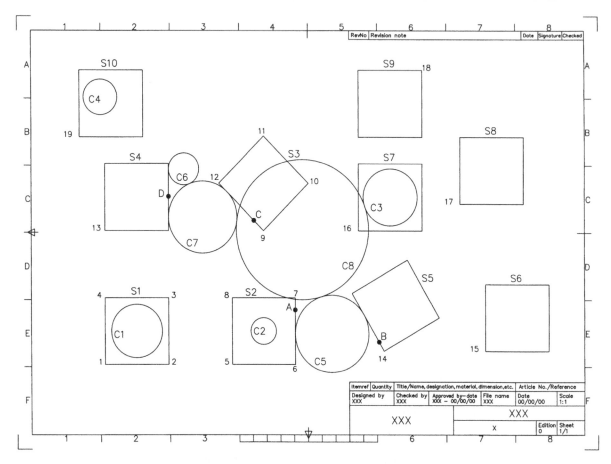

Figure 7.2   Line and circle creation by different options.

2   Select the **LINE icon** from the Draw toolbar and: *Figure 7.2 reference*
    prompt   Specify first point and enter **50,50 <R>**   point 1 start
    prompt   Specify next point and enter **100,50 <R>**   point 2
    prompt   Specify next point and enter **100,100 <R>**   point 3
    prompt   Specify next point and enter **50,100 <R>**   point 4
    prompt   Specify next point and enter **50,50 <R>**   point 1 end
    prompt   Specify next point and right-click
    prompt   shortcut menu displayed
    respond   **pick Enter** to end the line command

3   Square S1 will be created.

Relative (absolute) co-ordinate entry

1   The user specifies the X and Y co-ordinates of a point relative to the last point referenced.

2   The **@ symbol** is used for the relative/incremental entry.

3   From the menu bar select Draw-Line and enter the following X–Y co-ordinate pairs,
    remembering **<R>** after each entry.
    prompt   Specify first point and enter **150,50 <R>**   point 5 start
    prompt   Specify next point and enter **@50,0 <R>**   point 6
    prompt   Specify next point and enter **@0,50 <R>**   point 7
    prompt   Specify next point and enter **@−50,0 <R>**   point 8
    prompt   Specify next point and enter **@0,−50 <R>**   point 5 end
    prompt   Specify next point and right-click

prompt   shortcut menu displayed
respond   **pick Enter** to end the line command.

4   Square S2 will be created.

5   The @ symbol has the following effect:
  a) @50,0 is 50 units in the positive X direction and 0 units in the Y direction from the last point, which is 150,50
  b) @0,−50 is 0 units in the X direction and 50 units in the negative Y direction from the last referenced point
  c) thus an entry of @30,40 would be 30 in the positive X direction and 40 in the positive Y direction from the current cursor position
  d) similarly an entry of @ −80,−50 would be 80 units in the negative X direction and 50 units in the negative Y direction.

Relative (polar) co-ordinate entry

1   The user identifies the position of the next point by specifying the length of the line to be drawn and the angle of the line relative to the last point referenced.

2   The **@ symbol** is used to indicate relative entry and the **< symbol** used to indicate an angular entry.

3   Activate the LINE command (icon or menu bar) and enter the following co-ordinates:
  Prompt   Specify first point and enter **175,150 <R>**   point 9 start
  Prompt   Specify next point and enter **@50<45 <R>**   point 10
  Prompt   Specify next point and enter **@50<135 <R>**   point 11
  Prompt   Specify next point and enter **@50<225 <R>**   point 12
  Prompt   Specify next point and enter **C <R>** to close the square and end line command.

4   Square S3 will be drawn.

5   *Notes*
  a) The relative polar entries can be read as:
    1. @50<45 is 50 units at an angle of 45 degrees from the last point referenced which is 175,150
    2. @50<225 is 50 units at an angle of 225 degrees from the current cursor position.
  b) The entry **C <R>** is the **CLOSE** option and:
    1. closes the square, i.e. a line is drawn from the current screen position (point 12) to the start point 9
    2. ends the sequence, i.e. <R> not needed
    3. the close option works for any straight line shape.
  c) There is NO comma (,) with polar entries. This is a common mistake with new AutoCAD users, i.e.
    1. @50<45 is correct
    2. @50,<45 is wrong and gives the command line error: **Point or option keyword required**.

Grid and snap method

The grid and snap drawing aids can be set to any value suitable for current drawing requirements, so:

1   Set the grid and snap spacing to 25.

2   With the LINE command, draw a 50-unit square the start point being at 50,150 which is point 13 in Figure 7.2.

3   When the 50-unit square has been drawn (square S4) reset the grid and snap to original values.

Using polar tracking

1   Right-click **Polar from the Status bar**, pick Settings and:
    prompt   Drafting Settings dialogue box with Polar Tracking tab
                  active
    respond   a) Polar Tracking On (F10) active, i.e. tick in box
                  b) Polar Angle Settings: scroll at Incremental angle and pick 30
                  c) pick Options and:
                     prompt   Options dialogue box
                     respond   ensure the 3 Autotrack Settings are active then pick OK
    prompt   Drafting Settings dialogue box with Polar Tracking tab
                  active
    respond   pick OK.

2   Activate the LINE command and:
    a) enter 270,60 <R> at the command line as the line start point (point 14)
    b) move the cursor until the tooltip display is **Polar: ??? <30** and leave it
    c) enter 50 <R> from the keyboard
    d) move the cursor until the tooltip display is **Polar: ??? <120** and enter 50 <R>
    e) move the cursor until the display is **Polar: ??? <210** and enter 50 <R>
    f) finally enter C<R> to close the square (S5).

3   Now de-activate the Polar Tracking aid.

Moving the origin

1   The five squares created so far have been with the origin point (0,0) at the lower left
    corner of 'our drawing paper'.

2   Move the cursor onto this position (snap on) and the status bar will display 0.00,
    0.00, 0.00.

3   These are the X, Y and Z co-ordinates of the origin point.

4   At present we are drawing in 2D and thus the third co-ordinate will always be 0.00.

5   The origin can be moved to any point on the screen and we will reset it and draw
    another 50 unit square from this new origin position.

6   Menu bar with **View-Display-UCS Icon** and ensure that both On and Origin are
    active (tick at name).

7   Menu bar with **Tools-New UCS-Origin** and:
    prompt   Specify new origin point
    enter   **350,60 <R> <R>**
    and   the UCS icon will move to this position (if it does not then repeat
                  step 6).

8   Move the cursor (snap on) to the + in the icon and observe the status bar. The co-ordinates
    read:
    0.00, 0.00, 0.00, i.e. we have reset the origin point (point 15)

9   Draw a square (S6) of side 50 from this new origin using any of the methods previ-
    ously described.

10 At the command line enter **UCS <R>** and:

prompt   Enter an option [New/Move. . . .

enter   **P <R>** – the previous option

and   the UCS icon will be 'returned' to its original origin position at the lower left corner of our drawing paper.

Rectangles

1 Rectangular shapes (in our case, squares) can be created with a fillet and chamfer effect (later) with the user specifying:

a) two points on a diagonal of the rectangle

b) the area and the length/width of the rectangle.

2 From the menu bar select **Draw-Rectangle** and:

prompt   Specify first corner point or [Chamfer/Elevation/Fillet/ Thickness/Width]

enter   **250,150 <R>** – point 16

prompt   Specify other corner point or [Area/Dimensions/Rotation]

enter   **300,200 <R>**

and   square S7 created.

3 Select the rectangle icon from the Draw-toolbar and:

prompt   Specify first corner point or [Chamfer/Elevation/Fillet/ Thickness/Width]

enter   **380,170 <R>** – point 17

prompt   Specify other corner point or [Area/Dimensions/Rotation]

enter   **@ −50,50 <R>**

and   square S8 positioned.

4 Activate the rectangle command and:

prompt   Specify first corner point or [Chamfer/Elevation/Fillet/ Thickness/Width]

enter   **300,270 <R>** – point 18

prompt   Specify other corner point or [Area/Dimensions/Rotation]

enter   **A <R>** – the area option

prompt   Enter area of rectangle in current units

enter   **2500 <R>**

prompt   Calculate rectangle dimensions based on Length/Width

enter   **L <R>** – the length option

prompt   Enter rectangle length

enter   **−50 <R>** – square S9 created and note the 'direction' of the rectangle.

Using dynamic input

Before we can use the Dynamic Input drawing aid, it is necessary to set the options to our requirements, so:

1 Right-click **DYN from the Status bar**, pick Settings and:

prompt   The Drafting Settings dialogue box with the Dynamic Input tab active

respond   *a*) Enable Pointer Input active, i.e. tick

          b) pick Pointer Input Settings

prompt   Pointer Input Settings dialogue box

respond   *a*) Format: Cartesian and Absolute active

          b) Visibility: When a command asks for a point active

          c) pick OK.

> *prompt*   The Drafting Settings dialogue box with the Dynamic Input
>   tab active
> *respond*   pick OK.

2   Activate the line command and:
> *prompt*   Specify first point
> *and*   *a)* note the cursor display as it is moved about the screen
>   *b)* the tooltip displays two absolute co-ordinate value 'boxes'.
> *respond*   **enter 30 then press the TAB key**
> *and*   *a)* left tooltip box displays 30 with a locked icon, i.e. the X co-ordinate of
>   the line start point is fixed
>   *b)* right tooltip box (the Y co-ordinate value) now becomes active.
> *respond*   **enter 220 <R>** and the line start point is positioned (point 19)
> *prompt*   Specify next point
> *and*   tooltip displays two absolute co-ordinate values
> *respond*   *a)* **enter @50 and press the TAB key**
>   *b)* **enter 0 <R>**
> *prompt*   Specify next point
> *respond*   *a)* **enter @0 and press the TAB key**
>   *b)* **enter 50 <R>**
> *prompt*   Specify next point
> *enter*   **@−50, TAB, 0 <R>**
> *then*   close the square (S10).

3   At this stage save the drawing with the menu bar selection **File-Save As** and:
> *prompt*   Save Drawing As dialogue box
> *respond*   *a)* File type: AutoCAD 2004 Drawing (*.dwg)
>   *b)* scroll at Save in and pick C: drive
>   *c)* double left-click on MYCAD folder
>   *d)* enter file name as DEMODRG
>   *e)* pick Save.

4   The complete file name path for the saved drawing is **C:\MYCAD\DEMODRG**.

CIRCLES

1   AutoCAD 2006 allows circles to be created by six different methods.

2   The circle command can be activated by icon selection, menu bar selection or keyboard entry.

3   When creating circles, absolute co-ordinates are usually used to specify the circle centre, although the next chapter will introduce the user to the Object Snap modes. These Object Snap modes allow greater flexibility in selecting existing entities for reference.

4   To investigate how circles are created:
> *a)* continue from the line exercises or;
> *b)* open C:\MYCAD\DEMODRG
> *c)* refer to Figure 7.2.

Specifying a centre point and a radius

Select the **CIRCLE icon** from the Draw-toolbar and:
> *prompt*   Specify centre point for circle or [3P/2P/Ttr (tan tan
>   radius)]:
> *enter*   **75,75 <R>** – the circle centre point
> *prompt*   Specify radius of circle or [Diameter]

enter   **20 <R>** – the circle radius

and   circle C1 created.

Specifying a centre point and a diameter

From the menu bar select **Draw-Circle-Centre, Diameter** and:

prompt   Specify centre point for circle or [3P/2P/Ttr (tan tan radius)]:

enter   **175,75 <R>**

prompt   Specify diameter of circle

enter   **20 <R>** – circle C2.

Specifying two points on circle diameter

At the command line enter **CIRCLE <R>** and:

prompt   Specify centre point for circle or [3P/2P/Ttr (tan tan radius)]:

enter   **2P <R>** – the two point option

prompt   Specify first end point on circle's diameter

enter   **260,160 <R>**

prompt   Specify second end point on circle's diameter

enter   **290,190 <R>** – circle C3.

Specifying any three points on circle circumference

Menu bar selection with **Draw-Circle-3 Points** and:

prompt   Specify centre point for circle or [3P/2P/Ttr (tan tan radius)]:

respond   **pick any point within the top left square**

prompt   Specify second point on circle

respond   **pick another point within the top left square**

prompt   Specify third point on circle

respond   **drag out the circle and pick a point in the square** – circle C4.

Specifying two tangents and a radius

1   Menu bar with **Draw-Circle-Tan,Tan,Radius** and:

    prompt   Specify point on object for first tangent of circle

    respond   **move cursor to line A and leave for a second**

    and   *a*) a small marker is displayed

              b) tooltip with Deferred Tangent displayed.

    respond   **pick line A**, i.e. left-click on it

    prompt   Specify point on object for second tangent of circle

    respond   **pick line B**

    prompt   Specify radius of circle

    enter   **29 <R>**

    and   a circle (C5) is drawn tangential to the two selected lines.

2   At the command line enter **CIRCLE <R>** and:

    prompt   Specify point on object for first tangent of circle

    enter   **TTR <R>** – the tan,tan,radius option

    prompt   first tangent point prompt and: **pick line C**

    prompt   second tangent point prompt and: **pick line D**

prompt   radius prompt and enter **12 <R>**
and   a circle (C6) is drawn tangential to the two selected lines, line C being assumed extended.

Specifying three tangents

1   Menu bar with **Draw-Circle-Tan,Tan,Tan** and:
    prompt   Specify first point on circle and: **pick line C**
    prompt   Specify second point on circle and: **pick line D**
    prompt   Specify third point on circle and: **pick circle C6** – circle C7.

2   Activate the **Draw-Circle-Tan,Tan,Tan** sequence and:
    prompt   Specify first point on circle and: **pick circle C3**
    prompt   Specify second point on circle and: **pick circle C5**
    prompt   Specify third point on circle and: **pick circle C7** – circle C8.

3   Circles have been drawn tangentially to selected objects, these being:
    a) two lines and a circle
    b) three circles.

4   *Questions*
    a) How long would it take to draw a circle as a tangent to three other circles by conventional draughting methods, i.e. drawing board, T square, set squares, etc?
    b) Can a circle be drawn as a tangent to two circles and a line or to three inclined lines? Try these for yourself.

Saving the working drawing

1   Assuming that the LINE and CIRCLE commands have been entered correctly, your drawing should resemble Figure 7.2 (without the text) and is ready to be saved for future work.

2   From the menu bar select **File-Save As** and:
    prompt   Save Drawing As dialogue box
    respond   1. File type: AutoCAD 2004 Drawing (*.dwt)
                 2. Save in: MYCAD
                 3. File name: **DEMODRG**
                 4. **pick Save.**

Notes

1   *a*) we started the line and circle exercise by opening our A3PAPER template file
    b) we saved the completed work as a drawing file
    c) the original A3PAPER template file is thus unchanged
    d) this is the procedure for starting, completing and saving all future work.

2   With future work, when the user is asked to draw a line, a circle (or any other AutoCAD object), the method of creation is now at user discretion. Keyboard co-ordinates can be used, as can the dynamic input aid, polar tracking, etc. I will now assume that the user has this knowledge.

3   When positioning a LINE, mistakes can be made with co-ordinate entry. If the line 'does not appear to go in the direction it should', then either:
    a) enter U <R> from the keyboard to 'undo' the last line segment drawn, or
    b) right-click and pick undo from the pop-up menu
    c) this 'undo effect' can be used until all the line segments are erased.

4 The @ symbol very is useful if you want to 'get to the last point referenced on the screen'. Try the following:
   a) draw a line and cancel the command with a <RETURN>
   b) re-activate the line command and enter @ <R>
   c) the cursor 'snaps to' the endpoint of the drawn line.

5 After a command has been cancelled, a <RETURN> keyboard press will activate the last command.

6 A right-click on the mouse will activate a shortcut menu, allowing the last command to be activated.

Task

1 The two Tan,Tan,Tan circles have been created without anything being known about their centre points or radii.

2 From the menu bar select **Tools-Inquiry-List** and:

prompt	`Select objects`
respond	**pick the smaller TTT circle (C7)**
prompt	`1 found and Select objects`
respond	**right-click**
prompt	`AutoCAD Text` window with information about the circle.

3 Note the information then cancel the text window by picking the right (X) button from the title bar.

4 Repeat the Tools-Inquiry-List sequence for the larger TTT circle (C8).

5 The information for my two TTT circles is as follows:

	Smaller	*Larger*
Centre point	126.88,160.11	205.54,150,88
Radius	26.88	52.32
Circumference	168.88	328.74
Area	2269.56	8599.90

6 Could you calculate these figures manually as easily as has been demonstrated?

Assignment

This activity uses the LINE and CIRCLE commands and requires co-ordinate entry.

1 Close all existing drawings then open your A3PAPER template file.

2 Refer to *Activity 2* and draw the three shapes using co-ordinate input. Any entry method can be used but I would recommend:
   a) position the start points and circle centres with absolute entry
   b) use relative entry as much as possible.

3 When the drawing is complete, save it as **C:\MYCAD\ACT2**.

Chapter 8

Object snap

1   The lines and circles drawn so far have been created by co-ordinate input.

2   While this is the basic method of creating objects, it is often desirable to 'reference' existing objects already displayed on the screen, e.g. we may want to:
a) draw a circle with its centre at the midpoint of an existing line
b) draw a line, from a circle centre perpendicular to another line.

3   These types of operations are achieved using the **object snap modes** (referred to as **OSNAP**) and are one of the most useful (and powerful) draughting aids.

4   Object snap modes are used transparently (used whilst in a command) and can be activated:
a) from the Object Snap toolbar
b) by direct keyboard entry.

5   The toolbar method is quicker and easier to use and will be the only method which will be considered.

Getting ready

1   Open your **C:\MYCAD\DEMODRG** drawing of the squares and circles.

2   Referring to Figure 7.2, erase:
a) squares: S6, S7, S8
b) circles: C2, C3, C6, C7, C8.

3   Display the Draw, Modify and Object Snap toolbars and position them to suit.

4   Now refer to Figure 8.1.

Drawing line and circle objects using object snap

1   Several line segments will be created all using a different object snap mode, so activate the **LINE** command and:

prompt	Specify first point
respond	**pick the Snap to Nearest icon**
prompt	nea to
respond	*a*) move cursor to line AB
	b) note the coloured marker and tooltip display
	c) pick any point (left-click) on line AB
prompt	Specify next point
respond	**pick the Snap to Quadrant icon**
prompt	qua of
respond	**pick circle C1 at 'six o'clock position'**

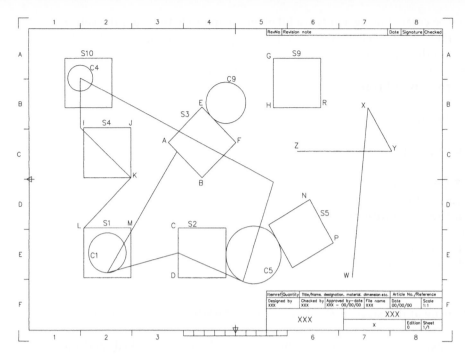

Figure 8.1 Using Object Snap to draw line and circle objects.

prompt	Specify next point
respond	**pick the Snap to Midpoint icon**
prompt	mid to
respond	**pick line CD**
prompt	Specify next point
respond	**pick the Snap to Tangent icon**
prompt	tan to
respond	**pick circle C5 on lower circumference**
prompt	Specify next point
respond	**pick the Snap to Apparent Intersect icon**
prompt	appint of
respond	**pick line EF**
prompt	and
respond	**pick line GH**
prompt	Specify next point
respond	**pick the Snap to Center icon**
prompt	qua of
respond	**pick circle C4**
prompt	Specify next point
respond	**pick the Snap to Perpendicular icon**
prompt	per to
respond	**pick line IJ**
prompt	Specify next point
respond	**pick Snap to Intersection icon**
prompt	Int of
respond	**pick point K**
prompt	Specify next point
respond	**pick Snap to Endpoint icon**
prompt	end of
respond	**pick line LM at 'L' end**
prompt	Specify next point
respond	**right-click and pick Enter** to end the line sequence.

2 A circle will now be created using a very useful object snap mode to obtain the centre point, and the radius will be determined by referencing an existing object. Select the **CIRCLE icon** from the Draw toolbar and:

prompt	Specify center point for circle or [3P/2P. . .
respond	**pick the Snap From icon**
prompt	from Base point
respond	**pick Midpoint icon**
prompt	mid of
respond	**pick line GH**
prompt	<Offset>
enter	**@−50,−20 <R>**
prompt	Specify radius of circle
respond	**pick the Perpendicular icon**
prompt	per to
respond	**pick line EF**
and	circle C9 created.

3 *Notes*

    a) save your drawing at this stage as **C:\MYCAD\DEMODRG** to update the existing drawing.

    b) the endpoint 'snapped to' depends on which part of the line is 'picked' and the coloured marker indicates this endpoint.

    c) a circle has four quadrants, these being at the 3,12,9,6 o'clock positions. The coloured marker indicates which quadrant will be snapped to.

The extension and parallel object snap modes

1 The object snap modes selected so far should have been self-explanatory to the user i.e. endpoint will snap to the end of a line, centre will snap to the centre of a circle, etc.

2 The extension and parallel object snap modes operate in a slightly different manner from those already used and:

    a) Extension: used with lines and arcs to give a temporary extension line as the cursor is passed over the endpoint of an object.

    b) Parallel: used with straight-line objects only and allows a vector to be drawn parallel to another object.

3 To demonstrate these two object snap modes continue with the DEMODRG, turn off the snap, activate the LINE command and:

prompt	Specify first point
respond	**pick the Snap to Extension icon**
prompt	ext of
respond	move cursor over point P then drag down-to-right in line with NP
and	*a*) highlighted extension line dragged out
	b) information displayed about distance from endpoint as a tooltip
respond	*a*) drag cursor until tooltip display is **Extension: 37.51<300.0** (or similar)
	b) enter 40
and	**point W identified**
prompt	Specify next point
respond	**pick the Snap to Extension icon**
prompt	ext of
respond	**move cursor over point R and drag to right**

and   *a*) highlighted extension line dragged out

                 b) information displayed as a tooltip

respond   *a*) move cursor until **Extension: 50.0<0.0** (or similar) is displayed

                 b) enter 50

and   **point X identified and line segment WX drawn**

prompt   Specify next point

respond   **pick the Snap to Parallel icon**

prompt   par to

respond   *a*) move cursor over line NP, leave for a few seconds and note the display

                 b) move cursor to right of current position until a highlighted line and tooltip displayed

                 c) enter 50

and   point Y identified and line XY drawn

prompt   Specify next point

respond   **pick the Snap to Parallel icon**

prompt   par to

respond   *a*) move cursor over line HR then move to left of point H

                 b) note tooltip display then enter 100

prompt   Specify next point

respond   **right-click and pick Enter** to end line sequence.

4 The line segments should be as displayed in Figure 8.1.

5 Save your layout as **C:\MYCAD\DEMODRG**, updating the existing DEMODRG.

Running object snap

1 Using the object snap icons from the toolbar will increase the speed of the draughting process, but it can still be 'tedious' to have to pick the icon every time an ENDpoint (for example) is required.

2 It is possible to 'preset' the object snap mode to ENDpoint, MIDpoint, CENter; etc. and this is called a running object snap.

3 Pre-setting the object snap does not preclude the user from selecting another mode, i.e. if you have set an ENDpoint running object snap, you can still pick the INTersection icon.

4 The running object snap can be set from the Drafting Settings dialogue box by:
     a) From the menu bar with **Tools-Drafting Settings** and picking the Object Snap tab
     b) Entering **OSNAP <R>** at the command line
     c) Picking the **Osnap Settings icon** from the Object Snap toolbar
     d) With a right-click on OSNAP in the Status bar and picking Settings

To demonstrate using a running object snap:

1 Select **Tools-Drafting Settings** from the menu bar and:

prompt   Drafting Settings dialogue box

respond   *a*) ensure the Object Snap tab is active

                 b) ensure Object Snap On (F3) is active (tick)

                 c) ensure Object Snap Tracking on is not active (no tick)

                 d) activate Endpoint and Midpoint by picking the appropriate box – tick means active

                 e) pick OK.

2 Now activate the LINE command and move the cursor cross-hairs onto any line and leave it.

3   A coloured marker will be displayed on the selected line.

4   Press the TAB key to cycle through the set running object snaps, i.e. the line should display the triangle and square coloured markers for the Endpoint and Midpoint object snap settings.

5   Cancel the line command with ESC.

AutoSnap and AutoTrack

1   The Object Snap tab of the Drafting Settings dialogue box allows the user to select **Options** which will display the Drafting tab of the Options dialogue box.

2   The user can control both the AutoSnap and the AutoTrack settings.

3   These can be defined as:
     a) AutoSnap: a visual aid for the user to see and use object snaps efficiently, i.e. marker and tooltip displayed.
     b) AutoTrack: an aid to the user to assist with drawing at specific angles, i.e. polar tracking.

4   The settings which can be altered with AutoSnap and AutoTrack are:
AutoSnap	*AutoTrack*
Marker	Display polar tracking vector
Magnet	Display full screen tracking vector
Display tool tip	Display AutoTrack tooltip
Display aperture box	Alter alignment point acquisition
Alter marker colour	Alter aperture size
Alter marker size	

5   The various terms should be self-explanatory:
     a) Marker: is the geometric shape displayed at a snap point
     b) Magnet: locks the aperture box onto the snap point
     c) Tool tip: is a flag describing the name of the snap location.

6   *Note*
     a) It is normal to have the marker, magnet and tool tip active (ticked)
     b) The colour of the marker and the aperture box sizes are at the user's discretion
     c) Having object snap tracking 'on' is also at user discretion.

Cancelling a running object snap

A running object snap can be left 'active' once it has been set, but this can cause problems if the user 'forgets' about it. The running snap can be cancelled:

1   Using the Object Snap settings dialogue box and selecting Clear All.

2   By entering **–OSNAP <R>** at the command line and:
     prompt   Enter list of object snap modes
     enter   **NONE <R>**.

3   *Notes*
     a) Selecting the Snap to None icon from the Object Snap toolbar will only turn off the object snap running modes for the next point selected.
     b) Using the Object Snap dialogue box is the recommended way of activating and de-activating object snaps.

Object snap tracking

Object snap tracking allows the user to 'acquire coordinate data' from the object snap modes which have been set. To demonstrate this drawing aid:

1   Right-click **OSNAP from the Status bar**, pick **Settings** and:
    prompt   Drafting Settings dialogue box with Object Snap tab active
    respond   a) Object Snap On active
                b) Endpoint object snap modes active
                c) Object Snap Tracking On active
                d) pick OK.

2   Activate the CIRCLE command and:
    a) move the cursor over point F until Endpoint marker (square) displayed
    b) move to right and a highlighted line is displayed
    c) move the cursor over point H until Endpoint marker displayed
    d) move vertically downward and display should be as Figure 8.2 – identifying the circle centre
    e) pick this point (left-click) then enter the radius to suit
    f) a very useful method of acquiring points?

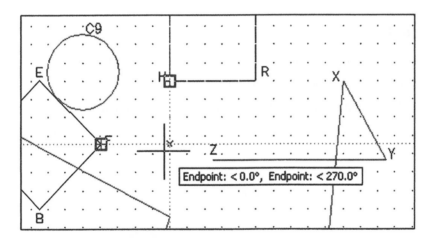

Figure 8.2   Using object snap tracking to acquire a circle centre.

This chapter is now complete. Ensure that you have saved your work as **C:\MYCAD\DEMODRG**.

Chapter 9

Creating arcs, donuts, ellipses, points and polygons

These drawings commands will be discussed in turn using DEMODRG. The commands can be activated from the toolbar, menu bar or by keyboard entry and both co-ordinate entry and referencing existing objects (using OSNAP) will be demonstrated.

Getting started

1   Open your C:\MYCAD\DEMODRG to display the squares, circles and object snap lines.

2   Erase the objects created during the object snap exercise.

3   Refer to Figure 9.1 and activated the Draw, Modify and Object Snap toolbars.

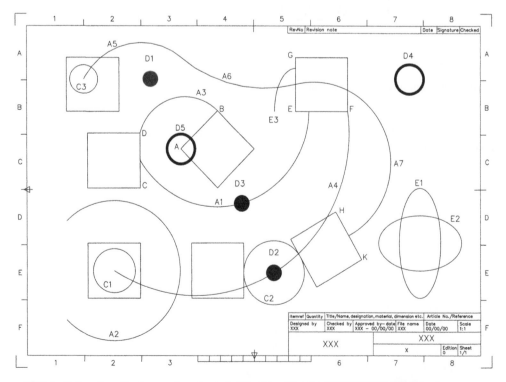

Figure 9.1   Creating arcs, donuts and ellipses using C:\MYCAD\WORKDRG.

Arcs

1   There are 10 different arc creation methods.

2   Arcs are normally drawn in an anti-clockwise direction with combinations of the arc start point, end point, centre point, radius, included angle, length of arc, etc.

3   We will investigate four different arc creation methods as well as continuous arcs. You can try the others for yourself.

4   **Center, Start, End**
From the menu bar select **Draw-Arc-Center, Start, End** and:

prompt	Specify center point of arc
respond	**Snap to Intersection icon and pick point B**
prompt	Specify start point of arc
respond	**Snap to Midpoint icon and pick line CD**
prompt	Specify end point of arc
respond	**Snap to Nearest icon and pick a point on line EF**
and	arc A1 created.

5   **Start, Center, Angle**
Menu bar with **Draw-Arc-Start, End, Angle** and:

prompt	Specify start point of arc
enter	**30,30 <R>** – absolute co-ordinate entry
prompt	Specify centre point of arc
respond	**Snap to Centre icon and pick circle C1**
prompt	Specify included angle
enter	**270 <R>**
and	arc A2 created
note	negative angle entries draw arcs in a clockwise direction.

6   **Start, End, Radius**
Menu bar again with **Draw-Arc-Start, End, Radius** and:

prompt	Start point and **snap to Intersection of point B**
prompt	End point and **snap to Intersection of point D**
prompt	Specify radius of arc
enter	**5 <R>**
and	command line displays *Invalid*, i.e. the arc radius is too small
now	repeat the Start, End, Radius and:

         a) intersection of point B as start point
         b) intersection of point D as end point
         c) enter 45 as the arc radius and arc A3 created.

7   **Three points (on arc circumference)**
Activate the **3 Points arc** command and:

prompt	Start point and **snap to Center of circle C1**
prompt	Second point and **snap to Center of circle C2**
prompt	End point and **snap to Intersection of point F**
and	arc A4 created.

8   **Continuous arcs**
Allows arc segments to be continually drawn from previous segments. To demonstrate:
a) Activate the **3 Points arc** command again and:

prompt	Start point and **Snap to Center of circle C3**
prompt	Second point and enter **80,280 <R>** – absolute co-ordinate entry
prompt	End point and enter **@60,−10 <R>** – relative co-ordinate entry
and	arc segment A5 drawn

b) Select from the menu bar **Draw-Arc-Continue** and:
>*prompt*   End point – and cursor snaps to end point of the last arc drawn
>*respond*   **Snap to midpoint of line EG** – arc A6

c) Repeat the **Arc-Continue** selection and:
>*prompt*   End point
>*respond*   **snap to Midpoint of line HK**
>*and*   arc A7 created.

Donut

1   A donut is a 'solid filled' circle or annulus (a washer shape).

2   The user specifies the inside and outside diametres and then selects the donut centre point.

3   Menu bar with **Draw-Donut** and:
>*prompt*   Specify inside diameter of donut and enter **0 <R>**
>*prompt*   Specify outside diameter of donut and enter **15 <R>**
>*prompt*   Specify center of donut and enter **110,250 <R>** – donut D1
>*prompt*   Specify center of donut
>*respond*   **Snap to Center of circle C2** – donut D2
>*prompt*   Specify center of donut
>*respond*   **Snap to Midpoint of arc A1** – donut D3
>*prompt*   Specify center of donut and **right-click**.

4   Repeat the donut command and:
>*prompt*   Specify inside diameter of donut and enter **25 <R>**
>*prompt*   Specify outside diameter of donut and enter **30 <R>**
>*prompt*   Specify center of donut and enter **360,250 <R>** – donut D4
>*prompt*   Specify center of donut
>*respond*   **Snap to Intersection of point A** – donut D5
>*prompt*   Specify center of donut and **right-click**.

5   *Note*: the donut command allows repetitive entries to be made by the user, while the circle command only allows one circle to be created per command.

Ellipse

1   Ellipses are created by the user specifying:
  a) either the ellipse centre and two axes endpoints
  b) or three points on the axes endpoints
  c) be aware that partial elliptical arcs can be created.

2   Select from the menu bar **Draw-Ellipse-Center** and:
>*prompt*   Specify center of ellipse and enter **370,100 <R>**
>*prompt*   Specify endpoint of axis and enter **370,150 <R>**
>*prompt*   Specify distance to other axis or [Rotation] and enter **390,100 <R>**
>*and*   ellipse E1 created

3   Select the **ELLIPSE icon** from the Draw toolbar and:
>*prompt*   Specify axis endpoint of ellipse or (Arc/Center)
>*enter*   **C <R>** – the center option
>*prompt*   Specify center of ellipse
>*respond*   **Snap to Center icon and pick the ellipse E1**
>*prompt*   Specify endpoint of axis and enter **@40,0 <R>**
>*prompt*   Specify distance to other axis and enter **@0,25 <R>** – ellipse E2

4 Menu bar with **Draw-Ellipse-Arc** and:

prompt   Specify axis endpoint of elliptical arc or [Center]
enter   C <R> – the center option
prompt   Specify center of elliptical arc
respond **Snap to Intersection of point E**
prompt   Specify endpoint of axis
enter   **@0,40 <R>**
prompt   Specify distance to other axis or [Rotation]
enter   **20<R>**
prompt   Specify start angle or [Parameter]
enter   **0 <R>**
prompt   Specify end angle
enter   **90 <R>** – elliptical arc E3.

6 *Note*
   a) At this stage your drawing should resemble Figure 9.1 without the text
   b) Save the layout as **C:\MYCAD\DEMODRG** for future recall if required. Remember that this will 'over-write' the existing C:\MYCAD\DEMODRG file
   c) Arcs, donuts and ellipses have centre points and quadrants which can be 'snapped to' with the object snap modes
   d) It is also possible to use the tangent snap icon and draw tangent lines etc between these objects. Try this for yourself.

FILL

1 Donuts are generally displayed on the screen 'solid' i.e. 'filled in'

2 This solid fill effect is controlled by the FILL system variable which can be activated from the menu bar or the command line

3 Still with the DEMODRG layout on the screen?

4 Menu bar with **Tools-Options** and:
prompt   Options dialogue box
respond **pick the Display tab**
then   1 Apply solid fill OFF i.e. no tick in box
           2 pick OK.

5 Menu bar with **View-Regen** and the donuts will be displayed without the fill effect

6 At the command line enter **FILL <R>**
prompt   Enter mode [ON/OFF] <OFF>
enter   **ON <R>**

7 At the command line enter **REGEN <R>** to 'refresh' the screen and display the donuts with the fill effect

8 *Note*: the fill effect also applies to polylines, which will be discussed in a later chapter.

Point

1 A point is an object whose size and appearance is controlled by the user

2 Erase the arcs, donuts and ellipses from DEMODRG and refer to Figure 9.2

3 From the Draw toolbar select the **POINT icon** and:
prompt   Current point modes: PDMODE=0 PDSIZE=0.00
           Specify a point

enter	**370,170 <R>** – point P1
prompt	Specify a point
respond	**snap to Center icon of circle C1** – point P2
prompt	Specify a point
respond	**snap to Midpoint icon of line AB** – point P3
prompt	Specify a point and **ESC** to end the command.

4 Three point objects will be displayed in red on the screen. You may have to toggle the grid off to 'see' these points properly.

5 From the menu bar select **Format-Point Style** and:

prompt	Point Style dialogue box
respond	1 point style: select the lower middle
	2 point size: 5%
	3 Set Size Relative to Screen: active – Figure 9.3
	4 pick OK.

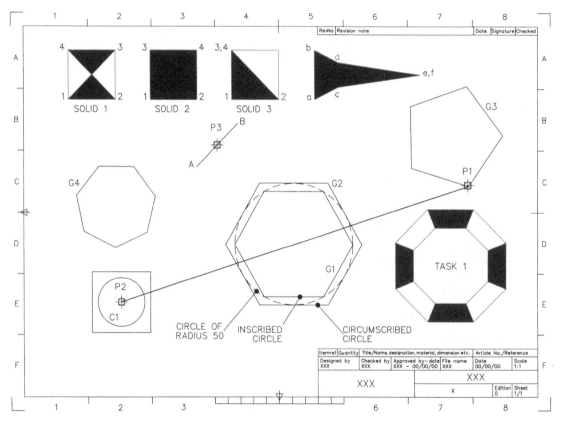

Figure 9.2   Point, polygon and 2D solid creation.

6 The screen will be displayed with the three entered points regenerated to this new point style.

7 Menu bar with **Draw-Point-Multiple Point** and:

prompt	Specify a point and enter **330,85 <R>**
prompt	Specify a point and enter **270,210 <R>**
prompt	Specify a point and ESC.

8 The screen will now display two additional points in the selected style.

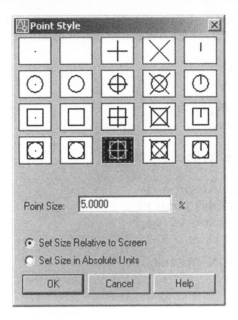

Figure 9.3 The Point Style dialogue box.

9 Select the **LINE icon** and:
    prompt   Specify first point
    respond   **Snap to Node icon and pick point P1**
    prompt   Specify next point
    respond   **Snap to Node icon and pick point P2**
    prompt   Specify next point and **right-click-Enter**.

Polygon

1 A polygon is a multi-sided figure, each side having the same length and can be drawn by the user specifying:
    a) a centre point and an inscribed or circumscribed radius
    b) the endpoints of an edge of the polygon.

2 Refer to Figure 9.2 and erase the lines and circles from DEMODRG not displayed then select the **POLYGON icon** from the Draw toolbar and:
    prompt   Enter number of sides<4>
    enter   **6 <R>**
    prompt   Specify center of polygon or [Edge]
    respond   **Snap to Midpoint icon and pick the line joining points P1 and P2**
    prompt   Enter an option [Inscribed in circle/Circumscribed about
              circle]<I>
    enter   **I <R>** – the inscribed (default) option
    prompt   Specify radius of circle
    enter   **50 <R>**
    and   polygon G1 created.

3 Repeat the **POLYGON icon** selection and:
    prompt   Enter number of sides<6> and enter **6 <R>**
    prompt   Specify center of polygon or [Edge]
    respond   **Snap to Midpoint icon and pick the line**
    prompt   Enter an option [Inscribed/Circumscribed...
    enter   **C <R>** – the circumscribed option
    prompt   Specify radius of circle and enter **50 <R>** – polygon G2.

4 The screen will display an inscribed and circumscribed circle drawn relative to a 50-radius circle as shown in Figure 9.2. These hexagonal polygons can be considered as equivalent to:
  a) inscribed: ACROSS CORNERS (A/C)
  b) circumscribed: ACROSS FLATS (A/F).

5 From the menu bar select **Draw-Polygon** and:
  prompt   Enter number of sides<6>
  enter   **5 <R>**
  prompt   Specify center of polygon or [Edge]
  enter   **E <R>** – the edge option
  prompt   Specify first endpoint of edge
  respond   **Snap to Node icon and pick point P1**
  prompt   Specify second endpoint of edge
  enter   **@50<55 <R>** – polygon G3.

6 At the command line enter **POLYGON <R>** and:
  a) Number of sides: 7
  b) Edge/Center: E
  c) First point: 70,120
  d) Second point: @30<25 – polygon G4.

7 A polygon is a POLYLINE type object and has the 'properties' of polylines. Polylines are discussed in a later chapter.

Solid (or more correctly 2D Solid)

1 A command which 'fills-in' lined shapes, the appearance of the final shape being determined by the pick point order.

2 With the snap on, draw three squares of side 40, towards the top part of the screen as Figure 9.2.

3 Menu bar with **Draw-Surfaces-2D Solid** and:
  prompt   Specify first point and **pick point 1 of SOLID 1**
  prompt   Specify second point and **pick point 2**
  prompt   Specify third point and **pick point 3**
  prompt   Specify fourth point and **pick point 4**
  prompt   Specify third point and right-click to end command.

4 At the command line enter **SOLID <R>** and:
  prompt   Specify first point and **pick point 1 of SOLID 2**
  prompt   Specify second point **and pick point 2**
  prompt   and pick points 3 and 4 in order displayed then right-click.

5 Activate the SOLID command and with SOLID 3 pick points 1–4 in the order given i.e. 3 and 4 are the same points.

6 The three squares demonstrate how 3 and 4-sided shapes can be solid filled.

7 Using the SOLID command and Snap on:
  First point   pick a point a
  Second point   pick a point b
  Third point   pick point c
  Fourth point   pick point d
  Third point   pick point e
  Fourth point   pick point f
  Third point   right-click

8 Step 7 demonstrates how continuous filled 2D shapes can be created.

9 *Task 1*
   a) Draw two concentric octagons (8 sides) centre point and circle radius to suit
   b) Use the SOLID command to produce the effect in Figure 9.2
   c) A running object snap to Endpoint will help.

10 *Task 2*
   What are the minimum and maximum number of sides allowed with the POLYGON command?

11 *Task 3*
   a) At the command line enter **FILL <R>** and:
      prompt   `Enter mode [ON/OFF]<ON>`
      enter   **OFF <R>**
   b) At command line enter **REGEN <R>**
   c) The solid fill effect is not displayed
   d) Turn FILL back on then REGEN the screen.

12 This exercise is complete and can be saved if required but will not be used again.

Assignments

Two activities have been included with this chapter. The procedure for completing these activities is:

1 Open your A3PAPER file – drawing or template?

2 Complete the drawing activity then save as C:\MYCAD\ACT3, etc.

Activity 3: Various designs

1 Draw the various shapes from arcs, donuts, ellipses, polygons and 2D solid.

2 No sizes are given, so use your imagination.

3 Setting the grid to 10 and the snap to 5 may help.

Activity 4: A backgammon board

1 This is a nice simple drawing to complete.

2 The filled triangles are created with the 2D SOLID command.

Chapter 10

Layers

1   All the objects that have been drawn so far have had a continuous linetype and no attempt has been made to introduce centre or hidden lines, or even colour.

2   AutoCAD has a facility called LAYERS which allows the user to assign different linetypes and colours to named layers. For example, a layer may for red continuous lines, another may be for green hidden lines and yet another for blue centre lines.

3   Layers can be used for specific drawing purposes, e.g. there may be a layer for dimensions, one for hatching, one for text, etc.

4   Individual layers can be 'switched' on/off by the user to mask out drawing objects which are not required.

5   The concept of layers can be imagined as a series of transparent overlays, each having its own linetype, colour and use. The overlay used for dimensioning could be switched off without affecting the other layers.

6   Figure 10.1 demonstrates the layer concept with:
    a) Five layers used to create a simple component with each 'part' of the component created on 'it's own' layer
    b) The layers 'laid on top of each other'. The effect is that the user 'sees' one component.

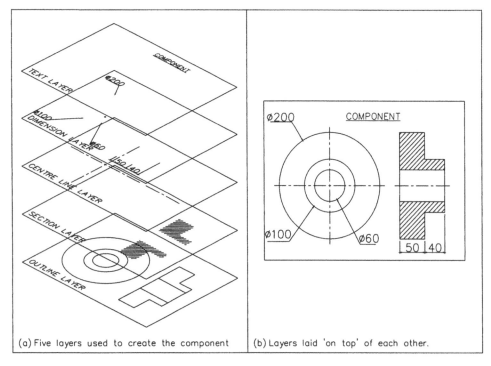

(a) Five layers used to create the component   (b) Layers laid 'on top' of each other.

Figure 10.1   Layer concepts.

7   The following points are worth noting when considering layers:
    a) all objects are drawn on layers
    b) layers should be used for each 'part' of a drawing, i.e. dimensions should not be on the same layer as centre lines (for example)
    c) new layers must be 'created' by the user, using the Layer Properties Manager dialogue box
    d) layers are one of the most important concept in AutoCAD
    e) layers are essential for good and efficient draughting.

8   *Notes*
    a) as layers are very important, and as the user must have a sound knowledge of how they are used, this chapter is therefore rather long (and perhaps boring). I make no apology for this, as all CAD operators must be able to use layers correctly
    b) I would recommend that this chapter be completed at 'the one sitting', if possible
    c) while layers were introduced to the user when we created the A3PAPER standard template file, no attempt was made to explain how they should be used.

Getting started

To demonstrate and investigate layers:

1   Open your C:\MYCAD\A3PAPER standard sheet and:
    a) layer OUT in red should be displayed in the Layers toolbar
    b) model space should be active.

2   Draw the following objects anywhere on the screen:
    a) a horizontal and vertical line each of length 200
    b) a circle of radius 75.

The Layer Properties Manager dialogue box

1   From the menu bar select **Format-Layer** and:
    prompt   `Layer Properties Manager dialogue box` similar to Figure 10.2
    respond   **Study the layout of the dialogue box and read the explanation following**
    Note: Much of the terminology in the explanation which follows may be entirely new to the reader, but should become apparent as the chapter progresses.

2   The Layer Properties Manager dialogue box has two distinct 'sections' these being:
    a) *The tree view pane*
        1. Displays a hierarchical list of layers and filters in the drawing
        2. The top node (**All**) displays all layers in the drawing
        3. Filters are displayed in alphabetical order
        4. A node is expanded to display a nested filter list
        5. The All Used Layers filter is read-only.
    b) *The list view pane*
        1. Displays layers and layer filters with their properties and descriptions
        2. If a layer filter is selected in the tree view, the list view displays only the layers in that layer filter
        3. The All filter in the tree view displays all layers and layer filters in the drawing
        4. When a layer property filter is selected and there are no layers that fit its definition, the list view is empty.

3   The icons displayed in the Layer Properties Manager dialogue box are:
    a) *New Property Filter icon*
        1. Displays the Layer Filter Properties dialogue box
        2. The user can create a layer filter based on one or more properties of the layers.

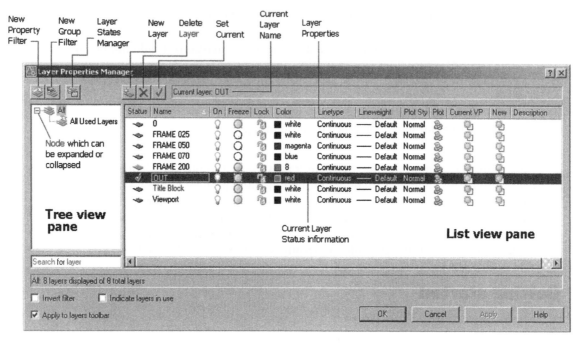

Figure 10.2   The Layer Properties Manager dialogue box.

b) *New Group Filter icon*
   1. Creates a layer filter created by the user.
c) *Layer States Manager icon*
   1. Displays the Layer States Manager
   2. The user can save the current property settings for layers in a named layer state and then restore those settings later.
d) *New Layer icon*
   1. Creates a new layer
   2. The list displays a layer named LAYER1
   3. The name is selected so that the user can enter a new layer name immediately
   4. The new layer inherits the properties of the currently selected layer in the layer list.
e) *Delete Layer icon*
   1. Marks selected layers for deletion
   2. Layers are deleted when the user clicks Apply or OK
   3. Only unreferenced layers can be deleted
   4. Referenced layers include layers 0 and DEFPOINTS, layers containing objects and the current layer.
f) *Set Current icon*
   1. Sets the selected layer as the current layer
   2. Objects that you create are drawn on the current layer.

4   Also displayed in the Layer Property Manager dialogue box are:
   a) *Current Layer name*
      1. Displays the name of the current layer
      2. This should be **OUT** at present.
   b) *Layer Properties information*:
      1. Status: indicates the item type, e.g. layer or filter
      2. Name: displays the layer or filter names. At present there are eight named layers
      3. On: turns highlighted layers on and off and yellow is on, blue is off
      4. Freeze: freezes or thaws a highlighted layer. Yellow is thawed, blue is frozen

5. Lock: locks or unlocks a highlighted layer. The icon display should be obvious to the user
6. Colour: allows the user to alter the colour of a highlighted layer
7. Linetype: allows the user to alter the linetype of a highlighted layer
8. Lineweight: allows the user to alter the lineweight of a highlighted layer
9. Plot Style: allows the plot style of highlighted layers to be altered
10. Plot: determines whether highlighted layers are plotted
11. Current VP: freezes selected layers from a layout tab only
12. New VP: freezes selected layers in new layout viewports from a layout tab only
13. Description: the user can enter a text description for the highlighted layer.

5   The eight layers displayed in the Layer Properties Manager dialogue box are:

Layer name	Comment
0	AutoCAD default layer – always available to the user
FRAME 025	Title block text details
FRAME 050	Title block text and drawing sheet grid marks
FRAME 070	Title block line spacers
FRANE 200	Drawing paper limits
OUT	Our created layer from the template exercise
Title Block	The actual title block layer
Viewport	The paper space viewport.

6   The user must realise that as we are using an AutoCAD template for our A3PAPER standard sheet, the named layers may seem meaningless at present. These layers will become obvious as we progress through the book.

7   All objects have so far, been created on layer OUT – the current layer. This is displayed in the Layers toolbar with the layer state icons at the top of the drawing area.

8   *Tasks*

    a) The layer OUT line should be highlighted in the Layer Properties Manager dialogue box. If it is not, move the pointing arrow onto name OUT and left-click

    b) Move the cursor along the highlighted area and pick the yellow On/Off icon and:

        prompt   AutoCAD layer message box – Figure 10.3
        respond   **pick No** from the message box
        and   icon now displayed in blue indicating off
        respond   **pick OK** from the Layer Properties Manager dialogue box
        and   the drawing screen will be returned and no objects are displayed. The lines and circle were drawn on layer OUT which has been turned off.

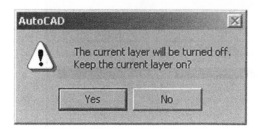

Figure 10.3   The AutoCAD current layer warning message box.

    c) Right-click the mouse button and pick **Repeat Layer** from the shortcut menu to re-activate the Layer Properties Manager dialogue box and:
      1. layer OUT highlighted in colour
      2. pick (left-click) the blue 'On' icon and it changes to yellow, i.e. layer is ON
      3. pick OK.

    d) The line and circle objects are again displayed as layer OUT is on.

Linetypes

AutoCAD allows the user to display objects with different linetypes, e.g. continuous, centre, hidden, dotted, etc. and all created objects have been displayed with a continuous linetype.

1 Activate the menu selection **Format-Layer** and:

prompt	Layer Properties Manager dialogue box
respond	**pick Continuous from layer OUT 'line'**
prompt	Select Linetype dialogue box
with	Loaded linetypes:

      Linetype   *Appearance*   *Description*
      Continuous   —————   Solid line

respond	**pick Load. . .**
prompt	Load or Reload Linetype dialogue box
with	*a*) Filename: acadiso.lin
	b) a list of all available linetypes in the named acadiso.lin file.
respond	1. scroll (at right) and pick CENTER2 – Figure 10.4(a)
	2. hold down the Ctrl key
	3. scroll and pick HIDDEN2
	4. pick OK from Load or Reload Linetypes dialogue box.
prompt	Select Linetype dialogue box – Figure 10.4(b)

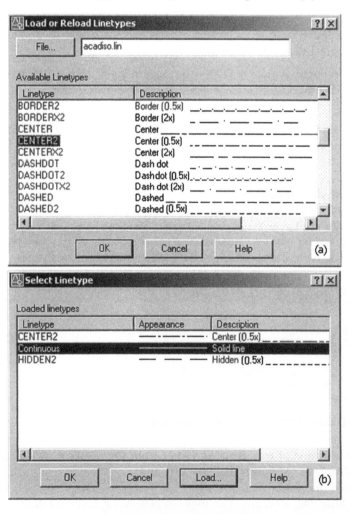

Figure 10.4 Loading linetypes: (a) selecting the CENTER2 linetype from the acadiso.lin file: (b) the CENTER2 and the HIDDEN2 linetype loaded into the current drawing.

with   Loaded Linetypes:

Linetype	Appearance	Description	
CENTER2	— — — – —	Center(0.5x)	— — — — — —
Continuous	——————	Solid line	
HIDDEN2	— — — —	Hidden(0.5x)	— — — — — —

Note: we have loaded the CENTER2 and HIDDEN linetypes from the acadiso.lin file into the Layer Properties Manager dialogue box and they can now be used in the current drawing

respond   1. pick CENTER2 and it becomes highlighted
          2. pick OK from Select Linetype dialogue box
prompt   `Layer Properties Manager dialogue box`
with   layer OUT having CENTER2 linetype
respond   pick OK from Layer Properties Manager dialogue box.

2   The drawing screen will display the three objects with a center linetype appearance. Remember that AutoCAD is an American package – hence center, and not centre.

3   *Note*: Although we have used the Layer Properties Manager dialogue box to 'load' the CENTER2 and HIDDEN2 linetypes, linetypes can also to loaded by selecting from the menu bar **Format-Linetype** and the Linetype Manager dialogue box will be displayed. The required linetypes are loaded 'into the current drawing' in the same manner as described.

Colour

Individual objects can be displayed on the screen in different colours, but I prefer to use layers for the colour effect. This is achieved by assigning a specific colour to a named layer and will be discussed later in the chapter. For this exercise:

1   Activate the Layer Properties Manager dialogue box. Layer OUT should be highlighted.

2   Left-click on 'red' and:
prompt   `Select Color dialogue box`
respond   1. pick the green square from the standard color bar
          2. pick OK

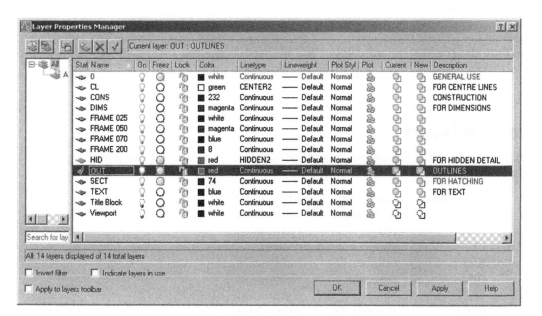

Figure 10.5   The Layer Properties Manager dialogue box after the new layer's have been 'made'.

prompt   Layer Properties Manager dialogue box
with   highlighted layer OUT line having color green – name and coloured square
respond   pick OK.

3   The screen will display the three objects with green centre lines.

4   Note the toolbars below the Standard toolbar:
    a) the Layer toolbar displays layer OUT with a green square
    b) the Properties toolbar displays:
        1. Colour Control: ByLayer is a green square
        2. Linetype Control: ByLayer is a center linetype.

5   *Tasks*
    By selecting the Layer Properties Manager icon from the Layers toolbar, use the Layer
    Properties Manager dialogue box to:
    a) set layer OUT linetype to Continuous
    b) set layer OUT colour to Red
    c) screen should display the original continuous linetype objects
    d) note the Layers and Properties toolbar.

6   *Note on colour*
    a) The default AutoCAD colour for objects is dependent on your screen configuration,
        but is generally:
        either   1. white background with black lines
        or   2. black background with white lines
    b) The white/black linetype can be confusing to new AutoCAD users
    c) All colours in AutoCAD are numbered. There are 255 colours available, but the
        seven standard colours are:
        1 red, 2 yellow, 3 green, 4 cyan, 5 blue, 6 magenta, 7 black/white
    d) The number or colour name can be used to select a colour. The numbers are asso-
        ciated with colour pen plotters
    e) The complete 'colour palette' can be activated:
        1. from the Layer Program Manager dialogue box by picking the colour square/name
            under Colour
        2. by selecting **Format-Color** from the menu bar
    f) The Select Color dialogue box allows the user further colour options, these being:
        1. Index Color: default
        2. True Color
        3. Colour Books
    g) Generally only the seven standard colours will be used in the but there may be
        the odd occasion when a colour from the default Index Colour palette will be
        selected.

Creating new layers

Layers should be made to suit individual or company requirements, but for our purposes the layers which will be made for all future drawing work are:

Usage	Layer name	Layer colour	Layer linetype
General	0	white/black	continuous
Outlines	OUT	red	continuous
Centre lines	CL	green	center2
Hidden detail	HID	number 12	hidden2
Dimensions	DIMS	magenta	continuous
Text	TEXT	blue	continuous
Hatching	SECT	number 74	continuous
Construction	CONS	to suit	continuous

New layers must be 'made' by the user. Layer 0 is the AutoCAD default layer and layer OUT was made in a previous chapter. This layer was created from the command line, but new layers can be created using the Layer Properties Manager dialogue box. This is the method we will now use, so:

1 Continue with the A3PAPER standard sheet displayed and erase the three objects.

2 Menu bar with **Format-Layer** and:

    prompt   Layer Properties Manager dialogue box
    with   1. eight listed layers
                2. layer OUT current
    respond   **pick the New Layer icon**
    and   Layer 1 added to layer list with the same properties as layer OUT, i.e. red colour and continuous linetype
    respond   **pick New another five times** until there are 14 listed layers in total:
                a) the original default layer 0
                b) the six AutoCAD template layers
                c) our previously created layer OUT
                d) the added new layers, named layer1–layer6, all colour red and continuous linetype.

3 *Naming the new layers*
    a) move pick arrow onto the **Layer1 name**, pick it and:
        1. it becomes highlighted
        2. a box effect is placed around the Layer1 name
    b) press the **F2 key**
    c) enter CL from the command line
    d) repeating steps (a)–(c), rename layer2–layer6 as follows:
        layer2: HID, layer3: DIMS, layer4: TEXT, layer5: SECT, layer6: CONS.

4 *Assigning linetypes to the new layers*
    a) pick the layer CL
    b) move pick arrow to **Linetype-Continuous** and pick it
        prompt   Select Linetype dialogue box
        with   CENTER2, Continuous, HIDDEN2 linetype names
        respond 1. pick CENTER2
               2. pick OK
        prompt   Layer Properties Manager dialogue box
        with   layer CL with CENTER2 linetype
    c) pick the layer HID
    d) move pick arrow to Linetype-Continuous and pick it
        prompt   Select Linetype dialogue box
        with   CENTER2, Continuous, HIDDEN2 linetype names
        respond 1. pick HIDDEN2
               2. pick OK
        prompt   Layer Properties Manager dialogue box
        with   layer HID with HIDDEN2 linetype.
    e) the remaining layers have all to have continuous linetype.

5 *Assigning colours to the new layers*
    a) highlight layer HID if not still highlighted
    b) move pick arrow to **Color-red** and pick it with a left-click
        prompt   Select Color dialogue box
        respond 1. at Color: red, enter 12
               2. pick OK
        prompt   Layer Properties Manager dialogue box
        with   layer HID with color number 12

    c) pick the other named layer lines and set the following colours:
      CL: green, DIMS: magenta, TEXT: blue, SECT: number 74, CONS: your own colour
      selection.

6   Adding a layer description
    a) select the layer 0 line
    b) move the arrow and pick under description to display a box effect
    c) press **F2** and enter **GENERAL USE** from the keyboard
    d) pick the other layer lines and enter a layer description based on the layer usage
    e) pick OK.

The current layer

1   The current layer is the one on which all objects are drawn.

2   The current layer name appears in the Layers toolbar which is generally docked
    below the Standard toolbar at the top of the screen.

3   The current layer is also named in the Layer Properties Manager dialogue box when
    it is activated and is displayed with a green status tick.

4   The current layer is 'set' by the user.

5   Activate the Layer Properties Manager dialogue box and:
    prompt   `Layer Properties Manager dialogue box`
    with   the layers created earlier displayed in numeric then alphabetical order
              i.e. 0, CL, CONS . . . TEXT, Title Block, Viewport as Figure 10.5
    respond   1. pick layer line CL – becomes highlighted
              2. pick the Set Current icon and the green status pick is placed at the CL
                 line layer
              3. pick OK.

6   The drawing screen will be returned, and the Layers toolbar will display CL with a
    green box and ByLayer will also display a green box.

7   The current layer can be set:
    a) from the Layer Properties Manager dialogue box as described in step 5
    b) from the Layers toolbar by scrolling at the right of the layer name and selecting
        the name of the layer to be current.

8   I generally start a drawing with layer OUT current, but this is a personal preference.
    Other users may want to start with layer CL or 0 as the current layer, but it does not
    matter, as long as the objects are eventually 'placed' on their correct layers.

9   Having created layers it is now possible to draw objects with different colours and
    linetypes, simply by altering the current layer. All future work should be completed
    with layers used correctly, i.e. if text is to be added to a drawing, then the TEXT layer
    should be current.

10 Now set OUT as the current layer.

Saving the layers to A3PAPER (template and drawing)

1 Ensure that layer OUT is current.

2 Erase the two lines and circle.

3 Menu bar with **File-Save As** and:
   prompt   Save Drawing As dialogue box
   respond 1. Files of type: scroll and pick AutoCAD Drawing Template (*.dwt)
              2. Save in: scroll and pick C:\MYCAD folder
              3. File name: A3PAPER – enter or pick
              4. pick Save.
   prompt   C:\MYCAD\A3PAPER.dwt already exists message
   respond **pick Yes**
   prompt   Template Description dialogue box
   enter   **text to suit then pick OK**.

4 Menu bar with **File-Save As** and:
   prompt   Save Drawing As dialogue box
   respond 1. Files of type: scroll and pick AutoCAD 2004 Drawing (*.dwg)
              2. Save in: scroll and pick C:\MYCAD folder
              3. File name: A3PAPER – enter or pick
              4. pick Save.
   prompt   C:\MYCAD\A3PAPER.dwt already exists message
   respond **pick Yes**.

5 The A3PAPER standard sheet has now been saved as a template file and a drawing file with:
   a) units set to metric
   b) sheet size A3
   c) grid, snap, etc. set as required
   d) several new layers
   e) a border and title box.

6 With the layers having been saved to the A3PAPER standard sheet, the layer creation process does not need to be undertaken every time a drawing is started. Additional layers can be added to the standard sheet at any time – the process is fairly easy?

Layer states

1 Layers can have different 'states', e.g., they could be:
   a) ON or OFF
   b) THAWED or FROZEN
   c) LOCKED or UNLOCKED.

2 The layer states are displayed both in the Layer Control box of the Properties toolbar as well as in the Layer Properties Manager dialogue box itself. In both, the layer states are displayed in icon form. These icon states are:
   a) yellow – ON, THAWED
   b) bluey grey – OFF, FROZEN
   c) lock and unlock should be obvious?

To demonstrate layer states:

1 Your A3PAPER standard sheet should be displayed with layer OUT current.

2 Using the seven created layers and layer 0:
   a) make each layer current in turn
   b) draw a 25-radius circle on each layer anywhere within the paper border
   c) make layer 0 current
   d) toggle the grid off to display eight coloured circles.

3 The green circle will be displayed with centre linetype and the hidden linetype circle displayed with colour number 12.

4   Activate the Layer Properties Manager dialogue box and:
    a) pick the yellow On icon of the CL layer line and:
        1. the line becomes highlighted
        2. the icon changes to blue.
    b) pick OK and the drawing screen will be returned with no green circle displayed as the CL layer has been turned off.

5   Layer Properties Manager dialogue box and:
    a) pick the yellow Freeze icon of DIMS layer line and:
        1. line becomes highlighted
        2. icon changes to blue.
    b) pick OK and drawing screen will display no magenta circle as the DIMS layer has been frozen.

6   Pick the scroll arrow from the Layers toolbar and:
    a) note CL and DIMS display information!
    b) pick Lock/Unlock icon on HID layer – note icon appearance
    c) left-click to side of pull-down menu
    d) hidden linetype circle still displayed although the HID layer has been locked.

7   From the Layers toolbar pull-down menu:
    a) pick the Freeze and Lock icons for the SECT layer
    b) left-click to side and no dark green circle, as layer SECT was frozen.

8   Using the pull down layer menu effect:
    a) turn off, freeze and lock the TEXT layer
    b) no blue circle as layer TEXT was turned off and frozen.

9   Activate the Layer Properties Manager dialogue box and:
    a) note the icon display for all the layers
    b) make layer OUT current
    c) pick Freeze icon for layer OUT
    d) Warning message – **Cannot freeze the current layer**
    e) pick OK
    f) turn layer OUT OFF
    g) Warning message:
        The current layer will be turned off
        Keep the current layer on?
    h) pick **No** from this message dialogue box
    i) pick OK from Layer Properties Manager dialogue box
    j) no red circle displayed as layer OUT has been turned off although it is the current layer.

10 The drawing area should now display:
    a) a black circle – drawn on layer 0
    b) a hidden linetype circle (colour 12) – on frozen layer HID
    c) a circle on layer CONS coloured to your own selection.

11 **Thus objects will not be displayed if their layer is OFF or FROZEN.**

12 Make layer 0 current.

13 *a*) erase the hidden linetype circle – you cannot and the prompt line displays:
        1 was on a locked layer
    b) cancel the Erase command.

14 *a*) Using the CEN object snap, draw a line from the centre of the hidden linetype circle to the centre of the black circle, i.e. you can reference the hidden linetype circle, although it is on a locked layer.
    b) erase the line.

15 *a*) make layer OUT current
    b) Warning message: **The current layer is turned off**
    c) pick OK
    c) draw a line from 50,50 to 200,200
    d) no line is displayed and the reason should be obvious?

16 Erase the coloured circle on layer CONS then activate the Layer Properties dialogue box.

17 Pick the **New Property Filter icon** and:
    prompt   Layer Filter Properties dialogue box
    respond   1. alter filter name to **TRY1**
                  2. pick the box at Status, scroll and select the layer in use (bluey) icon
                  3. pick the box at Name and * displayed – wildcard for all layer names
                  4. pick the box at On, scroll and select the On icon – dialogue box as Figure 10.6(a)
                  5. pick OK
    prompt   Layer Properties Manager dialogue box (part displayed in Figure 10.6(b)
    with   *a*) TRY1 added to the tree hierarchy side and only the filtered layers displayed which include the AutoCAD FRAME layers, etc.
                  b) Title block highlighted on the list side of the dialogue box – any idea why?

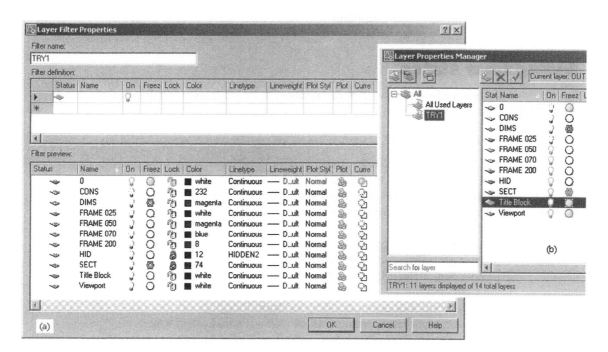

Figure 10.6   (a) The Layer Filter Properties dialogue box for TRY1: (b) The Layer Properties Manager dialogue box with TRY1 added.

18 Using the tree side of the Layer Properties Manager dialogue box, and excluding the AutoCAD layers:
    a) TRY1 is 'current,' i.e. highlighted with our five layers displayed
    b) pick All Used Layers and note the layers displayed
    c) pick All – any difference from (b)?
    d) collapse the tree hierarchy by picking the (−) at All
    e) expand the tree side by picking the (+) to display the other options
    f) pick the TRY1 filter to display the previous filtered layers

g) pick on the **Search for layer** line and:
  prompt   `*` displayed
  respond   **pick to right of * and backspace**
  and   `*` disappears and no layers displayed
  respond   1. enter OUT and no layer displayed – why?
              2. pick to right of OUT and backspace to remove the name
              3. enter HID and hidden line details displayed
              4. pick All from tree side then OK from dialogue box
  prompt   drawing screen with two circles and a line.

19 Using the Layer Properties Manager dialogue box:
  a) ensure all layers are: ON, THAWED and UNLOCKED
      1. On – three layers were turned off
      2. Thaw – three layers were frozen
      3. Unlock – three layers were locked
  b) pick OK
  c) 7 circles and 1 line displayed?
  d) remember that the CONS layer circle was erased.

20 The layer states can be activated from:
  a) the Layer Properties Manager dialogue box
  b) the Layers toolbar pull down menu.

21 This rather long exercise is now complete.

Saving layer states

Layer states can be saved so that the user can restore the original layer 'settings' at any time. To demonstrate this:

1 Erase the line to leave the circles.

2 At the command line enter LAYER <R> and:
  prompt   `Layer Properties Manager dialogue box`
  respond   **pick the Layer States Manager icon** from the tree side
  prompt   `Layer States Manager dialogue box`
  respond   **pick New**
  prompt   `New layer State to Save dialogue box`
  respond   1. New layer state name: enter **A3PAPER**
              2. Description: enter My Layer states for an A3-sized paper
              3. pick OK.
  prompt   `Layer States Manager dialogue box`
  respond   1. Layer setting to restore: make active (tick) the following states:
                  On/Off; Frozen/Thawed; Locked/Unlocked; Color; Linetype; Lineweight
              2. pick Close.
  prompt   `Layer Properties Manager dialogue box`
  respond   **pick OK**
  and   drawing screen returned with the 7 circles.

3 Scroll at layer information from the Layers toolbar and:
  a) turn off and lock all layers
  b) message about current layer displayed – pick OK
  c) screen display is blank – obviously?

4 Pick the **Layers Properties Manager icon** from the Layers toolbar and:
  prompt   `Layer Properties Manager dialogue box`
  with   all layers with off and locked icons

respond	**pick the Layer States Manager icon**
prompt	Layer States Manager dialogue box
respond	1. pick A3PAPER (probably highlighted)
	2. pick Restore
prompt	Layer Properties Manager dialogue box
with	all layers with on and unlocked icons
respond	**pick OK**.

5 The original screen objects will be displayed (I needed a screen regen).

6 *a*) It may be useful to have the saved A3PAPER layer states added to the A3PAPER standard sheet

   b) I will let you decide this for yourself.

The Layers toolbar

The Layers toolbar is generally displayed at all times and is usually docked below the Standard toolbar at the top of the drawing screen under the title bar. This toolbar gives information about layers and their states, and Figure 10.7 displays the toolbar details for our current drawing.

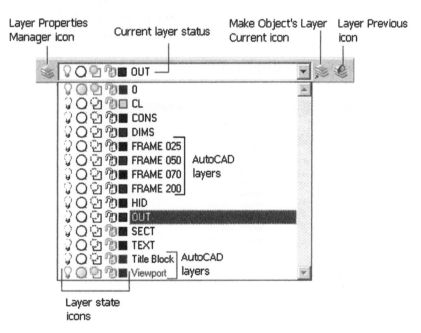

Figure 10.7 The Layers Properties toolbar.

The various 'parts' of the toolbar are:

a) *Layer Properties Manager icon*
Selecting this icon will display the Layer Properties Manager dialogue box. It is quicker alternative to the menu bar sequence **Format-Layers**.

b) *Make Object's Layer Current icon*
Select this icon and:

prompt	Select object whose layer will become current
respond	**pick the blue circle**
prompt	TEXT is now the current layer
and	**blue square and TEXT displayed in Layers toolbar**.

c) *Layer Previous icon*
Select this icon and:
prompt   Restored previous layer status
and   **Layer OUT will again be current**.

d) *Pull down layer information*
Allows the user to quickly set a new current layer, or activate one of the layer states, e.g. off, freeze, lock, etc.

The Properties toolbar

The Properties toolbar is also generally displayed at all times and is docked beside the Layers toolbar. Figure 10.8 displays the Properties toolbar for our A3PAPER drawing at this stage. The three main options of the Properties toolbar are:

a) *Color Control*
1. By scrolling at the colour control arrow, the user can select one of the standard colours.
2. By picking Select Color from the pull down menu, the Select Color dialogue box is displayed allowing access to all colours available in AutoCAD.
3. Selecting a colour by this method will allow objects to be drawn with that particular colour, irrespective of the current layer colour. This is **not recommended** but may be useful on certain occasions, i.e. coloured objects should (ideally) be created on their own layer.
4. It is recommended that **Color Control displays ByLayer**.

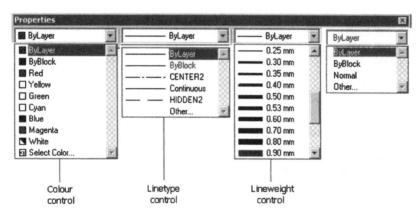

Figure 10.8   The Properties toolbar.

b) *Linetype Control*
1. Similar to colour control, the scroll arrow display all linetypes loaded in the current drawing
2. By picking Other the Linetype Manager dialogue box will be displayed, allowing the user to load any other linetype from a named file – usually **acadiso.lin**
3. By selecting a linetype from the pull down menu, the user can create objects with this linetype, independent of the current layer linetype. **This is not recommended**
4. Layers should be used to display a certain linetype, but it may be useful to have different linetypes displayed on the one layer occasionally
5. It is recommended that **Linetype Control displays ByLayer**.

c) *Lineweight Control*
1. Lineweight allows objects to be displayed (and plotted) with different thicknesses from 0 mm to 2.11 mm

2. The lineweight control scroll arrow allows the user to select the required thickness

3. We will be investigating this topic in greater detail in a later chapter, and will not discuss at this stage

4. **Lineweight Control should display ByLayer**.

d) *Plot Style Control*
Not discussed at this stage.

Renaming and deleting layers

Unwanted or wrongly named layers can easily be deleted or renamed in AutoCAD 2006. To investigate how this is achieved:

1 The A3PAPER drawing should still be displayed with 7 coloured circles (the circle on layer CONS was erased).

2 Erase any lines then activate the Layer Properties Manager dialogue box by picking the Layer icon from the Layers toolbar and:
 a) pick the New Layer icon twice to add two new layers to the list – Layer1 and Layer2
 b) rename Layer1 as NEW1 and Layer2 as NEW2
 c) pick OK from the Layer Properties Manager dialogue box.

3 Making each new layer current in turn, draw a circle anywhere on the screen.

4 Make layer OUT current.

5 *a*) activate the Layer Properties Manager dialogue box and:
   respond   pick NEW1 layer line then the Delete icon
   prompt   **AutoCAD message – The selected layer was not deleted**
 b) pick OK from this message box then pick OK from the Layer Properties Manager dialogue box.

8 *a*) erase the two added circles
 b) activate the Layer Properties Manager dialogue box
 c) pick NEW1 then Delete icon
 d) pick NEW2 then Delete icon
 e) an X is placed at Status for the NEW1 and NEW2 layers
 f) pick OK.

9 Re-activate the Layer Properties Manager dialogue box and layers NEW1 and NEW2 will not be listed.

This completes the chapter on layers. You should have the confidence and ability to create line and circle objects by various methods, e.g. co-ordinate entry, referencing existing objects dynamic input, etc. We are now ready to create a working drawing which will be used to introduce several new concepts, but as it is some time since you last completed a drawing on your own, I have included an assignment which is simple but interesting to complete.

Assignment

1 Ensure that your A3PAPER has been saved with the named layers from this chapter.

2 Complete activity 5 which requires you to:
 a) open your A3PAPER drawing
 b) complete the assignment
 c) save the completed work as C:\MYCAD\ACT5.

Chapter 11

Creating a working drawing

We will now create a drawing which will be used to investigate several new topics, e.g. using the modify commands, adding text and dimensions, etc. This working drawing will be created from lines and circles and is relatively easy to complete, so:

1 Open your A3PAPER standard file – drawing or template.

2 Refer to Figure 11.1 and:
   a) draw full size the component given
   b) ensure that layer OUT is current
   c) a start point is given – **USE THIS START POINT AS IT IS IMPORTANT FOR FUTURE WORK**
   d) *do not attempt to add the dimensions*

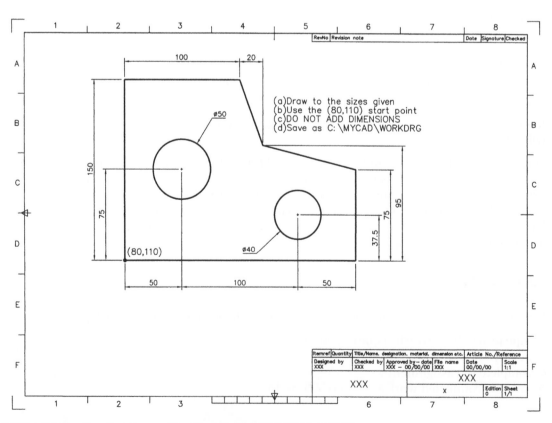

Figure 11.1   Creating the working drawing C:\MYCAD\WORKDRG.

e) use any entry method suitable but I would suggest:
1. absolute co-ordinates for the (80,110) start point then relative co-ordinates for the outline
2. absolute co-ordinates for the circle centres (some 'sums' needed).

3   When the drawing is complete, menu bar with **File-Save As** and:
prompt   Save Drawing As dialogue box
with   File name: A3PAPER
respond 1. ensure C:\MYCAD is current folder to Save in
           2. alter file name to: WORKDRG
           3. pick Save.

4   We have now opened our A3PAPER file (drawing or template), completed a drawing exercise and saved this drawing with a different name to that which was opened.

5   This is (at present) the method which will be used to complete all new drawing exercises.

6   Now continue to the next chapter.

7   *Note*: It should be apparent that the circles will be drawn without centre lines. This is deliberate, as it will allow us to investigate other CAD draughting techniques in a later chapter.

Chapter 12

Using the Modify commands 1

1   In this chapter we will investigate several modify commands which can be used to alter the appearance of and existing drawing.

2   The commands which will be investigated are fillet, chamfer, offset, extend, trim and change.

3   These commands can all be activated by icon and menu bar selection or by keyboard entry.

Getting started

1   Open the C:\MYCAD\WORKDRG created in the previous chapter or simply continue from the previous chapter.

2   Ensure model space is active, layer OUT is current and display the Draw and Modify toolbars.

3   Refer to Figure 12.1.

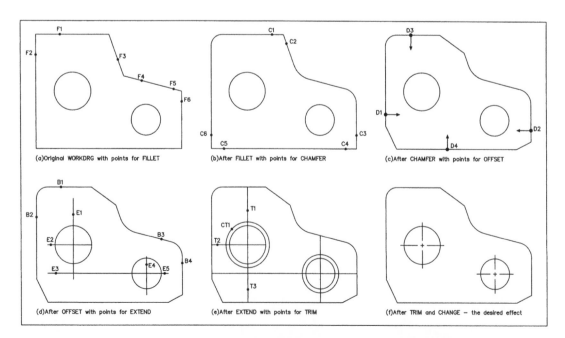

Figure 12.1   WORKDRG with the fillet, chamfer, offset, extend, trim and change commands.

Fillet

1   A fillet is a radius added to existing line/arc/circle objects.

2   The user:
    a) specifies the fillet radius
    b) selects the two objects which have to be filleted.

3   Select the **FILLET icon** from the Modify toolbar and:
    prompt   Current settings: MODE = TRIM, Radius = ??
                Select first object or [Undo/Polyline/Radius/Trim/
                Multiple]
    enter   **R <R>** – the radius option
    prompt   Specify fillet radius<??>
    enter   **15 <R>**
    prompt   Select first object or [Undo/Polyline/Radius/Trim/
                Multiple]
    respond   **pick line F1** – Figure 12.1(a)
    prompt   Specify second object
    respond   **pick line F2**.

4   The corner selected will be filleted with a radius of 15, the two 'unwanted line por-
    tions' will be erased and the command line will be returned.

5   From the menu bar select **Modify-Fillet** and:
    prompt   Current settings: MODE = TRIM, Radius = 15
                Select first object or [Polyline/Radius/Trim/mUltiple]
    enter   **R <R>** – the radius option
    prompt   Specify fillet radius<15.00>
    enter   **30 <R>**
    prompt   Select first object and **pick line F3**
    prompt   Select second object and **pick line F4**.

6   At the command line enter **FILLET <R>** and:
    a) set the fillet radius to 20
    b) fillet the corner picking the F5 and F6 lines.

7   The result of the fillet operations is displayed in Figure 12.1(b).

Chamfer

1   A chamfer is a straight 'cut corner' added to existing line objects.

2   The chamfer distances must be 'set' prior to selecting the object to be chamfered.

3   Select the CHAMFER icon from the Modify toolbar and:
    prompt   (TRIM mode) Current chamfer Dist1 = ?? Dist2 = ??
                Select first line or [Undo/Polyline/Distance/Angle/Trim/
                mEthod/Multiple]
    enter   **D <R>** – the Distance option
    prompt   Specify first chamfer distance<??>
    enter   **25 <R>**
    prompt   Specify second chamfer distance<25.00>
    enter   **25 <R>**
    prompt   Select first line and **pick line C1** – Figure 12.1(b)
    prompt   Select second line and **pick line C2**.

4   The selected corner will be chamfered, the unwanted line portions removed and the command line returned.

5   Menu bar selection with **Modify-Chamfer** and:

    prompt   (TRIM mode) Current chamfer Dist1 = 25, Dist2 = 25
               Select first line or [Undo/Polyline/Distance/Angle/Trim/
               mEthod/Multiple]
    enter   **D <R>**
    prompt   Specify first chamfer distance and enter: **10 <R>**
    prompt   Specify second chamfer distance and enter: **20 <R>**
    prompt   Select first line and **pick line C3**
    prompt   Select second line and **pick line C4**.

6   Note that the pick order is important when the chamfer distances are different. The first line picked will have the first set chamfer distance.

7   At the command line enter **CHAMFER <R>** and:
    a) set first chamfer distance: 15
    b) set second chamfer distance: 30
    c) chamfer the corner picking lines C5 and C6.

8   Figure 12.1(c) displays the result of the chamfer operations.

Offset

1   The offset command allows to user to draw objects parallel to other selected objects.

2   Lines, circles and arcs can all be offset.

3   The user specifies:
    a) an offset distance
    b) the side to offset of the selected object.

4   Refer to Figure 12.1(c), select the **OFFSET icon** from the Modify toolbar and:

    prompt   Specify offset distance or [Through/Erase/Layer]
    enter   **50 <R>** – the offset distance
    prompt   Select object to offset or <Exit/Undo>
    respond   **pick line D1**
    prompt   Specify point on side to offset or [Exit/Multiple/Undo]
    respond   **pick any point to right of line D1 as indicated**
    and   line D1 will be offset by 50 units to right
    prompt   Select object to offset, i.e. any more 50 offsets
    respond   **pick line D2**
    prompt   Specify a point on side to offset
    respond   **pick any point to left of line D2 as indicated**
    and   line D2 will be offset 50 units to left
    prompt   Select object to offset, i.e. any more 50 offsets
    enter   **E <R>** to exit (end) the command.

5   Menu bar with **Modify-Offset** and:

    prompt   Specify offset distance or [Through/Erase/Layer]<50.00>
    enter   **75 <R>**
    prompt   Select object to offset and **pick line D3**
    prompt   Specify a point on side to offset and **pick as indicated**
    then   right-click and pick Enter to end the command.

6   At the command line enter **OFFSET <R>** and:
    a) set an offset distance of 37.5
    b) offset line D4 as indicated.

7 We have now created lines through the two circle centres as Figure 12.1(d). Later in the chapter we will investigate how these lines can be modified to be 'real centre lines'.

Extend

1 This command will extend an object 'to a boundary edge', the user specifying:
   a) the actual boundary – an object
   b) the object which has to be extended.

2 Refer to Figure 12.1(d) and with SNAP OFF, select the EXTEND icon from the Modify toolbar and:
prompt	`Current settings: Projection = UCS, Edge = Extend`
	`Select boundary edges . . .`
	`Select objects or <select all>`
respond	**pick line B1**
prompt	`1 found`
	`Select objects, i.e. any more boundary edges`
respond	**pick line B2**
prompt	`1 found, 2 total`
	`Select objects`
respond	**right-click to end boundary edge selection**
prompt	`Select object to extend or shift-select to trim or [Fence/Crossing/Project/Edge/Undo]`
respond	**pick lines E1, E2 and E3 then right-click-Enter**
and	the three lines will be extended to the selected boundary edges.

3 From the menu bar select **Modify-Extend** and:
prompt	`Select objects, i.e. the boundary edges`
respond	**pick lines B3 and B4 then right-click**
prompt	`Select object to extend`
respond	**pick lines E4 and E5 then right-click and pick Enter.**

4 At the command line enter **EXTEND <R>** and extend the two vertical 'centre lines' to the lower horizontal outline.

5 When complete, the drawing should resemble Figure 12.1(e).

Trim

1 Allows the user to trim an object 'at a cutting edge', the user specifying:
   a) the cutting edge – an object
   b) the object to be trimmed.

2 Refer to Figure 12.1(e) and OFFSET the two circles for a distance of 5 'outwards' – easy?

3 Extend the top horizontal 'circle centre line' to the offset circle – should be obvious why?

4 Select the **TRIM icon** from the Modify toolbar and:
prompt	`Current settings: Projection = UCS, Edge = Extend`
	`Select cutting edges . . .`
	`Select objects or <select all>`
respond	**pick circle CT1 then right-click**
prompt	`Select object to trim or shift-select to extend or [Fence/Crossing/Project/Edge/erase/Undo]`
respond	**pick lines T1, T2 and T3 then right-click-Enter.**

5 From menu bar select **Modify-Trim** and:
   prompt   Select objects, i.e. the cutting edge
   respond **pick the other offset circle then right-click**
   prompt   Select object to trim
   respond **pick the four circle 'centre lines' then right-click**.

6 *a*) erase the two offset circles
   b) now have 'neat lines' through the circle centres as Figure 12.1(f)
   c) select the Save icon from the Standard toolbar to automatically update
       C:\MYCAD\WORKDRG.

Changing the offset centre lines

1 The WORKDRG drawing was saved with 'centre lines' obtained using the offset,
   extend and trim commands.

2 These lines pass through the two circle centres, but they are continuous lines and not
   centre lines.

3 We will now modify these lines to be centre lines using the CHANGE command.
   Changing the properties of an object will be fully investigated in a later chapter.

4 Continue with WORKDRG (or re-open if necessary).

5 At the command line enter **CHANGE <R>** and:
   prompt   Select objects
   respond **pick the four offset centre lines then right-click**
   prompt   Specify change point or [Properties]
   enter   **P <R>** – the properties option
   prompt   Enter property to change [Color/Elev/Layer/LType/
              ltScale/LWeight/Thickness/PLotstyle]
   enter   **LA <R>** – the layer option
   prompt   Enter new layer name<OUT>
   enter   **CL <R>**
   prompt   Enter property to change, i.e. any more changes?
   respond **right-click and Enter** to end command.

6 *a*) the four selected lines will be displayed as green centre lines, as they were changed
       to the CL layer
   b) this layer was made with centre linetype and colour green.

7 Although the changed lines are centre lines, their 'appearance' may not be ideal and
   an additional command is required to 'optimise' the centre line effect. This command
   is LTSCALE.

LTSCALE

1 LTSCALE is a system variable used to 'alter the appearance' of non-continuous lines
   on the screen.

2 It has a default value of 1.0.

3 The value is altered by the user to 'optimise' centre lines, hidden lines, etc.

4 At the command line enter **LTSCALE <R>** and:
   prompt   Enter new linetype scale factor<1.0000>
   enter   **0.6 <R>**.

5 The four centre lines may be 'better defined' for the user?

6 The value entered for LTSCALE depends on the type of lines being used in a drawing, and can be further refined – more on this in a later chapter.

7 *a*) investigate other LTSCALE values
   b) optimise the centre lines to your own requirements
   c) save the drawing as **C:\MYCAD\WORKDRG**, updating the original.

8 *Notes*
   a) LTSCALE must be entered from the command line.
   b) The LTSCALE system variable is **GLOBAL**. This means that when it's value is altered, all linetypes (centre, hidden, etc.) will automatically be altered to the new value. In a later chapter we will discover how this can be 'overcome'.
   c) The actual LTSCALE value entered varies greatly and some users may find that the 0.6 value is totally unsuitable and that a larger value (e.g. 12, 15, 25) is more suited to display the centre lines with a reasonable definition. This difference in the LTSCALE value is dependant on other factors which will not be discussed at this stage.

The command options

The fillet, chamfer and offset commands have options and which we will now investigate, so:

1 Begin a new metric drawing from scratch and refer to Figure 12.2.

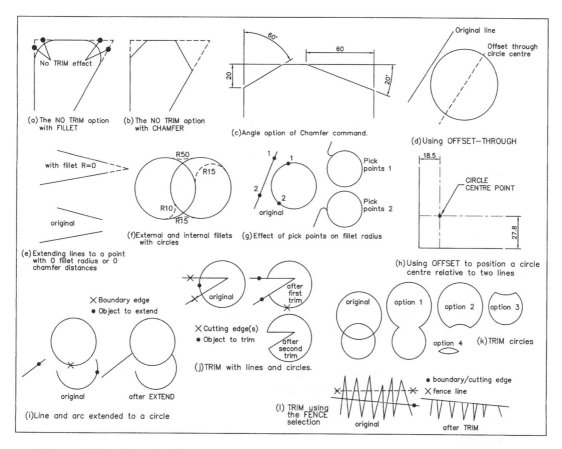

Figure 12.2 Using the fillet, chamfer, offset options and the trim and extend commands.

2　Fillet options: undo, polyline, radius, trim, multiple

a) Radius

The most common option and has been used several times

b) Polyline

Will be investigated during the polyline chapter

c) Trim

Allows the user to display the original 'fillet lines' if required. The option is obtained by entering **T <R>** at the prompt line after setting the radius value. The response is:

prompt	`Enter trim mode option [Trim/No Trim]<Trim>`
enter	*a*) **T <R>**: corner removed. This is the default
	b) **N <R>**: corners not removed
and	if the no trim (N) option is used, this effect will always be obtained until the user 'resets' the trim (T) option – Figure 12.2(a)

d) Multiple

Allows the command to be used repetitively with the same radius value. The user enters **M <R>** at the prompt line after the radius has be set

e) Undo

Will undo the last fillet operation if the multiple option is being used.

3　Chamfer options: undo, polyline, distance, angle, trim, method, multiple

a) Distance

The most common option and has been used several times

b) Polyline

As with the fillet command, will be discussed in a later chapter

c) Trim

Identical in operation to the fillet option – Figure 12.2(b)

d) Multiple and Undo

Identical in operation to the fillet option

e) Angle

Allows the user to specify a length and an angle from a selected line. When the chamfer command is activated, the response is:

prompt	`Select first line or [Undo/Polyline/Distance/Angle/Trim/mEthod/Multiple]`
enter	**A <R>** – the angle option
prompt	`Specify chamfer length on the first line`
enter	**20<R>**
prompt	`Specify chamfer angle from the first line`
enter	**60<R>**
and	the required lines can now be chamfered. Figure 12.2(c) displays:
	1. a length of 20 and angle of 60
	2. a length of 60 and angle of 20
note	the first selected line is used for the distance.

f) Method

Allows the user to toggle between the distance and the angle options. When **E <R>** is entered:

prompt	**Enter trim method [Distance/Angle]**
and	1. entering D sets the two distance method (default)
	2. entering A sets the length and angle method.

4　Error messages

The fillet and chamfer commands are generally used without any problems, but the following error messages (which should be self-evident to the user) may be displayed at the command prompt:

a) Radius is too large

b) Distance is too large

c) Chamfer requires two lines (not arc segments)

d) No valid fillet with radius??

e) Lines are parallel – this is a chamfer error

f) Cannot fillet an entity with itself.

5   Offset options: through, erase, layer

a) Through

A very useful option, as it allows an object to be offset through a specified point. When the offset command is activated, the response is:

prompt	`Specify offset distance or [Through/Erase/Layer]`
enter	**T <R>** – the through option
prompt	`Select object to offset or <Exit/Undo>`
respond	**pick the required object, e.g. a line**
prompt	`Specify through point or [Exit/Multiple/Undo]`
respond	**pick desired object, e.g. Snap to Center icon and pick a circle**
prompt	`Specify object to offset,` i.e. any more through offsets
respond	**<RETURN>** to end command
and	the line will be offset through the circle centre – Figure 12.2(d).

b) Erase

Allows the original source object to be erased after the offset operation. When offset activated:

prompt	`Specify offset distance or [Through/Erase/Layer]`
enter	**E <R>** – the erase option
prompt	`Erase source object after offsetting? <Yes/No>`
enter	**Y <R>**
prompt	`Specify offset distance`
and	command continues as 'normal'.

c) Layer

An offset object has the same layer as the source object and this may not be what is required (think back to offsetting the circle centre lines, when we used the CHANGE command to convert the four lines to centre lines). The layer option avoids the need for the CHANGE command. When offset is activated:

prompt	`Specify offset distance or [Through/Erase/Layer]`
enter	**L <R>** – the layer option
prompt	`Enter layer option for offset objects <Current/Source>`
enter	*a*) C <R> to place the offset object on the current layer
	b) S <R> to place the offset object on the original layer
prompt	`Specify offset distance`
and	command continues as 'normal'.

Additional exercises

To demonstrate additional use of the commands, try the examples which follow:

1   Fillet and chamfer with a zero radius

a) the commands can be used to extend inclined lines to a point

b) draw two inclined lines and set R = 0. Figure 12.2(e) demonstrates the effect.

2   Fillet with circles

The fillet command can be used to add external and internal fillets as Figure 12.2(f).

3   Fillet pick points

Adding a fillet to a line and a circle requires the pick points to be selected correctly. Figure 12.2(g) demonstrates the pick point effects.

4   Using offset to obtain a circle centre point

The offset command can be used to obtain a circle centre point and is one of the most common uses for the command. Figure 12.2(h) demonstrates offsets of 18.5 horizontally and 27.8 vertically to position the circle centre point.

Question: what about inclined line offsets?

5   Extending lines and arcs
    Lines and arcs can be extended to other objects, including circles – Figure 12.2(i).

6   Trim lines and circles
    Lines, circles and arcs can be trimmed to 'each other' as Figures 12.2(j) and (k). Can
    you obtain the various options in Figure 12.2(k)?

7   Trim/Extend with a fence selection
    When several objects have to be trimmed or extended, the fence selection option can
    be used. The effect is achieved by:
    a) activating the command
    b) selecting the boundary (extend) or cutting edge (trim)
    c) entering F <R> – the fence option
    d) draw the fence line then right-click
    e) the effect is displayed in Figure 12.2(l) for the TRIM command.

8   Trim/Extend toggle effect
    When the trim/extend command is activated and an object selected for the cutting
    edge/boundary, the prompt line will display:
    Select object to trim (extend) or shift-select to extend (trim)
    By using the shift key the user can 'toggle' between the two commands.

Match properties

This is a very useful 'tool' to the user, as it does exactly what it says – it matches
properties.

1   Have some objects drawn on the OUT and CL layers.

2   Select the **Match Properties icon** from the Standard toolbar and:
    prompt   Select source object
    respond   **pick any CL layer object**
    prompt   Select destination object(s) or [Settings]
    respond   **pick any OUT layer object**.

3   The selected objects will be displayed with green centre lines.

4   *Settings*
    The Match Properties command has a settings option, which allows the user to deter-
    mine which properties can be matched. To use this option:
    a) activate the Match Properties command
    b) pick any object as the source
    c) at the destination prompt, enter **S <R>** to display the Property Settings dialogue box
    d) generally the Basic and the Special Properties are active at all times but the user can
        alter as required
    e) cancel the dialogue box then ESC to end the command.

This completes the chapter.

Assignments

Three assignments for you to practice your skills with the OFFSET, TRIM and
EXTEND commands as well as FILLET and CHAMFER. The assignments should not
give you any problems. The procedure is:

1   Open your A3PAPER file (drawing or template).

2   Complete the drawing using layers correctly.

3   Save the completed drawing as C:\MYCAD\ACT??

4   Do NOT attempt to add dimensions.

Activity 6

Three interesting shapes to complete. One requires a bit of thought for the original shape, the other two require concentration to obtain the final shape. The commands to use are at your decision.

Activity 7

Two shapes to complete using the commands so far investigated. Make use of offset and trim as much as possible. No dimensions.

Activity 8

Three 'logo' type shapes. The CH is created from ellipses and lines. The circular shape uses fillets and the B shape should give you no trouble. Note that with these drawings I positioned the centre lines first.

Chapter 13

Text

1   Text should be added to a drawing whenever possible.

2   This text could simply be a title and date, but could also be a parts list, a company title block, notes on costing, customer data, etc.

3   AutoCAD 2006 allows the user to enter text:
   a) as short entries, i.e. a single line or a few lines
   b) as larger entries, i.e. several lines
   c) via a table.

In this chapter we will consider the above text types as well as text fonts and styles, so:

1   Open your WORKDRG saved drawing and refer to Figure 13.1.

2   The component should be positioned as displayed relative to the point (80,110).

3   Make layer TEXT current.

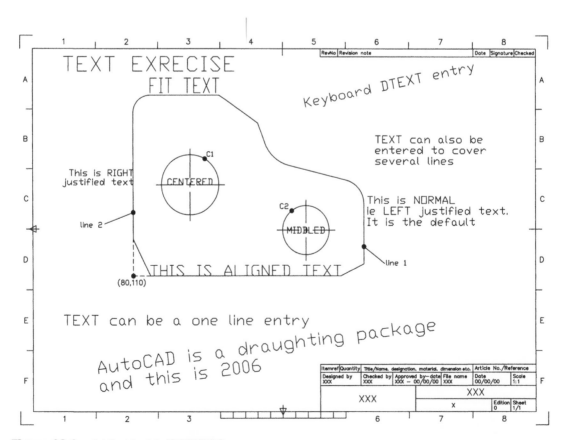

Figure 13.1   Adding text to WORKDRG.

Single line of text

1 Menu bar with **Draw-Text-Single Line Text** and:

   prompt   Specify start point of text or [Justify/Style]
   enter   **20,70 <R>**
   prompt   Specify height<??> and enter **8 <R>**
   prompt   Specify rotation angle of text<0.0> and enter **0 <R>**
   and   *a*) rectangular box with flashing cursor displayed
            b) this is the text start point at 20,70
   enter   **TEXT can be a one line entry <R>**
   and   cursor 'jumps' to start of next line i.e. any more lines of text
   respond **press the <RETURN> key** to end the command.

2 Repeat the **Draw-Text-Single Line Text sequence** with:
   a) start point: 20,280 <R>
   b) height: 12 <R>
   c) rotation: 0 <R>
   d) text: TEXT EXRECISE <R><R> – two returns (*Note*: **this spelling is deliberate**.)

3 At the command line enter **DTEXT <R>** and enter
   a) start point: 230,250
   b) height: 7.5
   c) rotation: 12
   d) text: Keyboard DTEXT entry <R><R>.

4 *Notes*
   a) the text is displayed on the screen as you enter it from the keyboard
   b) this is referred to as *dynamic* text.

Several lines of text

1 Select from the menu bar **Draw-Text-Single Line Text** and:

   prompt   Specify start point of text and enter **275,245 <R>**
   prompt   Specify height and enter **6 <R>**
   prompt   Specify rotation angle of text and enter **0 <R>**
   enter   **TEXT can also be <R>**
   enter   **entered to cover <R>**
   enter   **several lines <R>**
   enter   **<RETURN>**.

2 At the command line enter **TEXT <R>** and:

   prompt   Specify start point of text and enter **50,30 <R>**
   prompt   Specify height and enter **10 <R>**
   prompt   Specify rotation angle of text and enter **7 <R>**
   enter   **AutoCAD is a draughting package<R>**
   enter   **and this is 2006 <R>**
   enter   **<R>** – to end command.

Editing existing screen text

1 Hopefully text which has been entered on a drawing is correct, but there may be spelling mistakes and/or alterations to existing text may be required. Text can be edited as it is being entered from the keyboard if the user notices the mistake.

2 Screen text which needs to be edited requires a command.

3 From the menu bar select **Modify-Object-Text-Edit** and:

   prompt   `Select an annotation object or [Undo]`
   respond   **pick the TEXT EXRECISE item**
   and   `selected item is boxed and highlighted`
   respond   either   *a)* retype the highlighted phrase correctly then <R>
                 or   *b)* 1. left-click at right of the text item
                              2. backspace to remove error
                              3. retype correctly
                              4. <R>.
                  or   *c)* 1. move cursor to TEXT EXRE|CISE and left-click
                              2. backspace to give TEXT EX|CISE
                              3. move cursor to TEXT EXE|CISE and left-click
                              4. enter ER to give TEXT EXER|CISE
                              5. <R>.

4 The text item will now be displayed correctly.

5 *Note*: You could always erase the text item and enter it correctly?

6 *a)* AutoCAD has a built-in spell checker which 'uses' a dictionary to check the spelling.
   b) This dictionary can be changed to suit different languages and must be 'loaded' before it can be used.
   c) We will use a British English dictionary
   d) From the menu bar select **Tools-Spelling** and:

      prompt   `Select object`
      respond   **pick AutoCAD is a draughting package then right-click**
      prompt   `Check Spelling dialogue box`
      with   a named current dictionary
      respond   **pick Change Dictionaries**
      prompt   `Change Dictionaries dialogue box`
      respond   *a)* scroll at Main dictionary
                    b) pick British English (ise) – Figure 13.2(a)
                    c) pick Apply and Close.
      prompt   `Check Spelling dialogue box.`

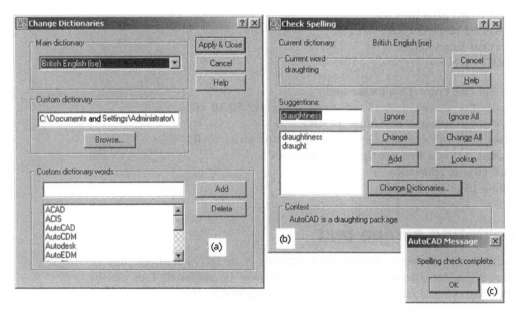

Figure 13.2   The Change Directories and Check Spelling dialogue boxes.

with	*a*) Current dictionary: British English (ise)
	b) Current word: draughting
	c) Suggestions: draughtiness – Figure 13.2(b).
respond	**pick Ignore**
prompt	AutoCAD Message box – Figure 13.2(c)
respond	**pick OK** – spell check is complete.

7 Save the screen layout if required, **BUT NOT AS** C:\MYCAD\WORKDRG.

Text justification

1 Text items added to a drawing can be 'justified', i.e. positioned to user specifications.

2 AutoCAD 2006 has several justification positions, these being:
a) six basic: left, align, fit, centre, middle, right.
b) nine additional: TL, TC, TR, ML, MC, MR, BL, BC, BR.

3 Continue with WORKDRG, layer TEXT current and toolbars Draw, Modify and Object Snap.

4 At the command line enter **TEXT <R>** and:
prompt	Specify start point of text or [Justify/Style]
respond	**Snap to Endpoint icon and pick line 1 'at top end'**
prompt	Specify height and enter **6 <R>**
prompt	Specify rotation angle of text and enter **0 <R>**
enter	**This is NORMAL <R>**
enter	**i.e. LEFT justified text. <R>**
enter	**It is the default <R>**
enter	**<RETURN>** to end command.

5 Menu bar with **Draw-Text-Single Line Text** and:
prompt	Specify start point of text or [Justify/Style]
enter	**J <R>** – the justify option
prompt	Enter an option [Align/Fit/Center/Middle/Right/TL/TC/TR/ML/MC/MR/BL/BC/BR]
enter	**R <R>** – the right justify option
prompt	Specify right endpoint of text baseline
respond	**Snap to Midpoint icon and pick line 2**
prompt	Specify height and enter **5<R>**
prompt	Specify rotation angle of text and enter **0 <R>**
enter	**This is RIGHT <R>**
enter	**justified text <R><R>**.

6 Repeat the single line text command and:
prompt	Specify start point of text or [Justify/Style]
enter	**J <R>**
prompt	Enter an option [Align/Fit/Center/Middle/Right/TL/TC/TR/ML/MC/MR/BL/BC/BR]
enter	**C <R>** – center option
prompt	Specify center point of text
respond	**Snap to Center icon and pick circle C1**
prompt	Specify height and enter **5 <R>**
prompt	Specify rotation angle of text and enter **0 <R>**
enter	**CENTERED <R><R>**.

7 Enter TEXT <R> at the command line then:
a) enter J <R> for justify
b) enter M <R> for middle option

c) Middle point: Snap to Center icon and pick circle C2

d) Height: 5 and Rotation: 0

e) Text: MIDDLED.

8 Activate the single line text command with the Fit justify option and:

prompt   `Specify first endpoint of text baseline`

respond   **Snap to Endpoint icon and pick left end of top horizontal line**

prompt   `Specify second endpoint of text baseline`

respond   **Snap to Endpoint icon and pick right end of top horizontal line**

prompt   `Specify height and enter` **12 <R>**

enter   **FIT TEXT <R><R>**

Note: this option has no rotation prompt.

9 With TEXT again:

a) select the Align justify option

b) pick left end of lower horizontal line as first text endpoint

c) pick right end of lower horizontal line as second text endpoint

d) text item: THIS IS ALIGNED TEXT

Note: This option has no height or rotation prompt.

10 Your drawing should now resemble Figure 13.1 and can be saved, but we will not use it again.

11 *a*) The text justification options are easy to use

b) Simply enter the appropriate letter for the justification option at the command line

c) The entered letters are:

A: Align	F: Fit	C: Center
M: Middle	R: Right	TL: Top left
MC: Middle center	BR: Bottom right	etc.

d) Try some of the other options in your own time.

Text fonts and styles

1 The words 'font' and 'style' are text terminology and can be explained as:

Font   *a*) defines the pattern which is used to draw characters, i.e. it is basically an alphabet 'appearance'.

       b) AutoCAD 2006 has over 100 fonts available to the user.

Style   defines the parameters used to draw the actual text characters, i.e. the width of the characters, the obliquing angle, whether the text is upside-down, backwards, etc.

2 *Notes*

a) Text fonts are 'part of' the AutoCAD package

b) Text styles are created by the user

c) Any text font can be used for many different styles

d) A text style uses only one font

e) If text fonts are to be used in a drawing, a text style MUST BE CREATED by the user

f) New text fonts can be created by the user, but this is outside the scope of this book

g) Text styles can be created:

     1. by keyboard entry

     2. via a dialogue box.

3 Close all existing drawings *then* open your A3PAPER template file with model space active.

4 At the command line enter **–STYLE <R>** and

prompt   `Enter name of text style or [?]<Standard>`

enter   **? <R>** – the 'query' option

prompt	Enter text style(s) to list<*>
enter	**<R>**
prompt	AutoCAD Text Window
with	Style name: "Standard" Font files: txt (or similar)
	Height: 0.00 Width Factor: 1.00 Obliquing angle: 0.0
	Generation: Normal
	Current test style: "Standard".

4   *a*) this is AutoCAD 2006's 'default' text style

      b) it has the text style name **STANDARD** and uses the text font 'txt'

      c) realise that your system may have a different text style name and font. If it does, do not worry. It will not affect this exercise.

5   Cancel the text window.

6   The entry –STYLE was to allow us to use the command line instead of a dialogue box.

Creating a text style from the keyboard

1   At the command line enter –**STYLE <R>** and:

prompt	Enter name of text style or [?]
enter	**ST1 <R>** – the style name
prompt	Specify full font name or font filename (TTF or SHX)
enter	**romans.shx <R>**
prompt	Specify height of text and enter **0 <R>**
prompt	Specify width factor and enter **1 <R>**
prompt	Specify obliquing angle and enter **0 <R>**
prompt	Display text backwards? and enter **N <R>**
prompt	Display text upside-down? and enter **N <R>**
prompt	Vertical? and enter **N <R>**
prompt	"ST1" is now the current style.

2   The above entries of height, width factor etc are the parameters which must be defined for every text style being created by the user.

Creating a text style from a dialogue box

1   From the menu bar select **Format-Text Style** and

prompt	Text Style dialogue box
with	*a*) ST1 as the Style Name as we have just used this style
	b) romans.shx as the Font Name
	c) Height: 0.0, Width Factor: 1.00, Oblique Angle: 0.0
respond	**pick New** and:
prompt	New Text Style dialogue box
respond	*a*) alter Style Name to ST2
	b) pick OK
prompt	Text Style dialogue box
with	**ST2 as the Style Name**
respond	*a*) pick the scroll arrow at right of romans.shx
	b) scroll and pick **italicc.shx**
	c) ensure that:
	Height: 0.0, Width Factor: 1.00, Oblique Angle: 0.0
	d) note the Preview box – Figure 13.3
	e) pick **Apply** then **Close**.

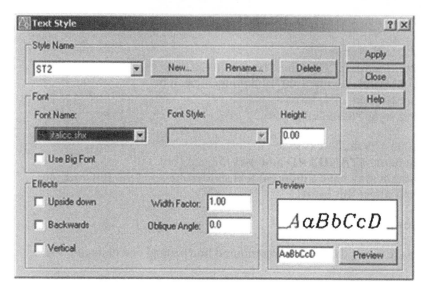

Figure 13.3 The Text Style dialogue box for created style ST2.

2 Now use the Text Style dialogue box to create another text style with:
   a) style name and font name: ST3 and Arial Black
   b) height, width factor and oblique angle: 0, 1, 0
   c) backwards, upside-down, vertical: all OFF.

3 *a*) using the Text Style dialogue box, scroll at Style Name, pick ST1 *then* close the dialogue box
   b) save A3PAPER file as both a template and drawing file, updated the existing file
   c) we have now saved the three created text style (ST1, ST2 and ST3) to our A3PAPER standard sheet.

Task

As an additional exercise, create the following text styles using the Text Style dialogue box:

Style name	Font name	Ht	Width factor	Oblique angle	*Effects* Back-wards	Upside-down	Vert'l	
ST4	gothice.shx	12	1	0	OFF	OFF	OFF	(blank OFF,
ST5	italict.shx	5	1	30	OFF	OFF	OFF	tick ON)
ST6	Romantic	10	1	0	OFF	ON	–	
ST7	scriptc.shx	5	1	−30	OFF	OFF	OFF	
ST8	monotxt.shx	6	1	0	OFF	OFF	ON	
ST9	Swis721 BdOulBT	12	1	0	OFF	OFF	OFF	
ST10	complex.shx	5	1	0	ON	OFF	OFF	
ST11	isoct.shx	5	1	0	ON	ON	–	
ST12	romand.shx	5	1	0	ON	ON	ON	

Using created text styles

1 Still have the blank A3PAPER standard sheet displayed?

2 Make layer TEXT current

3   At the command line enter **TEXT <R>** and:
   prompt   Specify start point of text or [Justify/Style]
   enter   **S <R>** – the style option
   prompt   Enter style name (or ?) – (ST1 should be default name?)
   enter   **standard** <R>
   prompt   Specify start point of text or [Justify/Style]
   enter   **110,275 <R>**
   prompt   Specify height and enter **8 <R>**
   prompt   Specify rotation angle of text and enter **0 <R>**
   enter   **AutoCAD 2006 <R><R>**.

4   Menu bar with **Draw-Text-Single Line Text** and
   prompt   Specify start point of text or [Justify/Style]
   enter   **S <R>** – the style option
   prompt   Enter style name (or ?) – Standard as default name?
   enter   **ST1 <R>**
   prompt   Specify start point of text or [Justify/Style]
   enter   **20,240 <R>**
   prompt   Specify height and enter **8 <R>**
   prompt   Specify rotation angle of text and enter **0 <R>**
   enter   **AutoCAD 2006 <R><R>**.

5   Using the step 4 procedure and the single line text command, add the text item
    AutoCAD 2006 using the following information:

Style	Start pt	Ht	Rot
ST1	20,240	8	0 – already entered
ST2	225,265	8	0
ST3	15,145	8	0
ST4	250,240	NA	0
ST5	25,175	NA	30
ST6	100,220	NA	0
ST7	145,195	NA	−30
ST8	215,235	NA	270 (default angle)
ST9	235,195	NA	0
ST10	365,165	NA	0
ST11	325,145	NA	0
ST12	400,140	NA	270 (default angle).

6   When completed, the screen should display 13 different text styles – the 12 created
    and the STANDARD default as Figure 13.4.

7   There is no need to save this drawing but:
   a) erase all the text from the screen
   b) save the 'blank' screen as **C:\MYCAD\STYLEX** – you are saving the created text
      styles for future use.

8   *Notes*
    Text styles and fonts can be confusing to new AutoCAD users due to the terminology,
    and by referring to Figure 13.4, the following may be of assistance:
   a) *Effects:*
      Three text style effects which can be 'set' are upside-down, vertical and backwards.
      These effects should be obvious to the user, and several of our created styles had
      these effects toggled on.
   b) *Width factor:*
      This parameter will 'stretch' the text characters and Figure13.4(a) displays an item
      of text with six width factors. The default width factor value is 1.

c) *Obliquing angle:*
A parameter which 'slopes' the text characters as Figure 13.4(b). The default value is 0.

d) *Height:*
When the text command was used with the created text styles, three styles prompted for a height, these being ST1, ST2 and ST3. The other text styles had a height value entered when the style was created – hence no height prompt. This also means that these text styles cannot be used at varying height values. The effect of differing height values is displayed in Figure 13.4(c).

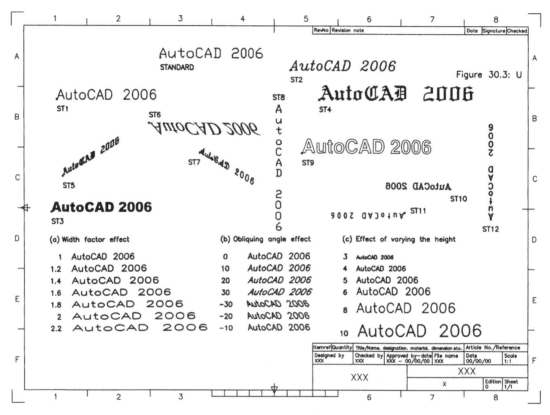

Figure 13.4   Using the created text styles.

e) *Recommendation:*
I would strongly recommend that if text styles are being created, the height be left at 0. This will allow the user to enter any text height at the prompt when the text command is used.

f) The text items displayed using styles ST5 and ST7 are interesting, these items having:
ST5   30 obliquing   30 rotation
ST7   −30 obliquing   −30 rotation
These styles give an 'isometric text' appearance.

Text control codes

1   When text is being added to a drawing, it may be necessary to underline the text item, or add diameter/degree symbol.

2   AutoCAD has several control codes which when used with the **Single Line Text** command will allow underscoring, overscoring and symbol insertion.

3 The available control codes are:
%%O toggles the OVERSCORE on/off
%%U toggles the UNDERSCORE on/off
%%D the DEGREE symbol for angle or temperature(°)
%%C the DIAMETER symbol (Ø)
%%P the PLUS/MINUS symbol (±)
%%% the PERCENTAGE symbol (%).

4 Close all existing drawings *then* open your A3PAPER standard sheet with layer TEXT current.

5 Refer to Figure 13.5, menu bar with **Draw-Text-Single Line Text** and:
prompt Specify start point of text or [Justify/Style]
enter **S <R>**
prompt Enter style name or [?]
enter **ST1 <R>**
prompt Specify start point of text and enter **25,250 <R>**
prompt Specify height and enter **15<R>**
prompt Specify rotation angle of text and enter **5 <R>**
enter **%%UAutoCAD 2006%%U <R><R>**.

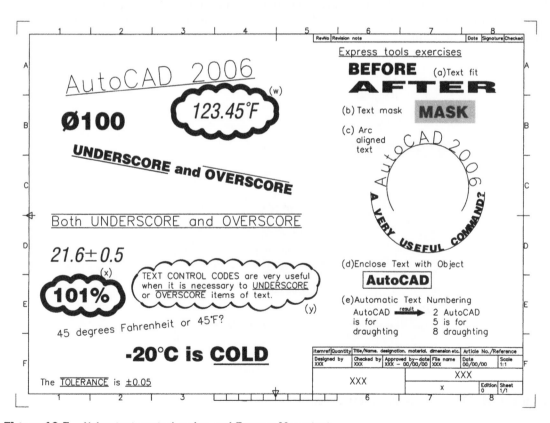

Figure 13.5 Using text control codes and Express Menu text.

6 At the command line enter **DTEXT <R>** and:
a) style: ST2
b) start point: 130,230
c) height, angle: 10,0
d) text: 123.45%%DF.

7 Activate the single line text command and:
   a) style: ST3
   b) start point: 30,200
   c) height, angle: 8, −10
   d) text: %%UUNDERSCORE%%U and %%OOVERSCORE%%O.

8 With the single line text command:
   a) style: ST3
   b) start point: 20,220
   c) height, angle: 15,0
   d) text: %%C100.

9 Refer to Figure 13.5 and add the other text items. The text style used is at your discretion.

10 *Note*: Later in the chapter another method of adding text symbols will be discussed.

Express menu text

If your 2006 package has been fully installed you will have access to the Express pull-down menu, which is located in the menu bar between Modify and Window. Figure 13.6 displays the Text options available to the user and we will now investigate several of these. If your system does not have the Express menu, then proceed to the Revision Cloud heading.

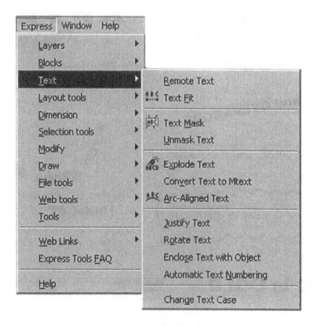

Figure 13.6 The Express Menu Text selections.

1 Refer to Figure 13.5, clear an area to the right of the sheet and attempt the following exercises.

2 *Text Fit*
   Allows the user to 'stretch' an existing item of text as Figure 13.5(a), the user:
   a) selects the item of text
   b) specifies the end point or the start and end points.

3 *Text Mask*

Will create a rectangular frame around a specified text item. To demonstrate the Figure 13.5(b) effect, create an item of text *then* menu bar with **Express-Text-Text Mask** and:

prompt	Select text object to mask or [Masktype/Offset]
enter	**M <R>** – masktype option
prompt	Specify entity type to use for mask
enter	**S <R>** – solid option
prompt	Select Color palette dialogue box
respond	**enter colour number 61 then pick OK**
prompt	Select text object to mask or [Masktype/Offset]
enter	**O <R>** – offset option
prompt	Enter offset factor relative to text height
enter	**0.4 <R>**
prompt	Select text object to mask or [Masktype/Offset]
respond	**pick required text item then right-click-enter**.

4 *Arc-aligned text*

Allows text to be aligned with an arc or trimmed circle as Figure 13.5(c). It will not work with full circles.

a) **Draw an arc** then menu bar with **Express-text-Arc-Aligned Text** and:

prompt	Select an Arc or an ArcAligned Text
respond	**pick your drawn arc**
prompt	ArcAlignedText Workshop - Create dialogue box
respond	1. style ST1 active – scroll and select if required
	2. Text: enter AutoCAD 2006
	3. C (center along the arc) active, i.e. 'pressed in'
	4. Text height: 10
	5. Width factor: 1
	6. Char spacing: 4
	7. Offset from arc: 0.5
	8. On convex side active
	9. Outward from center active – Figure 13.7
	10. pick OK

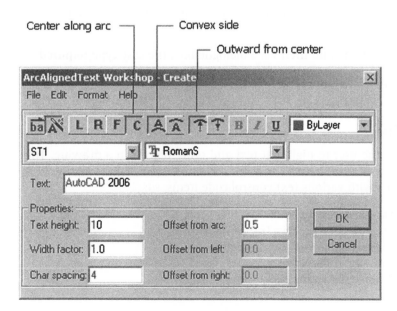

Figure 13.7 The ArcAlignedText – Create dialogue box.

and   the text item will be aligned along the outside of the drawn arc.

b) Repeat the arc aligned text command and align an item of text on the 'inside' of a drawn arc.

5   *Justify Text*
The user has access to the complete list of justify options when a text item is selected.

6   *Rotate Text*
Allows a selected text item to be rotated to a new entered value.

7   *Enclose Text with Object*
Creates a user-defined frame around a text item as Figure 13.5(d). To demonstrate this option, create an item of text, activate the command and:

prompt	Select objects
respond	**pick the text item then right-click**
prompt	Enter distance offset factor
enter	**0.35 <R>**
prompt	Enclose text with [Circles/Slots/Rectangles]
enter	**R <R>** – the rectangle option
prompt	Create rectangles of constant or variable size
enter	**C <R>** – constant size
prompt	Maintain constant rectangle [Width/Height/Both]
enter	**B <R>** – both option
and	the selected text item will be enclosed in a rectangle.

8   *Automatic Text Numbering*
Will number text items according to user specific entry as Figure 13.5(e). Enter a few lines of test, activate the command from the Express Tools menu and:

prompt	Select objects
respond	**window the text items then right-click**
prompt	Sort selected objects by [X/Y/Select-order]
enter	**Y <R>**
prompt	Specify starting number and increment [Start,increment]
enter	**2,3 <R>**
prompt	Placement of number in text
enter	**P <R>** – prefix option
and	the windowed text items will be numbered according to the specified entries.

9   *Change Text Case*
The user selects an item of text and then via a dialogue box, the case type required.

10   This completes the Express Tools demonstration.

Revision cloud

This feature will allow the user to highlight text and other objects in a drawing. The user creates a sequence of polyline arcs to form a cloud-shaped object. Figure 13.5 should still be displayed with various text control code items, so:

1   Menu bar with **Format-Lineweight** and:

prompt	Lineweight Settings dialogue box
respond	*a*) scroll at Lineweights and pick 2.00 mm
	b) units for listing: millimetres
	c) display lineweight active (tick)
	d) pick OK.

2   Draw a centred ellipse about the 123.45°F text item.

3   Menu bar with **Draw-Revision Cloud** and:
   prompt   Minimum arc length: 15 Maximum arc length: 15
                Specify start point or [Arc length/Object/Style]<Object>
   enter   **O <R>** – the object option
   prompt   Select object
   respond   **pick the ellipse**
   prompt   Reverse direction [Yes/No]
   enter   **N <R>** – Figure 13.5(w).

4   Repeat the **Draw-Revision Cloud** command and:
   prompt   Specify start point or [Arc length/Object]<Object>
   respond   *a*) Identify any suitable text item e.g. 101%
                b) pick a suitable start point.
   prompt   Guide crosshairs along cloud path
   respond   **move cursor around the text item to the start point**
   and   **revision cloud will be added** – Figure 13.5(x).

5   Revision cloud command again and:
   prompt   Specify start point or [Arc length/Object]<Object>
   enter   **S <R>** – the style option
   prompt   Select arc style [Normal/Calligraphy]<Normal>
   enter   **C <R>** – the calligraphy style
   prompt   Specify start point or [Arc length/Object]<Object>
   respond   *a*) Identify any suitable text item
                b) pick a suitable start point
   prompt   Guide crosshairs along cloud path
   respond   **move cursor around the text item to the start point**
   and   **revision cloud will be added** – Figure 13.5(y).

6   Try the revision cloud command with another item of text.

    This completes the text control codes and express text exercises. The drawing will not
    be used again.

Multiline text

1   Multiline allows large amounts of text to be added to a pre-determined area on the screen.

2   The added text can be edited.

Creating the multiline text

1   Close any existing drawing then open the STYLEX drawing, saved with the created
    text styles. The drawing area should be blank.

2   Display toolbars to suit, including the TEXT toolbar.

3   With layer TEXT current, select the **Multiline Text icon** from the Draw toolbar and:
   prompt   Specify first corner
   enter   **10,280 <R>**
   prompt   Specify opposite corner or [Height/Justify/Line spacing/
                Rotation/Style/Width]
   enter   **S <R>** then **ST1 <R>** – setting the text style

prompt	Specify opposite corner or [Height/Justify/Line spacing/ Rotation/Style/Width]
enter	**H <R>** then **5 <R>** – setting the height
prompt	Specify opposite corner or [Height/Justify/Line spacing/ Rotation/Style/Width]
enter	**125,195 <R>**
prompt	Multiline Text Editor dialogue box
with	three distinct 'sections':
	a) the Text Formatting toolbar, which can be positioned to suit user requirements
	b) the ruler with tab settings, etc.
	c) the text editor window
and	cursor positioned at top left of the 'window area'
respond	*a)* enter the text below in the text editor window
	b) INCLUDE the typing errors which are deliberate and have been underlined
	c) do not try to add the underline effect
	d) **DO NOT PRESS THE RETURN KEY AT ANY STAGE.**
enter	**CAD is a draughting <u>tol</u> with many benefits when compared to conventional draughting <u>techniches</u>. Some of these <u>benefitds</u> include <u>incresed</u> productivity, shorter lead <u>tines</u>, standardisation, <u>acuracy amd</u> rapid <u>resonse</u> to change.**
and	dialogue box as Figure 13.8
respond	**pick OK from the Text Formatting toolbar.**

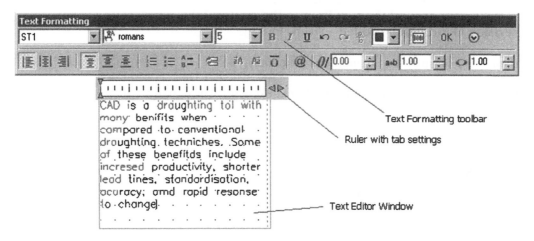

Figure 13.8 The Text Formatting toolbar and Multiline Text window.

4   Figure 13.9(a) displays the entered text.

5   *Notes*
   a) the text 'wraps around' the Text Editor window as it is entered from the keyboard
   b) the text is fitted into the width of the selected area of the screen, not the full rectangular area of the dialogue box
   c) Figure 13.9(a) displays a rectangular area equivalent to the 10,275 and 125,190 entered co-ordinates
   d) the width is determined by the entered coordinates
   e) the entered text in the dialogue box will not appear as it was entered from the keyboard. This is normal.

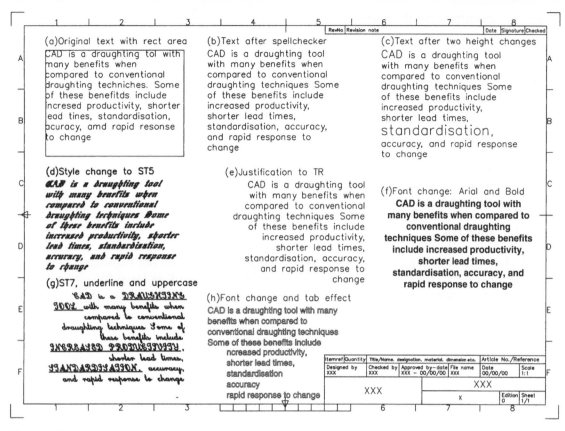

Figure 13.9   Editing multiline text.

Spellcheck

AutoCAD 2006 has a built-in spellchecker which has been used earlier in the chapter. The spellchecker can be activated:

a) from the menu bar with **Tools-Spelling**
b) by entering SPELL <R> at the command line.

1   Activate the spell check command and:

prompt	Select objects
respond	**pick any part of the entered text then right-click**
prompt	Check Spelling dialogue box
with	a) current dictionary: British English (ise)
	(if not, select Change Dictionaries, select as required then Apply & Close)
	b) current word: probably *draughting*
	c) suggestions: probably draughtiness, draught
	d) context: *CAD is a draughting tol with many benefits when compared to*
respond	**pick Ignore All**
and	we have agreed that **draughting** is the correct spelling
prompt	Check Spelling dialogue box
with	a) current word: *tol*
	b) suggestions: toll, tool, toil, toile, told, tolls....
respond	a) pick tool – becomes highlighted
	b) tool added to Suggestions box
	c) pick Change
prompt	Check Spelling dialogue box

with	*a*) current word: *techniches*
	b) suggestions: tech
respond	*a*) at suggestions, alter tech to techniques
	b) pick Change
prompt	Check Spelling dialogue box
respond	change the following as they appear:

original	*alter to*
benefitds	benefits
incresed	increased
acuracy	accuracy
amd	and (manual change required)
resonse	response

then	AutoCAD Message: Spelling check complete
respond	**pick OK**.

2   One of the original spelling mistakes was '**tines**' (times) and this was not highlighted with the spellcheck This means that the word 'tines' is a 'real word' as far as the dictionary used is concerned although it is wrong for our use. This can be a major problem with spellcheckers.

3   Enter **DDEDIT <R>** at the command line and:
   a) pick any part of the text
   b) the Multiline Text Editor dialogue box will be displayed and the word tines can be manually altered to times
   c) when the alteration is complete, pick OK.

4   The multiline text will be displayed with the correct spelling as Figure 13.9(b).

Editing multiline text

1   To investigate how multiline text can be edited, continue with the created multiline text and refer to Figure 13.9.

2   Select the **Edit (Text)** icon from the Text toolbar and:

prompt	Select an annotation object or [Undo]
respond	**pick the multiline text item**
prompt	Multiline Text Editor dialogue box
with	text displayed
respond	*a*) left-click and drag mouse over CAD
	b) alter height to 6
	c) left-click and drag over standardisation
	d) alter height to 8
	e) pick OK from the toolbar
	f) right-click-enter to end command
and	text displayed as Figure 13.9(c).

3   Menu bar with **Modify-Object-Text-Edit** and:

prompt	Select an annotation object
respond	**pick the text item**
prompt	Multiline Text Editor dialogue box
respond	*a*) scroll at Style, pick ST5 then pick OK from toolbar
	b) right-click-Enter to end command – Figure 13.9(d).

4   Edit text command, select the next multiline text then:

prompt	Multiline Text Editor dialogue box

respond	*a)* alter text style back to ST1
	b) **right-click in the text window**
prompt	`shortcut menu.`
respond	*a)* pick Justification
	b) pick Top Right
	c) pick OK then right-click-enter
and	text displayed as Figure 13.9(e).

5   Using the Edit Text command alter the multiline text item using the following information:

   a) style ST1, justification top center, highlight text and change font to Arial with Bold effect – Figure 13.9(f)

   b) style ST7, justification bottom right, highlight then underlined and UPPERCASE indicated text – Figure 13.9(g)

   c) style ST1, justification top left, height 5, highlighted and change font to SWIS721BdCnOulBT and produce tab effect – Figure 13.9(h).

6   This exercise is now complete and can be saved if required.

Text modifications

To investigate some of the other text modifications, refer to Figure 13.10 and restore the multiline text, i.e. with top left justification, style ST1 at height 5 – Figure 13.10(a).

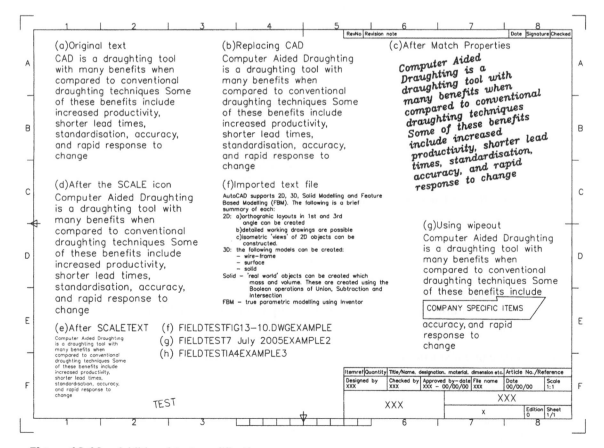

Figure 13.10   Additional text modifications.

1 Select the **Find (and Replace icon)** from the Text toolbar and:

prompt   Find and Replace dialogue box

respond   *a*) Find text string: enter CAD

                      b) Replace with: Computer Aided Draughting

                      c) Search in: Entire drawing

                      d) pick Select objects icon.

prompt   Select objects

respond   **pick the multiline text item then right-click**

prompt   Find and Replace dialogue box

with   Search in: Current selection

respond   **pick Find**

prompt   Find and Replace dialogue box as Figure 13.11

with   *a*) Object Type: Mtext

                      b) Context: CAD is a draughting tool . . .

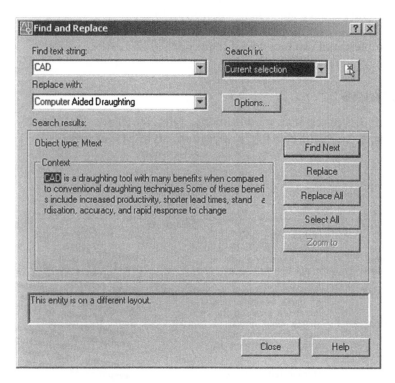

Figure 13.11   The Find and Replace dialogue box.

respond   **pick Replace**

prompt   Find and Replace dialogue box

with   No more occurrences found

respond   **pick Close**

and   text item displayed with CAD replaced by Computer Aided Draughting –
                      Figure 13.10(b).

2 With the single line text icon from the text toolbar, create an item of text using:

    a) style: ST2

    b) start point: pick to suit

    c) height: 5.25

    d) rotation angle: 8

    e) text: TEST.

3   Select the **Match Properties icon** and:
  prompt   `Select source objects`
  respond   **pick the TEST text item**
  prompt   `Select destination object`
  respond   **pick the multiline text then right-click-enter**
  and   the multiline text will be displayed with the same text 'settings' as the
              TEST text – Figure 13.10(c).

4   Undo the match properties effect to restore the Computer Aided Draughting multiline
     text.

5   Select the **Scale (Text) icon** from the Text toolbar and:
  prompt   `Select objects`
  respond   **pick the multiline text item then right-click**
  prompt   `Enter a base point option for scaling [Existing/Left/`
              `Center/Middle/Right...`
  respond   **<RETURN>** i.e. accept the existing default
  prompt   `Specify new height or [Match object/Scale factor]`
  enter   **M <R>** – the match object option
  prompt   `Select a text object with the desired height`
  respond   **pick TEST item of text**
  and   the selected multiline text will be displayed at a height of 5.25 –
              Figure 13.10(d)
  note   the modified multiline text does not have the style or rotation angle as the
              TEST item, only the height.

7   Undo the scale text effect then enter **SCALETEXT <R>** at the command line and:
  prompt   `Select objects`
  respond   **pick the multiline text item then right-click**
  prompt   `Enter a base point option for scaling [Existing/Left/`
              `Center/Middle...`
  respond   **<RETURN>** i.e. accept the existing default
  prompt   `Specify new height or [Match object/Scale factor]`
  enter   **S <R>** – the scale factor option
  prompt   `Specify scale factor`
  enter   **0.5 <R>**
  and   selected text item scaled as Figure 13.10(e).

Importing text files into AutoCAD

Text files can be imported into AutoCAD from other application packages using
the Multiline Text Editor dialogue box. To demonstrate the concept we will use a
text editor and write a new item of text, save it and then import it into our existing
drawing.

1   Save the existing layout as a precaution

2   Select   *a*) **Start** from the Windows taskbar
  then   *b*) **Programs-Applications(or Accessories)-Notepad** and:
  prompt   `Blank Notepad screen displayed`
  enter   the following lines of text as displayed and:
                 a) tab out the spacing to suit
                 b) enter a <R> at end of each line.

AutoCAD supports 2D, 3D, Solid Modelling and Feature Based Modelling (FBM). The following is a brief summary of each:

2D: *a*) orthographic layouts in 1st and 3rd angle can be created
    b) detailed working drawings are possible
    c) isometric 'views' of 2D objects can be constructed.
3D: the following models can be created:
    – wire-frame
    – surface
    – solid
Solid – 'real world' objects can be created which mass and volume. These are created using the Boolean operations of Union, Subtraction and Intersection
FBM – true parametric modelling using Inventor.

3   When the text has been entered as above, menu bar with **File-Save As** and:
    prompt   Save As dialogue box
    respond   *a*) scroll at Save in and pick C:\MYCAD
                b) enter File name as MYTEST
                c) note type: Text Document (*.txt)
                d) pick Save
    then   Minimise Notepad (left button from title bar) to return to the AutoCAD screen.

4   Layer TEXT still current?

5   Activate the Multiline Text command and:
    respond   **pick two suitable points of your own** and create a reasonably sized rectangle
    prompt   Multiline Text Editor dialogue box
    respond   **right-click in blank text area**
    prompt   shortcut menu
    respond   **pick Import text**
    prompt   Select File dialogue box
    respond   *a*) scroll at Look in and pick C:\MYCAD
                b) pick MYTEST
                c) pick Open
    prompt   Multiline Text Editor dialogue box
    with   **imported text displayed**, but it may not appear as you would expect
    respond   *a*) highlight all the imported text
                b) scroll at the text styles and select ST1
                c) change text height to 3
                d) right-click in text window and set a top left justification
                e) pick OK from the Text Formatting toolbar.

6   The imported text will be displayed as Figure 13.10(f).

7   This part of exercise is now complete and can be saved, but remember that Notepad may still be open.

Wipeout

1   Wipeout is a draw command which can be used to cover existing objects with a blank area.

2   This blank area can be used:
  a) to allow notes to be added
  b) mask out existing objects.

3   We will demonstrate the command by masking out an area of text.

4   With snap off, menu bar with **Draw-Wipeout** and:
  prompt   Specify first point or [Frames/Polyline]
  respond   **pick a suitable point in the displayed multiline text**
  prompt   Specify next point
  respond   **pick other points then right-click-enter**.

5   The wipeout area can be used as required by the user as Figure 13.10(g).

Fields

1   A field is text that contains instructions to display data expected to change during the 'life' of a drawing.

2   When a field is updated, the latest data is displayed.

3   Fields can be inserted in any kind of text (except tolerances), including table text and attributes.

4   Clear an area of the screen, menu bar with **Draw-Text-Single Line Text** and:
  prompt   Specify start point of text or [Justify/Style]
  enter   **S <R>** then **ST1 <R>** – to set the text style
  prompt   Specify start point of text or [Justify/Style]
  respond   **pick a point to suit**
  prompt   Specify height and enter **6 <R>**
  prompt   Specify rotation angle of text and enter **0 <R>**
  enter   **FIELDTEST**
  then   **right-click**
  prompt   shortcut menu
  respond   **pick Insert Field**
  prompt   Field dialogue box
  respond   1. Field names: scroll and pick Filename
                   2. Uppercase active
                   3. Filename only active
                   4. Display file extension active – Figure 13.12
                   5. pick OK
  and   filename and extension added to text item
  respond   continue with text entry and enter **EXAMPLE then <R> and <R>**.

5   The text item will be displayed as **FIELDTESTDrawing file nameEXAMPLE**.

6   When this file name is altered, this item of text will also be altered.

7   Figure 13.10 displays three different added field text items:
  drawing file name, date and paper size

8   This short exercise is now complete.

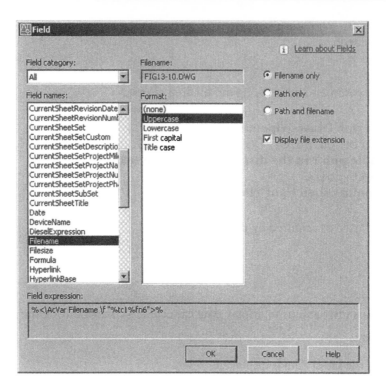

Figure 13.12 The Field dialogue box with Filename selected.

Assignments

Two assignments requiring text to be added to drawings are included.

Activity 9

Four simple components for you to complete. Text should be added, but not dimensions. The procedure is as always:

1 Open A3PAPER standard sheet

2 Complete the drawings using layers correctly

3 Save as C:\MYCAD\ACT9.

Activity 10: *Club cards*

Using your C:\MYCAD\STYLEX saved drawing, create several club cards of your own design, and:
a) All cards have to display the welcome symbol, created from lines (and an arc) using a 5 × 5 grid
b) You can use the saved text styles, or create other styles as appropriate.

Chapter 14

Dimensioning

1   AutoCAD has both automatic and associative dimensioning, the terms meaning:
   a) Automatic: when an object to be dimensioned is selected, the actual dimension text, arrows, extension lines, etc. are all added in one operation.
   b) Associative: the arrows, extension lines, dimension text, etc. which 'make up' a dimension are treated as a single object.

2   AutoCAD allows different 'types' of dimensions to be added to drawings. These are displayed in Figure 14.1 and can be categorised as:
   a) Linear: horizontal, vertical and aligned
   b) Baseline and Continue
   c) Ordinate: both X-datum and Y-datum
   d) Angular
   e) Radial: diameter and radius
   f) Leader: taking the dimension text 'outside' the object.

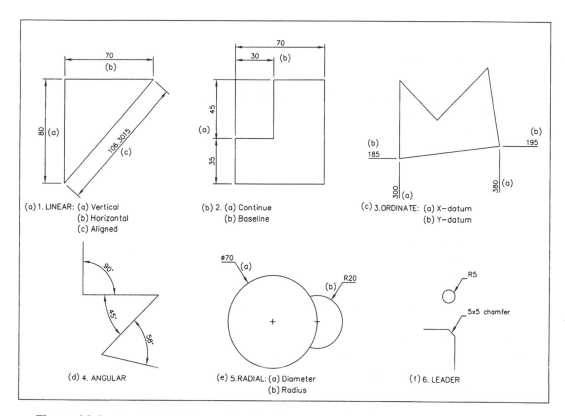

Figure 14.1   The AutoCAD dimension 'types'.

Dimension terminology

1 All dimensions used with AutoCAD objects have a terminology associated with them and it is important that the user has an understanding of this terminology.

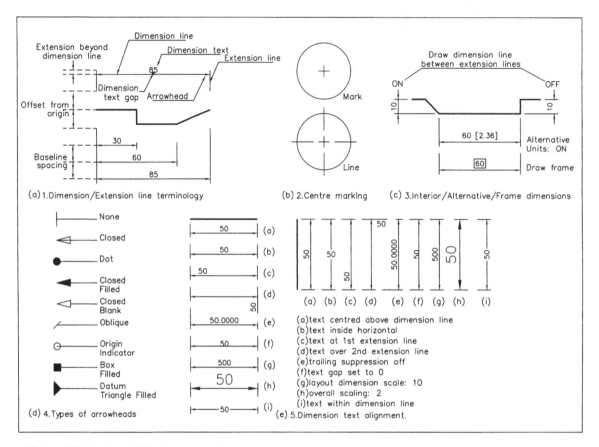

Figure 14.2   AutoCAD dimension line terminology.

2 The dimension terms used in AutoCAD, displayed in Figure 14.2, are:
   a) The dimension and extension lines
      These two features consist of:
      1. the dimension line
         – the actual line
         – the dimension text
         – arrowheads
         – extension lines
      2. the extension line
         – an origin offset from the object
         – an extension beyond the line
         – spacing (for baseline)
   b) Centre marking
      Circular features can be displayed with either a mark, a line or nothing.
   c) Dimension text
      It is possible for the dimension text to:
      1. have interior dimension line drawn or not drawn
      2. display alternative units (i.e. imperial)
      3. have a frame drawn around it.
   d) Arrowheads
      A selection of available arrowheads is displayed.

e) Dimension text alignment

The dimension text can be aligned relative to the dimension line by altering certain dimension variables. A selection of dimension text positions is displayed.

Dimension styles

1 Dimension styles allow the user to set dimension variables to individual/company requirements.

2 This permits various styles to be customised and saved for different clients.

3 To demonstrate how a dimension style is 'set and saved', we will create a new dimension style called A3DIM, use it with WORKDRG and then save it to our standard sheet.

4 *Notes*
   a) The exercise which follows will display several new dialogue boxes and certain settings will be altered within these boxes. It is important for the user to become familiar with these Dimension Style dialogue boxes, as a good knowledge of their use is essential if different dimension styles have to be used.
   b) The settings used in the exercise are my own, designed for our A3PAPER standard sheet.
   c) You can alter the settings to your own values at this stage.

Getting started

1 Close any existing drawing.

2 Open C:\MYCAD\A3PAPER to display your standard sheet.

3 Display the Draw, Modify, Dimension and Object Snap toolbars.

Creating the dimension style A3DIM

1 *Either*   menu bar with **Dimension-Dimension Style**
  or   **Dimension Style icon** from Dimension toolbar.
  prompt   Dimension Style Manager dialogue box
  with   a) Current Dimstyle: ISO-25 or similar
                b) Several other named styles for selection may be displayed
                c) Preview of the current dimension style
                d) Description of current dimension style.
  respond   **pick New**
  prompt   Create New Dimension Style dialogue box
  respond   a) alter New Style Name: A3DIM
                b) Start with: ISO-25 or similar
                c) Use for: All dimensions – Figure 14.3
                d) Pick Continue

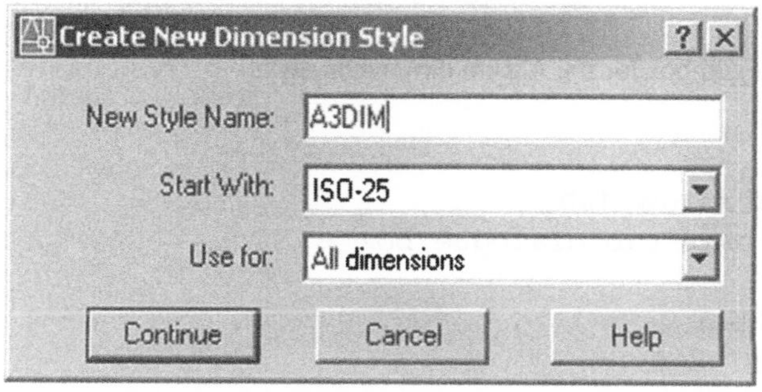

Figure 14.3 The Create New Dimension Style dialogue box.

prompt New Dimension Style: A3DIM dialogue box
with seven tab options for selection
respond **pick the tab name which follows**.

2 *Lines*
 respond **pick Lines tab**
 prompt Lines tab dialogue box
 alter *a*) Dimension Lines
   1. Color: ByBlock
   2. Linetype: ByBlock
   3. Lineweight: ByBlock
   4. Baseline spacing: 10
   5. Suppress: both not active, i.e. blank boxes.
  b) Extension lines
   1. Color: ByBlock
   2. Linetype extension lines: ByBlock
   3. Lineweight: ByBlock
   4. Suppress: both not active
   5. Extend beyond dim lines: 2.5
   6. Offset from origin: 2.5
   7. Fixed length extension lines: not active
 and dialogue box as Figure 14.4.

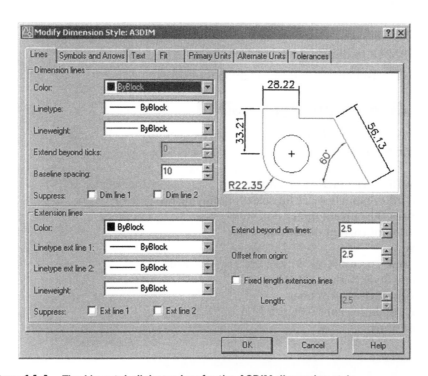

Figure 14.4 The Lines tab dialogue box for the A3DIM dimension style.

3 *Symbols and Arrows*
 respond **pick Symbols and Arrows tab**
 prompt Symbols and Arrows tab dialogue box
 alter *a*) Arrowheads
   1. First: Closed filled
   2. Second: Closed filled
   3. Leader: Closed filled
   4. Arrow size: 4.

    b) Centre marks
      1. Mark active, i.e. black dot
      2. Size: 2.
    c) Arc length symbol
      1. Above dimension text: active.
    d) Radius dimension jog
      1. Jog angle: 45
  and   dialogue box similar to Figure 14.5.

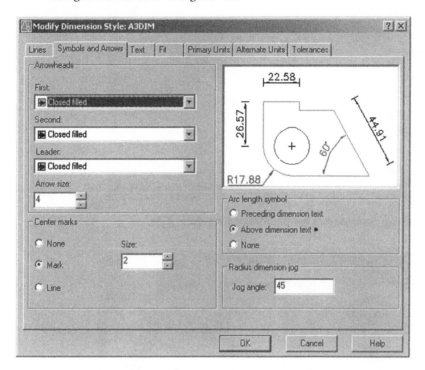

Figure 14.5 The Symbols and Arrows tab dialogue box for the A3DIM dimension style.

4 *Text*
  respond **pick Text tab**
  prompt Text tab dialogue box
  alter   *a*) Text appearance
      1. Text style: scroll and pick ST1
      2. Text color: by Block
      3. Fill color: none
      4. Text height: 5
      5. Draw frame around text: not active.
    b) Text placement
      1. Vertical: Above
      2. Horizontal: Centred
      3. Offset from dim line: 1.5.
    c) Text alignment
      1. ISO Standard: active
  and   dialogue box similar to Figure 14.6.

5 *Fit*
  respond **pick Fit tab**
  prompt Fit tab dialogue box
  alter   *a*) Fit options
      1. Either the text or arrows (best fit): active
      2. Suppress arrows if they don't fit inside extension lines: active.

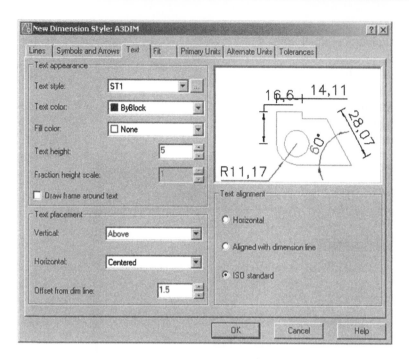

Figure 14.6 The Text tab dialogue box for the A3DIM dimension style.

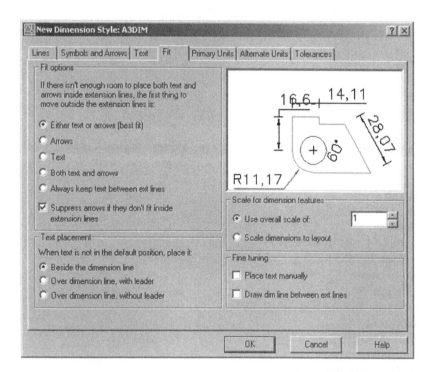

Figure 14.7 The Fit tab dialogue box for the A3DIM dimension style.

    b) Text placement
       1. Beside the dimension line: active.
    c) Scale for dimension features
       1. Use overall scale of: 1.
    d) Fine tuning
       1. Place text manually: not active
       2. Draw dim line between ext lines: not active

and   dialogue box similar to Figure 14.7.

6 *Primary Units* (Figure 14.8)

respond **pick Primary Units tab**

prompt Primary Units tab dialogue box

alter *a*) Linear Dimensions

    1. Unit Format: scroll and pick Decimal if required

    2. Precision: 0.00

    3. Decimal separator: '.' Period

    4. Round off: 0

    5. Prefix and Suffix: both blank

    6. Measurement Scale:

        a) Scale factor: 1

        b) Apply to layout dimension only: not active

    7. Zero Suppression:

        a) Leading: not active

        b) Trailing: active, i.e. tick in box

b) Angular Dimensions

    1. Units Format: Decimal Degrees

    2. Precision: 0.0

    3. Zero Suppression:

        a) Leading: not active

        b) Trailing active

and dialogue box similar Figure 18.5.

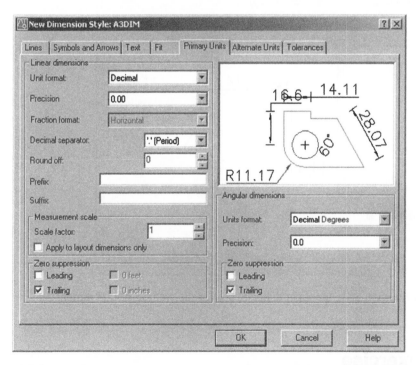

Figure 14.8 The Primary Units tab dialogue box for the A3DIM dimension style.

7 *Alternate Units*

respond **pick Alternate Units tab**

prompt Alternate Units tab dialogue box

respond Display alternate units: not active, i.e. no tick (no alterations to this dialogue box).

8 *Tolerances*

respond **pick Tolerances tab**

prompt	`Tolerances tab dialogue box`
respond	*a*) Tolerance Format
	1. Method: None (no alterations to this dialogue box).

9 *Continue*

respond	**pick OK**
prompt	`Dimension Style Manager dialogue box`
with	*a*) A3DIM added to styles list
	b) Preview of the A3DIM style
	c) Description of the A3DIM style
respond	*a*) pick A3DIM – becomes highlighted
	b) pick Set Current – note description
	c) scroll at List and pick: Styles in use
	d) dialogue box similar to Figure 14.9
	e) pick Close
	f) *Note*: if you have been altering other dimension styles during this process, the AutoCAD Alert message about over-rides may be displayed. If it is, respond as you require.

10   *a*) Menu bar with **File-Save** to automatically update our A3PAPER standard sheet

     b) Remember to save as both a drawing file and a template file.

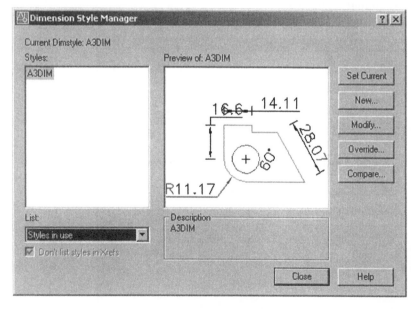

Figure 14.9   The Dimension Style Manager dialogue box for the created A3DIM dimension style.

Dimension exercise

1 Now that we have created a customised dimension style we want to use it and investigate the various dimension types which are available.

2 The WORKDRG will be used for this demonstration, but there is a problem: the A3DIM dimension style has been saved to our A3PAPER standard sheet and is therefore not available for use with the WORDRG drawing.

3 How can this be resolved? Using the AutoCAD Design Centre is the answer.

4 Open your **C:\MYCAD\WORKDRG** to display the component from the fillet/chamfer/offset exercise.

5 Menu bar with **Tools-Design Centre** and:
prompt Design Centre palette
Note: there will be a later chapter dedicated to the Design Centre
respond *a*) scroll at right side until MYCAD folder displayed
b) pick the + at MYCAD to 'expand' the folder
c) pick the + at A3PAPER to 'expand' the file
d) right-click on Dimstyles and pick Explore
e) on right side, right-click A3DIM and pick Add Dimstyle(s)
f) on left side, right-click on Textstyles and pick Explore
g) on right side, right-click on ST1 and pick Add Text Style(s)
h) repeat step 7 for styles ST2 and ST3
i) cancel the Design Centre – top right X
j) menu bar with **Dimension-Dimension Style** and:
prompt Dimension Style Manager dialogue box
with A3DIM style listed
respond 1. pick A3DIM
2. pick Set Current
3. pick Close
and we are now ready to dimension our WORKDRG with the A3DIM dimension style.

Linear dimension

1 Make layer DIMS current and refer to Figure 14.10.

2 Select the **LINEAR DIMENSION icon** from the Dimension toolbar and:
prompt Specify first extension line origin or select object
respond **Endpoint icon and pick line LD1**
prompt Specify second extension line origin
respond **Endpoint icon and pick the other end of line LD1**
prompt Specify dimension line location or
[Mtext/Text/Angle/Horizontal/Vertical/Rotated]
respond **pick any suitable point below line LD1.**

3 From the menu bar select **Dimension-Linear** and:
prompt Specify first extension line origin or <select object>
respond **Endpoint icon and pick line LD2**
prompt Specify second extension line origin
respond **Endpoint icon and pick the other end of line LD2**
prompt Specify dimension line location or ...
respond **pick a suitable point to the right of line LD2.**

Baseline dimension

1 Select the **LINEAR DIMENSION icon** and:
prompt Specify first extension line origin
respond **Endpoint icon and pick point P1**
prompt Specify second extension line origin
respond **Endpoint icon and pick point P2**
prompt Specify dimension line location
respond **pick any suitable point to left.**

2 Select the **BASELINE icon** from the Dimension toolbar and:
prompt Specify a second extension line origin
respond **Endpoint icon and pick point P3**
prompt Specify a second extension line origin
respond **Endpoint icon and pick point P4**

prompt   Specify a second extension line origin
respond   **press the ESC key** to end command.

3   The menu bar selection **Dimension-Baseline** could have been selected for step 2.

Continue dimension

1   Select the **LINEAR DIMENSION icon** and:
prompt   Specify first extension line origin
respond   **Endpoint icon and pick point P4**
prompt   Specify second extension line origin
respond   **Endpoint icon and pick point P5**
prompt   Specify dimension line location
respond   **pick any point above the line.**

2   Menu bar with **Dimension-Continue** and:
prompt   Specify a second extension line origin
respond   **Endpoint icon and pick point P6**
prompt   Specify a second extension line origin
respond   **Endpoint icon and pick point P7**
prompt   Specify a second extension line origin
respond   **press ESC.**

3   The Continue-Dimension icon could have been selected for step 2.

Diameter dimension

Select the **DIAMETER icon** from the Dimension toolbar and:
prompt   Select arc or circle
respond   **pick the larger circle**
prompt   Dimension text = 50
            Specify dimension line location or [Mtext/Text/Angle]
respond   **drag out and pick a suitable point.**

Radius dimension

Select the **RADIUS icon** and:
prompt   Select arc or circle
respond   **pick the right arc at point P7**
prompt   Dimension text = 20
            Specify dimension line location or [Mtext/Text/Angle]
respond   **drag out and pick a suitable point.**

Angular dimension

Select the **ANGULAR icon** and:
prompt   Select arc, circle, line or <specify vertex>
respond   **pick line LD1**
prompt   Select second line
respond   **pick line P1-P2**
prompt   Specify dimension arc line location or [Mtext/Text/Angle]
respond   **drag out and pick a point to suit.**

Aligned dimension

Select the **ALIGNED icon** and:
prompt   Specify first extension line origin
respond   **Endpoint icon and pick point P5**

prompt   Specify second extension line origin
respond **Endpoint icon and pick point P6**
prompt   Specify dimension line location or [Mtext/Text/Angle]
respond **pick any point to suit.**

Leader dimension

Menu bar with **Dimension-Leader** and:
prompt   Specify first leader point or [Settings]
respond **Nearest icon and pick any point on smaller circle**
prompt   Specify next point
respond **drag to a suitable point and pick**
prompt   Specify next point
respond **right-click**
prompt   Specify text width<0>
respond **right-click**
prompt   Enter first line of annotation text
enter   **Circle with <R>**
prompt   Enter next line of annotation text
enter   **a radius of 20<R>**
prompt   Enter next line of annotation text
respond **right-click to end leader dimension command.**

Jogged dimension

Menu bar with **Dimension-Jogged** and:
prompt   Select arc or circle
respond **pick the smaller circle**
prompt   Specify centre location over-ride
respond **pick a suitable point**
prompt   Specify dimension line location or [Mtext/Text/Angle]
respond **pick point to position dimension text**
prompt   Specify jog location
respond **move cursor and place 'jog' position to suit**
and   command line returned.

Arc length

Menu bar with **Dimension-Arc Length** and:
prompt   Select arc or polyline arc segment
respond **pick arc AR1**
prompt   Specify arc length dimension location or [Mtext/Text/
           Angle/Partial]
respond **pick position to suit**
and   command line returned.

Notes
1   At this stage your drawing should resemble Figure 14.10 and can be saved.

2   Object snap is used extensively when dimensioning. This is one time when a running Object Snap (e.g. Endpoint) will assist, but remember to cancel the running object snap!

3   *a*) From the menu bar select **Format-Layer** to display the Layer Properties Manager dialogue box
    b) Note the layer **Defpoints** – we did not create this layer
    c) It is automatically made by AutoCAD any time a dimension is added to a drawing

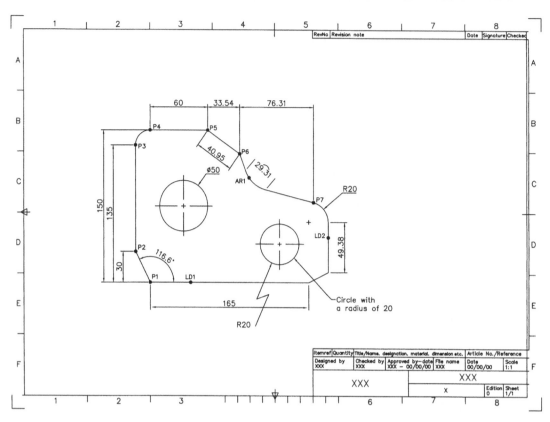

Figure 14.10 Dimensioned WORKDRG using dimension style A3DIM.

d) This layer can be turned off or frozen but cannot be deleted. **It is best left untouched**.

4 The A3DIM style can be modified at any time. To demonstrate this:
 a) Dimensioned WORKDRG should still be displayed
 b) Menu bar with Dimension-Dimension Style and:

> *prompt*   Dimension Style Manager dialogue box
> *with*   style A3DIM highlighted, i.e. current
> *respond*   1. pick Modify to display the seven tab selection dialogue boxes
>   2. activate the Text tab
>   3. text style: scroll and pick ST3
>   4. activate the Symbols and Arrows tab
>   5. arrowheads: scroll and select Dot small for first, second and leader
>   6. set arrowsize to 10
>   7. pick OK
> *prompt*   Dimension Style Manager dialogue box
> *respond*   pick Close
> *and*   the dimensioned component will be displayed with the dimension text with style ST3 and dots will have replaced the arrowheads.

 c) Do not save these changes to A3DIM.

Ordinate dimensions

This type of dimensioning is very popular with many companies and will now be investigated.

1 *a*) The dimensioned WORKDRG should still be displayed
 b) Erase all objects from the screen and refer to Figure 14.11.

2 With layer OUT current, draw the following objects using absolute co-ordinate entry:

LINE		*CIRCLE*	
First point:	60,60	centre:	100,110
Next point:	170,70	radius:	20
Next point:	130,150		
Next point:	40,160		
Next point:	close.		

3 Make the DIMS layer current.

4 Select the **ORDINATE icon** from the Dimension toolbar and:

prompt	Specify feature location
respond	**pick point A** – snap on helps
prompt	Specify leader endpoint or [Xdatum/Ydatum/Mtext/Text/Angle]
enter	**X <R>** – the Xdatum option
prompt	Specify leader endpoint or [Xdatum/Ydatum/Mtext/Text/Angle]
enter	**@0,−10 <R>**
prompt	Dimension text=60
and	command line returned.

5 Menu bar with **Dimension-Ordinate** and:

prompt	Specify feature location
respond	**pick point A**
prompt	Specify leader endpoint or [Xdatum/Ydatum/Mtext/Text/Angle]
enter	**Y <R>**
prompt	Specify leader endpoint or [Xdatum/Ydatum/Mtext/Text/Angle]
enter	**@−10,0 <R>**
prompt	Dimension text=60
and	command line returned.

6 Repeat the Ordinate dimension command and:
   a) select point B as the feature location
   b) with snap on, drag vertically downwards and click to suit.

7 With the Ordinate dimension active:
   a) select point B as the feature location
   b) drag horizontally to the right and click to suit to position the ordinate dimension.

8 Now add ordinate dimensions to the points C and D of the outline shape and point E, the circle centre. The displayed dimensions should be the same as your co-ordinate entries from step 2.

9 The result should be similar to Figure 14.11(a).

Quick dimensioning

1 AutoCAD 2006 has a quick dimensioning option.

2 To demonstrate the topic, refer to Figure 14.11(b) and draw a shape consisting of lines and arcs/circles.

3 Layer DIMS should still be current.

4 Menu bar with **Dimension-Quick Dimension** and:

prompt	Select geometry to dimension

respond	**window the complete shape then right-click**
prompt	Specify dimension line position or [Continuous/
	Staggered/Baseline/Ordinate/radius/Diameter ...
enter	B <R> – the baseline option
prompt	Specify dimension line position or ...
respond	**pick any point to the right of the shape**.

5 *a*) all vertical dimensions will be displayed from the horizontal baseline of the shape
     as Figure 14.11(b)
   b) the user can now delete any dimensions which are unwanted.

6 Erase all the dimensions.

7 Select the **QUICK DIMENSION icon** from the Dimension toolbar and:

prompt	Select geometry to dimension
respond	**window the shape then right-click**
prompt	Specify dimension line position or [Continuous/
	Staggered/Baseline ...
enter	**R <R>** – the radius option
prompt	Specify dimension line location
respond	**pick any point to suit**.

8 All radius dimensions for the component will be displayed as Figure 14.11(c).

9 Erase the radius dimensions then use the Quick dimension command to add:
   a) staggered dimensions to the component as Figure 14.11(d)
   b) ordinate dimensions as Figure 14.11(e).

10 If Quick Dimension is being used, then it may be necessary to use the command sev-
    eral times to obtain the required dimension effect. QDIM is another method of adding
    dimensions to a drawing and it is the user decision as to whether to use quick dimen-
    sions or 'ordinary' dimensions.

11 The command line entry to activate the Quick Dimension command is QDIM.

Dimension options

When using the dimension commands, the user may be aware of various options
when the prompts are displayed. To investigate these options:

1 Continue with the current drawing – Figure 14.11.

2 Make layer OUT current and draw six horizontal lines of length 60 at the top of the
   screen.

3 Make layer DIMS current.

4 *The RETURN option*
   Select the **LINEAR DIMENSION icon** and:

prompt	Specify first extension line origin or <select object>
respond	**press the RETURN/ENTER key**
prompt	Select object to dimension
respond	**pick the first line**
prompt	Specify dimension line location
respond	**pick below the line** – Figure 14.11(q).

5 *The ANGLE option*
   Select the **LINEAR DIMENSION icon**, press RETURN, pick the second line and:

prompt	Specify dimension line location
	[Mtext/Text/Angle/Horizontal/Vertical/Rotated]

enter	**A <R>** – the angle option
prompt	`Specify angle of dimension text`
enter	**45 <R>**
prompt	`Specify dimension line location`
respond	**pick a point to suit** – Figure 14.11(r).

6 *The ROTATED option*
LINEAR dimension icon, right-click, pick third line and:

prompt	`Specify dimension line location`
	`[Mtext/Text/Angle/Horizontal/Vertical/Rotated]`
enter	**R <R>** – the rotated option
prompt	`Specify angle of dimension line <0>`
enter	**15 <R>**
prompt	`Specify dimension line location`
respond	**pick to suit** – Figure 14.11(s).

7 *The TEXT option*
LINEAR dimension icon, right-click, pick the fourth line and:

prompt	`Specify dimension line location`
	`[Mtext/Text/Angle/Horizontal/Vertical/Rotated]`
enter	**T <R>** – the text option
prompt	`Enter dimension text<60>`
enter	**THIS IS 60 <R>**
prompt	`Specify dimension line location`
respond	**pick to suit** – Figure 14.11(t).

8 *Dimensioning with keyboard entry*
At the command line enter **DIM <R>** and:

prompt	`Dim`
enter	HOR <R> – horizontal dimension
prompt	`Specify first extension line origin`
respond	**right-click and pick the next line**
prompt	`Specify dimension line location`
respond	**pick to suit**
prompt	`Enter dimension text<60>`
enter	**SIXTY <R>** – Figure 14.11(u)
prompt	`Dim returned at command line`
respond	**Press ESC to end command**.

9 *Dimension cheating* – not recommended
Enter **DIM <R>** at the command line and:

prompt	`Dim`
enter	**HOR <R>** – horizontal dimension
prompt	`Specify first extension line origin`
respond	**right-click and pick the last line**
prompt	`Specify dimension line location`
respond	**pick to suit**
prompt	`Enter dimension text<60>`
enter	**2006 <R>** – Figure 14.11(v)
prompt	`Dim returned`
respond	**ESC to end command**.

10 *a*) Your screen should resemble Figure 14.11
b) The exercise is now complete and does not need to be saved.

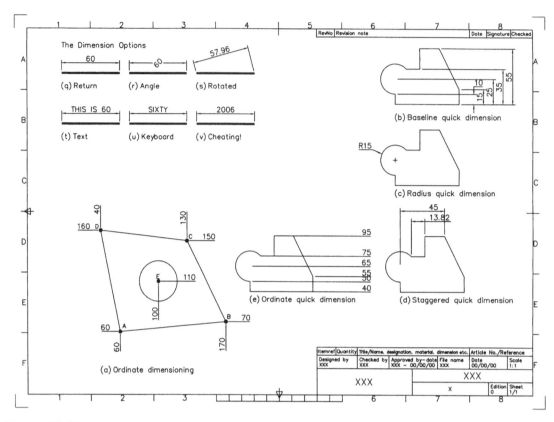

Figure 14.11 Dimensions: ordinate, quick and options.

Assignments

As dimensioning is an important concept, I have included three activities which will give you practice with:

1 using the standard sheet with layers and the draw commands

2 adding suitable text

3 adding dimensions with the A3DIM dimension style.

In each activity the procedure is the same:

1 Open the A3PAPER standard drawing sheet

2 Using layers correctly, complete the drawings

3 Save the completed work as C:\MYCAD\ACT11, etc.

Activity 11

Two simple components to be drawn and dimensioned. The sizes are more awkward than usual.

Activity 12

Two components created mainly from circles and arcs. Use offset as much as possible. The signal arm is interesting and more difficult than you would expect.

Activity 13

A component which is much easier to complete than it would appear. Offset and fillet will assist.

Chapter 15

The MODIFY commands 2

1   The draw and modify commands are probably the most commonly used of all the AutoCAD commands and we have already used several of each.

2   The modify commands discussed previously have been Erase, Offset, Trim, Extend, Fillet and Chamfer.

3   In this chapter, we will use C:\MYCAD\WORKDRG to investigate several other modify commands as well as some additional selection set options.

Getting ready

1   Open C:\MYCAD\WORKDRG to display the dimensioned red component with green centre lines.

2   *a*) Erase all dimensions
    b) Layer OUT current, with the Draw, Modify and Object Snap toolbars displayed
    c) Freeze layer CL – you will find out why shortly
    d) Menu bar with **View-Zoom-All**.

Copy

1   Allows objects to be copied to other parts of the screen.

2   The command can be used for single or multiple copies of selected objects.

3   The user specifies a:
    a) start (base) point
    b) displacement (or second point).

4   Select the **COPY icon** from the Modify toolbar, refer to Figure 15.1 and:
prompt	Select objects
enter	**C <R>** – the crossing selection set option
prompt	Specify first corner
respond	**pick a point P1**
prompt	Specify opposite corner
respond	**pick a point P2**
but	**DO NOT RIGHT-CLICK YET**
prompt	**11 found – note objects not highlighted**
then	Select objects, **i.e. any more objects to be included in the selection**
enter	**A <R>** – the add selection set option
prompt	Select objects
respond	**pick objects C1, C2 and C3 then right-click**
and	note command line as each object is selected
prompt	Specify base point or [Displacement]
enter	**80,110 <R>** – note copy image as mouse moved
prompt	Specify second point or <use first point as displacement>
enter	**300,300 <R>**

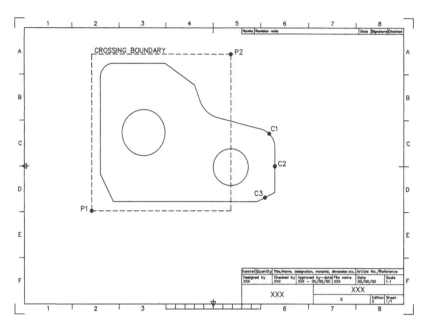

Figure 15.1 WORKDRG with selection information for the COPY command.

prompt	Specify second point or [Exit/Undo]
enter	**@220,0**
prompt	Specify second point or [Exit/Undo]
enter	**@0,−200**
prompt	Specify second point or [Exit/Undo]
enter	**@185<110**
prompt	Specify second point or [Exit/Undo]
respond	**right-click-enter**.

5 The original component will be copied to other parts of the screen and will not all be visible.

6 *a*) Don't panic!
   b) Select from the menu bar **View-Zoom-All** to 'see' the complete copied effect –
      Figure 15.2.
   c) You may have to reposition your toolbars?

Move

1 Allows selected objects to be moved to other parts of the screen.

2 The user defines the start and end points of the move by:
   a) co-ordinate entry
   b) picking points on the screen
   c) referencing existing entities.

3 Refer to Figure 15.2 and erase the shapes indicated.

4 Select the **MOVE icon** from the Modify toolbar:

prompt	Select objects
enter	**W <R>** – the window selection set option
prompt	Specify first corner
respond	**pick a point P1**
prompt	Specify opposite corner
respond	**pick a point P2**
but	**DO NOT RIGHT-CLICK YET**

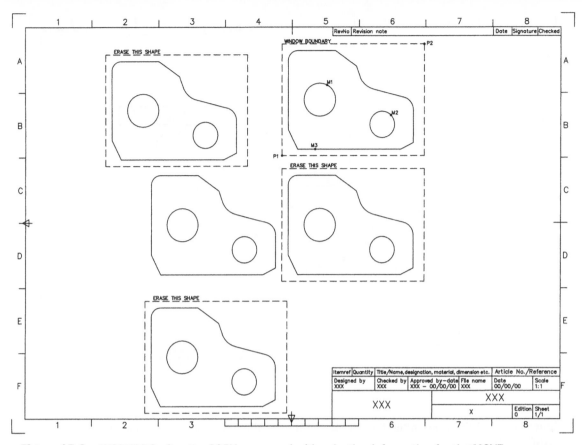

Figure 15.2 WORKDRG after the COPY command with selection information for the MOVE command.

prompt	**14 found**
and	Select objects
enter	**R <R>** – the remove selection set option
prompt	Remove objects
respond	**pick circles M1 and M2 then right-click**
and	note command line as each object is selected
prompt	Specify base point or [Displacement]
respond	**Endpoint icon and pick line M3 at end indicated** (note image as cursor is moved)
prompt	Specify second point or <use first point as displacement>
enter	**@–85,–70 <R>**.

5   The result will be Figure 15.3, i.e. the red outline shape is moved, but the two circles do not move due to the Remove option.

6   *Task*

    a) with the Layer Properties Manager dialogue box, THAW layer CL

    b) the centre lines are still in their original positions, i.e. they have not been copied or moved

    c) this can be a problem with objects which are on frozen or off layers.

7   Now:

    a) freeze layer CL

    b) erase the copied-moved objects to leave the original WORKDRG

    c) View-Zoom-All to 'restore' the original screen.

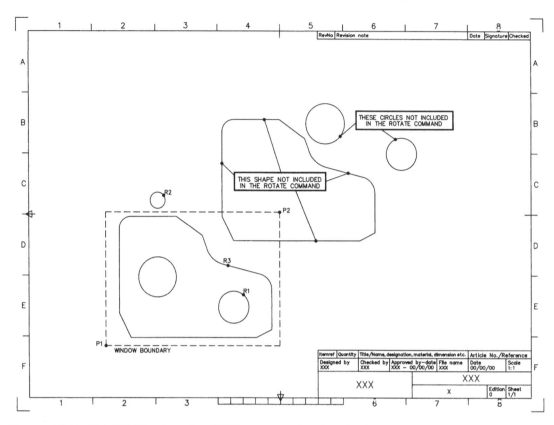

Figure 15.3 WORKDRG after the MOVE command with selection information for the ROTATE command.

Rotate

1   Selected objects can be 'turned' about a designated point in either a clockwise (−ve angle) or counter-clockwise (+ve angle) direction.

2   The user specifies:
    a) a base point for the rotation
    b) an angle of rotation.

3   The base point can be:
    a) selected as a point on the screen
    b) entered as a co-ordinate value
    c) referenced to existing objects.

4   Refer to Figure 15.3 and draw a circle of radius 10 with the centre point at 130,280.

5   Menu bar with **Modify-Rotate** and:

prompt	Current positive angle in UCS: ANGDIR=counterclockwise: ANGBASE=0.0
then	Select objects
respond	**window from P1 to P2 – but NO right-click yet**
prompt	14 found
then	Select objects
enter	**R <R> –** the remove option
prompt	Remove objects
respond	**pick circle R1**
prompt	1 found, 1 removed, 13 total
then	Remove objects
enter	**A <R> –** the add option

prompt	Select objects
respond	**pick circle R2**
prompt	1 found, 14 total
then	Select objects
respond	**right-click** to end selection sequence
prompt	Specify base point
respond	**Endpoint icon and pick line R3** – at 'upper end'
prompt	Specify rotation angle or [Copy/Reference]
enter	**−90 <R>**.

6 *a*) the selected objects will be rotated as displayed in Figure 15.4
   b) use your scroll bars to 'position' WORKDRG to your requirements.

7 Note that two selection set options were used in this single rotate sequence:
   a) R – removed a circle from the selection set
   b) A – added a circle to the selection set.

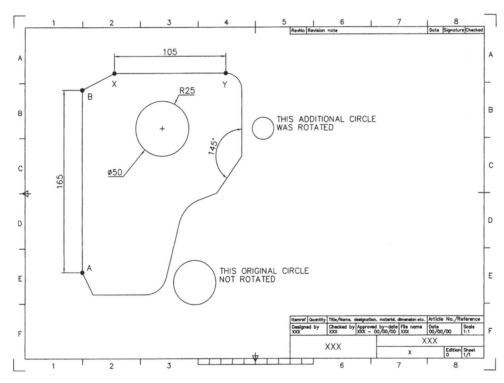

Figure 15.4 WORKDRG after the ROTATE command with information for the SCALE command.

Scale

1 The scale command allows selected objects or complete 'shapes' to be increased/decreased in size.

2 The user specifies:
   a) a base point
   b) the actual scale factor.

3 Refer to Figure 15.4 and erase the two circles 'outside the shape'.

4 Make layer DIMS current and with the Dimension command (icon or menu bar):
   a) linear dimension line AB
   b) diameter dimension the circle
   c) angular dimension as indicated.

5   At the command line enter **DIM <R>** and:
    prompt   Dim
    enter   **HOR <R>**
    prompt   Specify first extension line origin
    respond   **Endpoint icon and pick line at X**
    prompt   Specify second extension line origin
    respond   **Endpoint icon and pick line at Y**
    prompt   Specify dimension line location
    respond   **pick above line**
    prompt   Enter dimension text<105>
    enter   **105 <R>**
    prompt   Dim
    enter   **RAD <R>**
    prompt   Select arc or circle
    respond   **pick the circle**
    prompt   Enter dimension text<25>
    enter   **R25 <R>**
    prompt   Specify dimension text location
    respond   **pick to suit**
    prompt   Dim and press **ESC** to end sequence.

6   Make layer OUT current.

7   Select the **SCALE icon** from the Modify toolbar and:
    prompt   Select objects
    respond   **window the complete shape with dimensions then right-click**
    prompt   Specify base point
    respond   **Endpoint of line AB at 'B' end**
    prompt   Specify scale factor or [Reference]
    enter   **0.5 <R>**.

8   The complete shape will be scaled as Figure 15.5.

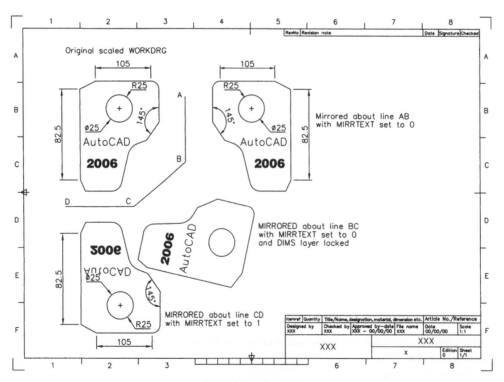

Figure 15.5   WORKDRG after the SCALE and MIRROR commands.

9 Note the dimensions:
   a) the vertical dimension of 165 is now 82.5
   b) the diameter value of 50 is now 25
   c) the horizontal dimension of 105 is still 105
   d) the radius of 25 is still 25.

10 *Questions*:
   a) Why have two dimensions been scaled by 0.5 and two have not?
       Answer: 1. the scaled dimensions are those which used the icon or menu bar selection
               2. the dimensions not scaled were dimensioned with the command line DIM.
   b) Which of the resultant dimensions are correct?
       Is it the 82.5 and diameter 25, or the 105 and R25?
       Answer: I will let you reason this one for yourself, but I can assure you that it causes
               a great deal of debate.

Mirror

1 Allows objects to be mirror imaged about a line (real or imaginary) designated by the user.

2 The command has an option for deleting the original set of objects – the source objects.

3 Refer to Figure 15.5 and with layer OUT current, draw the following three line segments:
   a) first point 110,215   point D
   b) next point 175,215   point C
   c) next point 225,255   point B
   d) next point 225,315   point A

4 With layer TEXT current, add the following two text items:

Style	Position	ht	rot	Item
ST2	centred on 150,270	7	0	AutoCAD
ST3	centred on 145,250	8	0	2006

5 At the command line enter **MIRRTEXT <R>**
   prompt   Enter new value for MIRRTEXT<?>
   enter   **0 <R>**.

6 Make layer OUT current and select the **MIRROR icon** from the Modify toolbar and:
   prompt   Select objects
   respond   **window the shape, text and dimensions then right-click**
   prompt   Specify first point of mirror line
   respond   **Endpoint icon and pick point A**
   prompt   Specify second point of mirror line
   respond   **Endpoint icon and pick point B**
   prompt   Erase source objects [Yes/No]<N>
   enter   **N <R>**.

7 The selected objects are mirrored about the line AB. The dimensions may surprise the user?

8 The system variable MIRRTEXT (step 5) determines the appearance of text during the mirror command and:
   a) if MIRRTEXT is set to 1 then text items are mirrored
   b) if MIRRTEXT is 0 then text items are not mirrored.

9 Set the MIRRTEXT system variable to 1 then menu bar with **Modify-Mirror** and:
   prompt   Select objects
   enter   **P <R>** – previous selection set option

prompt	`20 found`
then	`Select objects`
respond	**right-click** to end selection
prompt	`Specify first point of mirror line` and **pick endpoint of point C**
prompt	`Specify second point of mirror line` and **pick endpoint of point D**
prompt	`Delete source objects` and enter **N <R>**.

10 The shape is mirrored. What about the text and dimensions?

11 *a*) Set MIRRTEXT to 0
    b) Lock layer DIMS.

12 At the command line enter **MIRROR <R>** and:
    a) select objects: enter P <R><R> – yes two returns – why?
    b) first point – pick endpoint of C
    c) second point – pick endpoint of B
    d) delete – N.

13 The final result should be similar Figure 15.5.

14 *Task*: Before leaving the exercise, thaw layer CL. The circle centre lines are still in their original positions.

Copy and reference options

1 The rotate and scale commands have a copy and reference option and we will use the rotate option to demonstrate these options.

2 Close any existing drawing (no save) and open the original WORKDRG. Refer to Figure 15.6.

3 Draw a 30-radius circle with centre at 200,180.

4 Activate the **rotate** command, window the shape (not the circle) right-click and:

prompt	`Specify base point`
respond	**Endpoint icon and pick as indicated**
prompt	`Specify rotation angle or [Reference]`
enter	**R <R>** – the reference option
prompt	`Specify the reference angle` and enter **70 <R>**
prompt	`Specify the new angle or [Points]` and enter **90 <R>**
and	the selected shape will be rotated through an angle of 20 degrees as Figure 15.6(a).

5 This 20 degree angle is equivalent to (new entered angle value – reference angle value).

6 Undo the rotate effect to restore the original WORKDRG.

7 With the rotate command:
    a) objects: window the WORKDRG shape then right-click
    b) base point: endpoint icon and pick as indicated
    c) rotation angle: quadrant icon and pick point A on circle
    d) the selected shape will be rotated about the base point as Figure 15.6(b)
    e) the actual angle of rotation is from the horizontal to a line from the base point through the selected quadrant of the circle. I dimensioned this angle and it was 36.6 degrees.

8 Undo the rotate effect.

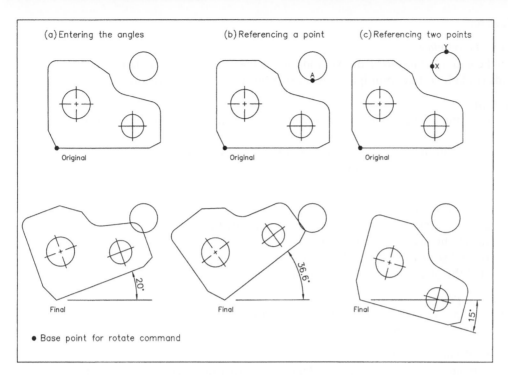

Figure 15.6   Using the REFERENCE option of the ROTATE command with WORKDRG.

9 With the rotate command again:
   a) objects: window the shape and right-click
   b) base point: endpoint icon and pick as indicated
   c) rotation angle: enter **R <R>**, the reference option and:
prompt	Specify the reference angle
respond	**quadrant icon and pick point X as indicated**
prompt	Specify second point
respond	**quadrant icon and pick point Y as indicated**
prompt	Specify new angle
enter	**30 <R>**
   d) the selected shape will be rotated about the base point as Figure 15.6(c)
   e) the actual rotation angle is equivalent to −15 degrees (new–reference)
   f) I will let you work out how this (new entered angle value – reference angle value) gives −75.

10 This exercise does not need to be saved.

Assignments

1 Several assignments have been included for this chapter as the Modify commands are used extensively in CAD.

2 Each assignment activity should be completed on the A3PAPER standard drawing sheet with layers used correctly.

3 The components do not need to be dimensioned, but you could attempt some dimensioning.

4 Remember to save your completed drawings as C:\MYCAD\ACT?? etc.

5 Use your discretion for sizes which are not given.

The activities are:

1   *Activity 14: A vent cover plate*
    a) The original component has to be drawn and then copied into the rectangular plate.
    b) Commands could be rotate, multiple copy, mirror but that is your decision.

2   *Activity 15: Designs*
    a) You have to create the basic shapes then multiple copy them into the geometric outline.
    b) Relatively simple, but involves extensive use of the TRIM command.
    c) The geometric shapes are to your own sizes.

3   *Activity 16: A template*
    a) This is easier to complete than you may think.
    b) Draw the quarter template using the sizes given then MIRROR.
    c) The text item is to be placed at your discretion.
    d) Multiple copy and scale by half, third and quarter.
    e) Additional: add the given dimensions to a quarter template.

4   *Activity 17: An Aztec pattern*
    a) With grid and snap set to 5, create the basic shape (or design your own).
    b) The shape is to 'fit into' a 75-sided square.

5   *Activity 18: Two well known objects*
    a) 1. With the simple steps given, the SUNGLASSES are easier than you would expect.
       2. Really a drawing exercise, the only modify command being mirror to get the complete effect.
       3. Some sizes of your own in step 3.
       4. Add the arc aligned text.
    b) 5. The screwdriver can be completed as a half shape then mirrored.
       6. Use offset as much as possible.
       7. Add the dimensions to the completed shape and check dimensions A. It should be 21.54.

Chapter 16

Grips

1 Grips are an aid to the draughting process and offer the user access to a limited number of modify commands.

2 *a*) In an earlier chapter we 'turned grips off' with the command line entry of GRIPS 0.
  b) This was to allow the user to become reasonably proficient with using the draw and modify commands.
  c) Now that this has been achieved, we will investigate how grips are used.

Toggling grips on/off

Grips can be toggled on/off using:

a) the Selection tab from the Options dialogue box
b) command line entry.

1 Open your A3PAPER standard sheet with layer OUT current.

2 From the menu bar select **Tools-Options** and:
  prompt Options dialogue box
  respond **pick Selection tab**
  prompt Selection tab dialogue box
  with five distinct sections, two of which refer to grips:
    a) Grip Size
    b) Grips
  respond refer to Figure 16.1 and:
    a) set grip size to suit
    b) set the three grip colours to:
      1. unselected grip: blue
      2. selected grip: red
      3. hover grip: green
    c) Enable grips: active, i.e. tick in box
    d) Enable grips within blocks: not active
    e) Enable grip tips: not active
    f) Object selection limit: 100
    g) pick Apply then OK
  and the drawing screen will be returned, with the grip box 'attached' to cursor cross-hairs.

3 At the command line enter **GRIPS <R>** and:
  prompt Enter new value for GRIPS<1>
  respond **<RETURN>** i.e. leave the 1 value.

4 The command entry to toggle grips on/off is GRIPS set to 1 are ON, GRIPS set to 0 are OFF.

5 *Notes*
  a) the grip box attached to the cross-hairs should not be confused with the pick box used with modify commands. Although similar in appearance they are entirely different.
  b) when any command is activated (e.g. LINE), the grip box will disappear from the cross-hairs, and re-appear when the command is terminated.

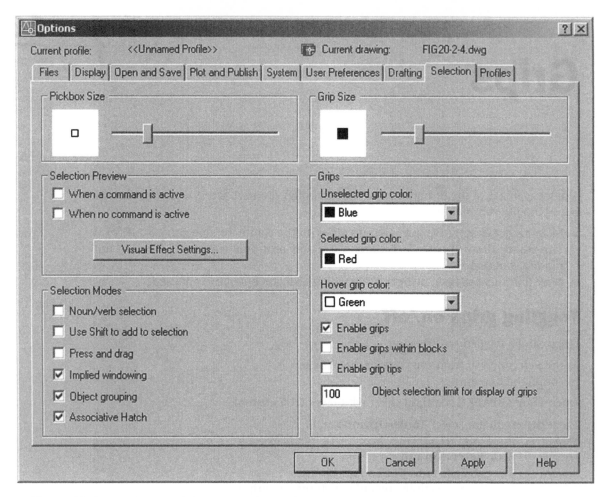

Figure 16.1   The Options dialogue box with the Selection tab for the grip settings.

What do grips do and how do they work?

1   Grips are small solid filled squares displayed at strategic points on objects selected by the user.

2   Grips provide the user with five modify commands: stretch, move, rotate, scale and mirror.

3   Grips work in the 'opposite sense' from normal command selection:
   a) the usual sequence is to activate the command then select the objects, e.g. MOVE, then pick the object to be moved
   b) with grips, the user selects the objects and then activates one of the five grip options.

4   To demonstrate how grips work:
   a) Draw a line, circle, arc and text item anywhere on the screen
   b) Ensure grips are on
   c) Refer to Figure 16.2 and move the cursor to each object, 'pick them' with the grip box and:
      1. solid blue grip boxes appear at each object 'snap point' and the selected object is highlighted as Figure 16.2(a)
      2. move the grip box on the cursor cross-hairs over any solid blue box and leave for a second. The solid blue box will change to a green solid box. This is the **hover grip**

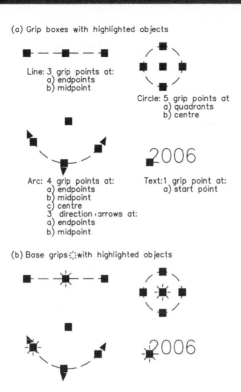

Figure 16.2 The grip types.

3. move the grip box to any one of the solid blue boxes and 'pick it'. The solid blue box changes to a solid red box and the selected object is still highlighted. The red box is the **BASE GRIP** – Figure 16.2(b)
4. press the ESC key – solid blue grips with highlighted object
5. ESC again to cancel the grips operation and restore the selected objects to their original appearance.

Grip exercise

This demonstration is relatively simple but rather long. It is advisable to work through the exercise without missing out any of the steps:

1　Erase all objects from the screen, or re-open A:A3PAPER.

2　*a*) Refer to Figure 16.3 and draw the original shape using the sizes given as Figure 16.3(a).
　　b) Make the lower left corner at the point (100,100) and ensure grips are on.

3　*a*) Move the cursor to the circle and pick it, then move to the right vertical line and pick it.
　　b) Solid blue grip boxes appear and the two objects are highlighted as Figure 16.3(b).

4　*a*) Move the cursor grip box over the grip box at the circle centre and the green hover grip will be displayed.
　　b) Left click, i.e. pick the circle centre grip.
　　c) The selected box will be displayed in red as it is now the base grip as Figure 16.3(c).

5　Observe the command line:
　　prompt　　`** STRETCH **`
　　　　　　　　`Specify stretch point or [Base point/Copy/Undo/eXit]`
　　respond　　**with a <RETURN>**

```
prompt      ** MOVE **
            Specify move point or [Base point/Copy/Undo/eXit]
enter       @50<25<R>
```

6 The following should have happened:
a) the circle and line are moved
b) the command prompt line is returned
c) the grips are still active
d) there is no base grip – Figure 16.3(d).

7 Move the cursor and pick the text item to add it to the grip selection – Figure 16.3(e).

8 Make the left grip box of the text item the base grip, by moving the cursor pickbox onto it, left-clicking as Figure 16.3(f) and:

```
prompt      ** STRETCH **
            Specify move point or [Base point/Copy/Undo/eXit]
respond     with a <RETURN>
prompt      ** MOVE **
            Specify move point or [Base point/Copy/Undo/eXit]
respond     <RETURN>
prompt      ** ROTATE **
            Specify move point or [Base point/Copy/Undo/Reference/
            eXit]
enter       −25 <R>.
```

9 The circle, line and text item will be rotated and the grips are still active – Figure 16.3(g).

10 Make the same text item grip box the base grip (easy!) and:

```
prompt      ** STRETCH **
            Specify move point or [Base point/Copy/Undo/eXit]
enter       SC <R> – the scale grip option
prompt      ** SCALE **
            Specify move point or [Base point/Copy/Undo/Reference/
            eXit]
enter       0.75 <R>.
```

11 The three objects are scaled and the grips are still active as Figure 16.3(h).

12 Make the lowest box on the line the base grip and:

```
prompt      ** STRETCH **
enter       MI <R> – the mirror option
prompt      ** MIRROR **
            Specify second point or [Base point/Copy/Undo/eXit]
enter       B <R> – the base point option
prompt      Specify base point
respond     Midpoint icon and pick the original horizontal line
prompt      ** MIRROR **
            Specify second point or [Base point/Copy/Undo/eXit]
respond     Midpoint icon and pick the arc.
```

13 The three objects are mirrored about the selected 'line' and the grips are still active as Figure 16.3(i).

14 Press ESC – removes the grips and ends the sequence.

15 The exercise is now complete. Do not exit yet.

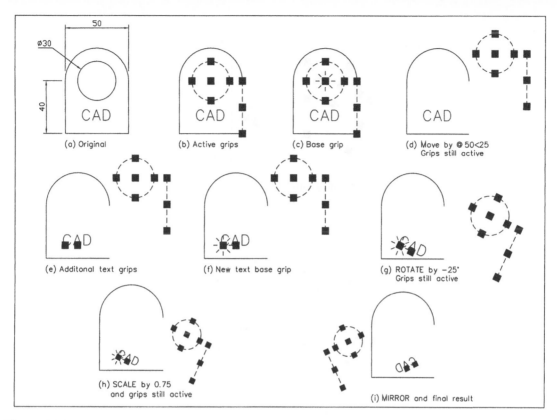

Figure 16.3 Grip exercise 1.

Selection with grips

1 Selecting individual objects for use with grips can be tedious.

2 It is possible to select a window/crossing option when grips are on.

3 Your screen should still display the line, circle and text item after the grips exercise has been completed?

4 *a*) Refer to Figure 16.4(a), move the cursor and pick a first point 'roughly' where indicated.
 b) Move the cursor down and to the right, pick a second point and all complete objects within the window will display solid blue grip boxes with highlighted objects.

5 ESC to cancel the grip effect.

6 *a*) Refer to Figure 16.4(b), move the cursor and pick a first point 'roughly' where indicated.
 b) Move the cursor upwards and to the left and pick a second point.
 c) All objects within or which cross the boundary will display solid blue grip boxes with highlighted objects.

7 ESC to cancel grip selection.

8 The effect can be summarised as:
 a) Window effect to the right of first pick.
 b) Crossing effect to the left of the first pick.

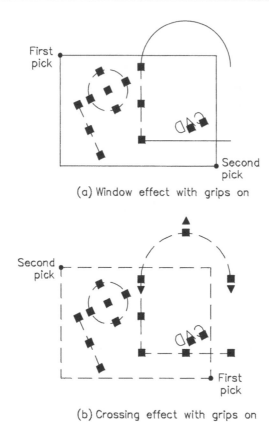

(a) Window effect with grips on

(b) Crossing effect with grips on

Figure 16.4 Window/crossing selection with grips.

Figure 16.5 The grips shortcut menu.

The grips shortcut menu

1 When grips are being used and the base grip has been selected, a right-click will display the grips shortcut menu as Figure 16.5.

2 This allows the user access to the grip options.

Assignment

Activity 19 involves the re-positioning of a robotic arm and using the arm to re-position a component:

1 Refer to Activity 19(a) and:
   a) create the robotic arm in the original position using the sizes given
   b) use your discretion for sizes not given
   c) position the two pallets (A and B) and the component C.

2 Refer to Activity 19(b) and:
   Fig(a): the original positions
   Fig(b): the upper and lower arms MIRRORED about a vertical line through the large lower arm circle with the hot grip as the large circle centre
   Fig(c): the upper arm MOVED, STRETCHED and ROTATED onto the component
   Fig(d): the upper arm ROTATED to a horizontal position and STRETCHED with component attached
   Fig(e): the upper arm STRETCHED and ROTATED to place the component on pallet B
   Fig(f): the robot home position with the lower arm vertical and the upper arm horizontal.

3 Save when complete.

This completes the grip chapter.

Chapter 17

Viewing drawings

1   Viewing a drawing is important as the user:
    a) may want to enlarge a certain part of the screen for more detailed work
    b) return the screen to a previous display.

2   AutoCAD allows several methods of altering the screen display including:
    a) the scroll bars
    b) the Pan and Zoom commands
    c) the Aerial view option
    d) Realtime pan and zoom.

3   In this chapter we will investigate several of these view options.

Getting ready

1   Open WORKDRG and:
    a) erase any dimensions
    b) Grips on or off – your decision.

2   With layer TEXT current, menu bar with **Draw-Text-Single Line Text** and:
    a) start point: **centred** on 120,175
    b) height: 0.1 – yes 0.1
    c) rotation: 0
    d) text item: AutoCAD.

3   *a*) with layer OUT current, draw a circle centred on 119.98,175.035 and radius 0.01
    b) the awkward co-ordinate entries are deliberate.

4   These two objects cannot yet be 'seen' on the screen.

5   Ensure the Zoom toolbar is displayed and positioned to suit.

Pan

1   Allows the graphics screen to be 'moved', the movement being controlled by the user:
    a) with co-ordinate entry
    b) by selecting points on the screen
    c) using the mouse tracker wheel (if available).

2   From the Standard toolbar select the **PAN REALTIME icon** and:
    prompt   Press Esc or Enter to exit, or right-click to display
               shortcut menu
    and   cursor changes to a hand
    respond *a*) hold down the left button of mouse
               b) move mouse and complete drawing moves
               c) note that scroll bars also move
               d) move image roughly back to original position

　　　　e) right-click and:
　　　　　　1. pop-up shortcut menu displayed
　　　　　　2. Pan is active – tick
　　　　　　3. pick Exit.

3　Menu bar with **View-Pan-Point** and:
　　prompt　　Specify base point or displacement and enter **0,0 <R>**
　　prompt　　Specify second point and enter **500,500 <R>**.

4　No drawing on screen – don't panic!

5　Use PAN REALTIME to pan down and to the left, and 'restore' the drawing roughly in its original position, then right-click and exit.

6　Menu bar with **View-Zoom-All** to restore original display.

Zoom

1　This is one of the most important and widely used of all the AutoCAD commands.

2　It allows parts of a drawing to be magnified/enlarged on the screen.

3　The command has several options, and it is these options which will now be investigated.

4　To assist us in investigating the zoom options:
　　a) use the COPY command, window the shape and copy the component from 80,110 to 240,320
　　b) refer to Figure 17.1.

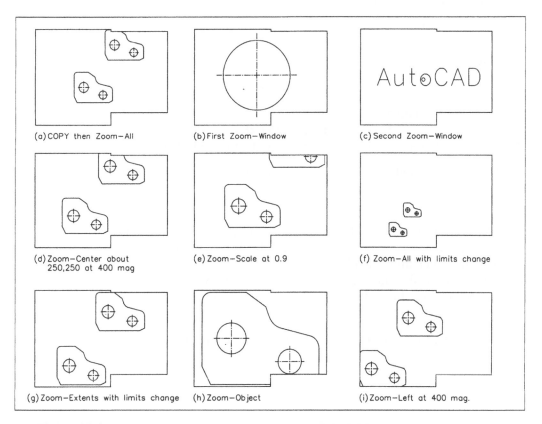

Figure 17.1　Using the various zoom options with WORKDRG.

Zoom All

1 Displays a complete drawing including any part of the drawing which is 'off' the current screen.

2 From menu bar select **View-Zoom-All**.

3 The two components are displayed as Figure 17.1(a).

Zoom Window

1 Perhaps the most useful of the zoom options.

2 It allows areas of a drawing to be 'enlarged' for clarity or more accurate work.

3 Select the **ZOOM WINDOW icon** from the Zoom toolbar and:
   prompt   Specify first corner
   respond   **window the original left circle** – Figure 17.1(b).

4 At the command line enter **ZOOM <R>** and:
   prompt   Specify corner of window, enter a scale factor and list of zoom options
   enter   **W <R>** – the window option
   prompt   Specify first corner and enter **119.6,174.9 <R>**
   prompt   Specify opposite corner and enter **120.4,175.2 <R>**.

5 The text item and circle will now be displayed – Figure 17.1(c).

Zoom Previous

1 Restores the drawing screen to the display before the last view command.

2 Menu bar with **View-Zoom-Previous** – restores Figure 17.1(b).

3 At command line enter **ZOOM <R>** then **P <R>** – restores Figure 17.1(a).

Zoom Center

1 Allows a drawing to be centred about a user-defined centre point.

2 Select the **ZOOM CENTER icon** and:
   prompt   Specify center point and enter **250,250 <R>**
   prompt   Enter magnification or height<?> and enter **400 <R>**.

3 The complete drawing is centred on the drawing screen about the entered point – Figure 17.1(d).

4 *Notes*
   a) the 'size' of the displayed drawing depends on the magnification/height value entered by the user and is relative to the displayed **default <?> value** with:
      1. a value less than the default – magnifies drawing on screen
      2. a value greater than the default – reduces the size of the drawing on the screen.
   b) toggle the grid ON and note the grid effect with this centre option, then toggle the grid off.

5 Menu bar with View-Zoom-Center and enter the following centre points, all with 1000 magnification:
   a) 200,200
   b) 0,0
   c) 500,500.

6 Now Zoom-Previous four times to restore Figure 17.1(a).

Zoom Scale

1 Centres a drawing on the screen at a scale factor entered by the user.

2 It is similar (and easier?) than the Zoom-Center option.

3 Select the **ZOOM SCALE icon** and:
prompt   Enter a scale factor (nX or nXP)
enter   **0.9 <R>**.

4 The drawing is displayed centred and scaled – Figure 17.1(e).

5 Menu bar with **View-Zoom-Scale** and enter a scale factor of 0.25.

6 Zoom to a scale factor of 1.5.

7 Zoom-Previous three times to restore Figure 17.1(a).

Zoom Extents

1 Zooms the drawing to extent of the current limits.

2 At command line enter **LIMITS <R>** and:
prompt   Specify lower left corner and enter **0,0 <R>**
prompt   Specify upper right corner and enter **1000,1000 <R>**.

3 Menu bar with **View-Zoom-All** – Figure 17.1(f).

4 Select the **ZOOM EXTENTS icon** – Figure 17.1(g).

5 Menu bar with **Format-Drawing Limits** and set the limits back to 0,0 and 420,297.

6 Zoom-All to restore Figure 17.1(a).

Zoom Object

1 Zooms selected objects as large as possible in the centre of the drawing.

2 Select the ZOOM OBJECT icon and:
prompt   Select objects
respond   window the original component then right-click – Figure 17.1(h).

3 Zoom-Previous to restore Figure 17.1(a).

Zoom Dynamic

1 This option will not be discussed.

2 The other zoom options and the realtime pan and zoom should be sufficient for all users needs?

Zoom Realtime

1 Menu bar with **View-Zoom-Realtime** and:
prompt   Press ESC or ENTER to exit, or right-click to display
        shortcut menu
and   cursor changes to a magnifying glass with a + and −
respond   a) hold down left button on mouse and move upwards to give a magnifi-
        cation effect

  b) move downwards to give a decrease in size effect

  c) left–right movement – no effect

  d) right-click to display pop-up menu with Zoom active

  e) pick Exit.

2 Menu bar with **View-Zoom-All** to display Figure 17.1(a).

3 The Zoom Realtime icon is in the Standard Toolbar.

Zoom Left

1 This option is not available as an icon and does not appear as a command line option.

2 It is a zoom option from previous AutoCAD Releases and can still be activated.

3 At the command line enter **ZOOM <R>** and:
   prompt   `Specify corner of window, enter a scale factor...`
   enter   **L <R>** – the left option
   prompt   `Lower left corner point and enter` **80,110 <R>**
   prompt   `Enter magnification or height and enter` **400 <R>**.

4 Complete drawing is moved to left of drawing screen and displayed at the entered magnification – Figure 17.1(i).

5 Zoom-Previous to restore Figure 17.1(a).

Aerial view

1 The aerial view is a navigation tool – an AutoCAD expression.

2 It allows the user to pan and zoom a drawing interactively.

3 This view option will be investigated when we construct a large-scale drawing.

Transparent zoom

1 A transparent command is one which can be activated while using another command.

2 The zoom command has this facility.

3 Restore the original Figure 17.1(a) – zoom all?

4 Select the **LINE icon** and:
   prompt   `Specify first point`
   enter   **80,110 <R>**
   prompt   `Specify next point`
   enter   **'ZOOM <R>** – the transparent zoom command
   prompt   `the zoom options now available at command line`
   enter   **W <R>** – the window zoom option
   prompt   `Specify corner of window and enter` **119.5,174.5 <R>**
   prompt   `Specify opposite corner and enter` **@1,1 <R>**
   prompt   `Specify next point`
   respond   with Snap off, **centre icon and pick the circle created at the start of the chapter** (which should now be displayed)
   prompt   `Specify next point`
   enter   **'Z <R> then A <R>** – the zoom all option
   prompt   `Specify next point`
   respond   **right-click-enter** to end line sequence.

5　The transparent command was activated by entering 'ZOOM <R> or 'Z <R> at the command line.

6　*a*) Only certain commands have transparency, generally those which alter a drawing display
　　b) GRID, SNAP and ZOOM.

7　Do not save this drawing modification.

8　*Notes*
　　a) in this exercise we entered Z <R> to activate the zoom command
　　b) this is an example of using an AutoCAD Alias or Abbreviation
　　c) many of the AutoCAD commands have an alias, which allow the user to activate the command from the keyboard by entering one or two letters
　　d) typical aliases are:
　　　L for line, C for circle, E for erase, M for move, CP for copy, MI for mirror, RO for rotate.

This chapter is now complete.

Chapter 18

Hatching

1   AutoCAD 2006 allows the user to add four 'types' of hatching:
    a) predefined, i.e. AutoCAD's stored hatch patterns
    b) user-defined
    c) custom – not considered in this book
    d) gradient.

2   When applying hatching (section detail) the user has two methods of defining the hatch pattern boundary:
    a) by selecting objects which make the boundary
    b) by picking a point within the boundary.

3   Hatch terminology includes:
    a) Boundary: the objects (lines, circles, arcs, etc.) which define the area to be hatched
    b) Pattern: the appearance of the hatching
    c) Island: an enclosed area within an existing hatch boundary
    d) Associativity: when the hatch area is altered, the hatching will 'fill the new area'.

4   The hatch command can be activated from the command line, menu bar or icon selection. In this chapter we will only consider the menu bar and icon methods. The user can investigate – **HATCH** at the command line for themselves.

Getting ready

1   Open the A3PAPER standard sheet and draw a 50-unit square on layer OUT, then multiple copy it to several other parts of the screen.

2   Refer to Figure 18.1 and add the other lines within the required squares (any size).

3   Note that I have included additional squares to indicate appropriate object selection and to demonstrate the before and after effect.

4   Make layer SECT current.

User-defined hatching

1   Menu bar with **Draw-Hatch** and:
    prompt   Hatch and Gradient dialogue box
    with   two tab selections:
               a) Hatch, with the following sections:
                    1. Type and pattern
                    2. Angle and scale
                    3. Hatch origin
                    4. Boundaries
                    5. Options

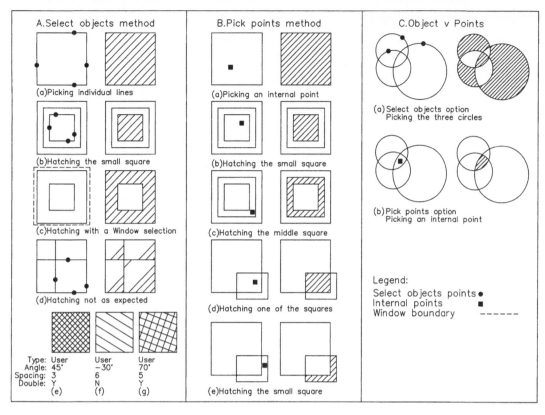

Figure 18.1   User-defined hatching. Both select objects and pick points methods.

6. More options (>) with:
   a) Islands
   b) Boundary retention
   c) Boundary set
   d) Gap tolerance
   e) Inherit options.
b) Gradient, with similar sections to the Hatch tab but includes sections for:
   1. Colour
   2. Orientation

respond   *a*) Set the following parameters
   1. Type: scroll and pick User-defined
   2. Angle: 45
   3. Double: not active, i.e. blank
   4. Spacing: 3
   5. Options: Associative active
   6. Island detection: not active
   7. Retain boundaries: active with Object type: Region – Figure 18.2

respond	*b*) **pick Add: Select objects**
and	dialogue box disappears
prompt	`Select objects or [pick internal point/remove Boundaries]`
respond	**pick the four lines of the first square then right-click-Enter**
prompt	`Hatch and Gradient dialogue box`
respond	**pick Preview**
and	drawing screen displayed with hatching added
respond	either: 1. ENTER or right-click if hatching correct
	or: 2. ESC to return to dialogue box and reset as required
and	hatching added as Figure 18.1A(a).

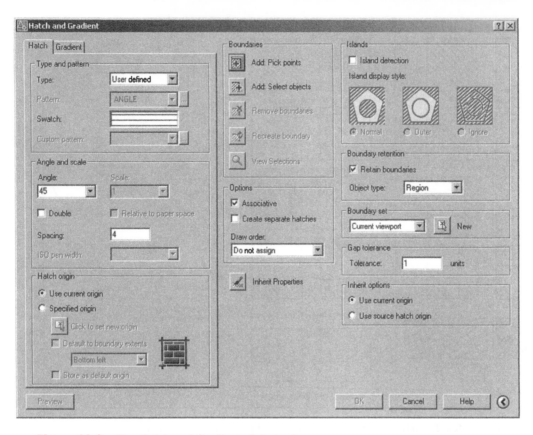

Figure 18.2 The Hatch and Gradient dialogue box.

2 Figure 18.1A(b)
Select the **HATCH icon** from the Draw toolbar and:
prompt   Hatch and Gradient dialogue box – with settings from step 1?
respond   **pick Add: Select objects**
prompt   Select objects
respond   **pick the four lines of the second square the right-click-Enter**
prompt   Hatch and Gradient dialogue box
respond   **Preview and right-click.**

3 Figure 18.1A(c)
With **HATCH <R>** at the command line:
prompt   Hatch and Gradient dialogue box – with settings from step 1?
respond   **pick Add: Select objects**
prompt   Select objects
enter   **W <R>** – the window selection set option
prompt   Specify first corner
respond   **window the third square then right-click-Enter**
prompt   Hatch and Gradient dialogue box
respond   **Preview and right-click.**

4 Figure 18.1A(d)
Repeat the HATCH command and using the Hatch and Gradient dialogue box and with the same hatch parameters as before:
a) select objects
b) pick the four lines of the fourth square
c) preview, etc.

5   Using the Hatch and Gradient dialogue box, with the Select objects window method, alter the following hatch parameters:

	Figure 18.1A(e)	*Figure 18.1A(f)*	*Figure 18.1A(g)*
Type	User-defined	User-defined	User-defined
Angle	45	−30	70
Spacing	3	6	5
Double	Y	N	Y

6   Refer to Figure 18.1(B) and with layer Sect still current, pick the **HATCH icon** and:
   prompt   Hatch and Gradient dialogue box
   respond   *a)* check/alter:
   　　　　　　　1. Type: User-defined
   　　　　　　　2. Angle: 45
   　　　　　　　3. Spacing: 3
   　　　　　　b) **pick Add: Pick points**
   prompt   Pick internal point or [Select objects/remove Boundary]
   respond   **pick any point within the next suitable square**
   prompt   Various prompts at command line
   then   Pick internal point, i.e. any more areas to be hatched
   respond   **right-click and Enter**
   prompt   Hatch and Gradient dialogue box
   respond   **Preview-right click**
   and   hatching added to square as Figure 18.1B(a).

7   Using the **HATCH icon** with the **Add: Pick points** option from the Hatch and Gradient dialogue box, add hatching using the internal points indicated in Figure 18.1B.

Select Objects vs Pick Points

1   With two options available for hatching, new users to AutoCAD may be confused as to whether they should Select Objects or Pick Points.

2   In general, the Pick Points option is the simpler to use, and will allow complex shapes to be hatched with a single pick within the area to be hatched.

3   To demonstrate the effect, refer to Figure 18.1(C):
   a) draw two sets of three intersecting circles – any size
   b) we want to hatch the intersecting area of the three circles.

4   Select the HATCH icon and from the Hatch and Gradient dialogue box:
   a) set: User-defined, Angle of 45, Spacing of 2
   b) pick the Select Objects option
   c) pick the three circles
   d) preview-right click-OK and hatching as Figure 18.1C(a).

5   Select the HATCH icon again and:
   a) pick Pick Points option
   b) pick any point within the area to be hatched
   c) preview-right click-OK and hatching as Figure 18.1C(b).

6   Figure 18.1 is now complete and can be saved if required. We will not use it again.

Hatch style

1   AutoCAD has a hatch style (called Island detection) option which allows the user to control three 'variants' of the hatch command.

2 Open your A3PAPER standard sheet, refer to Figure 18.3(A) and:
   a) with layer OUT current, draw a 60-sided square with smaller squares 'inside'. Snap on helps
   b) copy the squares to two other areas of the screen.

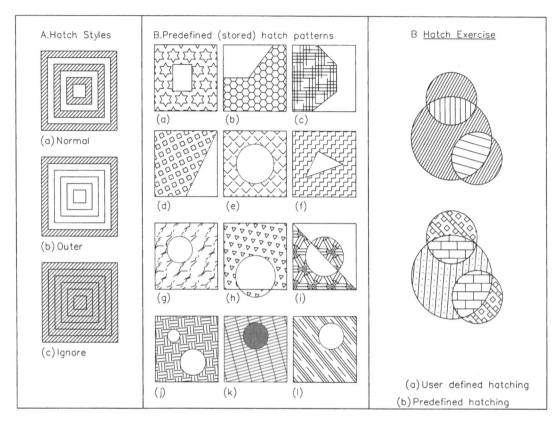

Figure 18.3 Hatch styles, predefined patterns and the hatch exercise.

3 With layer SECT current, select the **HATCH icon** and:
   prompt   Hatch and Gradient dialogue box
   respond   *a*) set: user-defined, angle 45, spacing 2
               b) more options arrow selected
               c) Island detection active, i.e. tick
               d) Island display style: normal active, i.e. dot
               e) pick Add: Select objects
   prompt   Select objects or [pick internal point/remove Boundaries]
               at the command line
   respond   **window the first set of squares then right-click**
   prompt   shortcut menu as Figure 18.4
   respond   *a*) pick Preview
               b) right-click
   and   hatching added as Figure 18.3A(a).

4 Repeat the HATCH icon and:
   a) alter Island detection style to Outer
   b) pick Add: Select objects and window the second set of squares
   c) preview, etc. – hatching as Figure 18.3A(b).

5 Using the Hatch and Gradient dialogue box, set the Island detection to Ignore, window the third set of squares to give hatching as Figure 18.3A(c).

Figure 18.4 Shortcut menu.

Predefined hatch patterns

1 AutoCAD 2006 has several stored hatch patterns, accessed from the Hatch and Gradient dialogue box.

2 The user specifies the:
   a) pattern name
   b) scale for the pattern
   c) angle of the pattern.

3 Refer to Figure 18.3(B) and:
   a) draw a 50-unit square
   b) multiple copy the square 11 times
   c) add other lines and circles as displayed.

4 Select the **HATCH icon** and:
   prompt   Hatch and Gradient dialogue box
   respond   *a*) Ensure More Options active
            b) Island detection style: Normal
            c) Retain boundaries: not active
            d) Pattern type: Predefined
            e) Pattern: pick the [. . .] button.
   prompt   Hatch Pattern Palette dialogue box with four tab selections:
            ANSI, ISO, Other Predefined, Custom
   respond   *a*) ensure Other Predefined tab active (should be)
            b) scroll until LINE to ZIGZAG displayed
            c) pick STARS (Figure 18.5) then OK.
   prompt   Hatch and Gradient dialogue box
   with   *a*) Pattern: STARS
            b) Swatch: display of stars hatch pattern
   respond   *a*) Angle: 0
            b) Scale: 1
            c) pick Add: Pick points
            d) select any internal point in first square
            e) right-click and pick Preview from pop-up menu
            f) right-click to accept the hatch pattern
   and   Hatching added as Figure 18.3B(a).

4 Repeat the HATCH icon selection, and with the same procedure as step 3, add hatching to the other squares with the pick points method and using the following

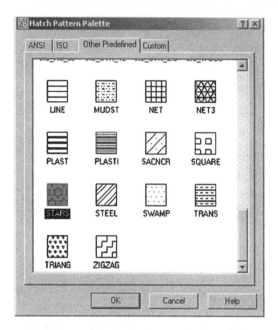

Figure 18.5   The Hatch Pattern Palette dialogue box with STARS selected.

hatch pattern names, scales and angles:

Figure 18.3(B)	*Pattern*	*Scale*	*Angle*	*Figure 18.3(B)*	*Pattern*	*Scale*	*Angle*
b	HONEY	1	0	h	TRIANG	0.7	−10
c	HOUND	1.5	0	i	ESCHER	0.5	0
d	SQUARE	1	25	j	EARTH	1	5
e	ANGLE	1	45	k	SOLID		
f	ZIGZAG	1	0	k	GRATE	3	15 i.e. two hatches for
g	GRAVEL	0.75	0				Figure 18.3B(k)

5   Finally enter – **HATCH <R>** and the command line and:
   prompt   Specify internal point or [Properties/Select objects ...
   enter   P <R> – the properties option
   prompt   Enter a pattern name or [?/Solid/User-defined]
   enter   **CLAY <R>**
   prompt   Specify a scale for the pattern and enter: **2 <R>**
   prompt   Specify an angle for the pattern and enter: **−50 <R>**
   prompt   Specify internal point or [Properties/Select objects . . .
   respond **pick point in next square then <R>** – Figure 18.3B(i).

6   The predefined hatch exercise is now complete and should be saved as it will be used again.

Hatch exercise

1   Refer to Figure 18.3(C) and draw two sets of circles to your own sizes on layer OUT.

2   With layer SECT current, add the following hatching using the pick points method:

   User-defined   *Predefined*
   a) angle: 60, spacing: 2   *a*) BRICK, scale: 1, angle: 0
   b) angle: 90, spacing: 4   *b*) SACNCR, scale: 2, angle: 40
   c) angle: −15, spacing: 6   *c*) BOX, scale: 0.5, angle: 45

3   This exercise is now complete and the drawing can be saved if required.

Associative hatching

1   AutoCAD 2006 has associative hatching.

2   This means that if the hatch boundary is altered, the hatching within the boundary will be 'regenerated' to fill the new boundary limits.

3   Associative hatching is applicable to both user-defined and predefined hatch patterns, irrespective of whether the select objects or pick points option was used.

4   To demonstrate the concept:
    a) continue from the previous exercise or re-open the predefined hatch squares exercise
    b) refer to Figure 18.5 and erase squares which are not displayed
    c) note that I have included two sets of each squares to demonstrate the 'before and after' effect
    d) make layer SECT current
    e) zoom in on each square to make selection easier.

5   Select the MOVE icon and:
    a) window the small square in the first square
    b) move it to another position in the square
    c) hatching 'changes' – Figure 18.6A(a) i.e. it is associative.

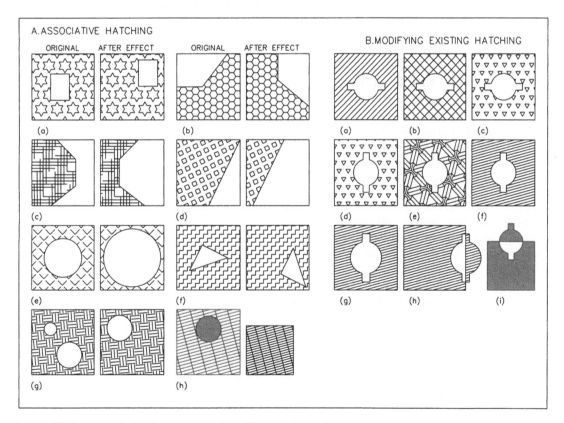

Figure 18.6   Associative hatching and modifying existing hatching.

6   Figure 18.6 displays some associative hatching effects, these being:

Figure 18.6A	*Effect*
b	rotating the two lines by 90 degrees about 'the centre of the square'
c	mirroring the three lines about midpoints of top and bottom lines and erasing source objects

d	moving the line in a horizontal direction
e	scaling the circle by factor of 1.5
f	rotating and moving the triangle
g	erasing the smaller circle and moving the larger circle
h	erasing the solid hatch and scaling the hatched square

7 *a*) Try the above operations and some other modify commands of your own.
   b) You may find that associated hatching does not always work.

8 This exercise is complete and can be saved.

Modifying hatching

1 Hatching which has been added to an object can be modified.

2 The angle, spacing, scale and pattern can be altered if required.

3 Open A3PAPER and refer to Figure 18.6(B). Display the MODIFY II toolbar.

4 On layer OUT, draw a square with a circular and linear shape (any size).

5 *a*) With layer SECT current, add hatching to the square.
   b) The hatching should be: user-defined, angle 45, spacing 3 – Figure 18.6B(a).
   c) Ensure that Associative hatching is active.

6 Menu bar with **Modify-Object-Hatch** and:
   prompt   Select hatch object
   respond **pick the added hatching**
   prompt   Hatch Edit dialogue box
   with   User-defined active
   respond *a*) alter angle to −45
            b) alter spacing to 5
            c) double active
            d) pick OK – Figure 18.6B(b).

7 Select the **EDIT HATCH icon** from the Modify II toolbar and:
   prompt   Select hatch object
   respond **pick the altered hatching**
   prompt   Hatch Edit dialogue box
   respond *a*) Type: Predefined
            b) Pattern: pick the [. . .] button
            c) scroll and pick TRIANG then OK
            d) scale: 0.75 and angle: 0
            e) pick OK – Figure 18.6B(c).

8 *a*) Rotate the circular and linear objects inside the square by 90 degrees.
   b) Use the circular object centre as the base point – Figure 18.6B(d).

9 Using the Edit Hatch icon:
   a) Type: Predefined ESCHER
   b) Scale of 0.6 and angle of 15 – Figure 18.6B(e).

10 Edit the hatching to:
   User-defined with angle of 0 and spacing of 2, double not active – Figure 18.6B(f).

11 *a*) Scale the circular and linear objects inside the square by 1.25
   b) Use a window selection – Figure 18.6B(g).

12 Move the circular and linear objects onto a side of the square – Figure 18.6B(h).

13 Finally, select the circular and linear objects and:
   a) Scale the complete component by 0.75.
   b) Move the linear and circular objects as shown.
   c) Change hatch pattern to SOLID – Figure 18.6B(i).

14 This exercise is now complete and can be saved if required.

Text and hatching

1 Text which is placed in an area to be hatched can be displayed with a 'clear border' around it.

2 *a*) Clear a space and refer to Figure 18.7(A).
   b) Draw four 50 diameter circles on layer OUT.
   c) Add any suitable text to each circle using layer TEXT (style, height, etc. your own).

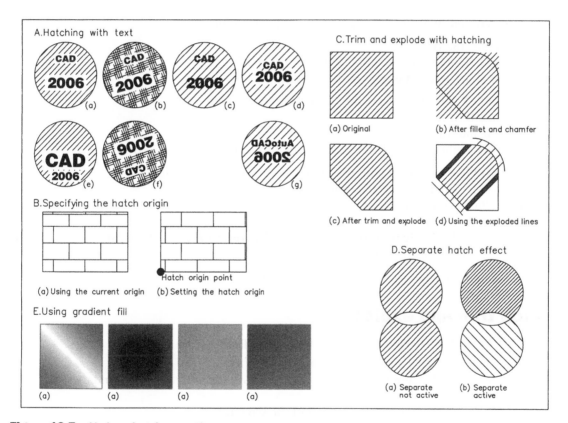

Figure 18.7 Various hatch operations.

3 With layer SECT current, use the HATCH icon with:
   a) type: User-defined, angle of 45 and spacing of 3
   b) pick points method and pick any point in first circle
   c) right click-preview-right click – Figure 18.7A(a).

4 Use the HATCH icon with:
   a) type: Predefined pattern HOUND, scale: 1, angle: 20
   b) pick points method and pick any point inside second circle
   c) right click-preview-right click – Figure 18.7A(b).

5 HATCH icon using:
   a) user-defined at 45 with 4 spacing
   b) **select objects method and pick third circle only**
   c) preview, etc. – Figure 18.7A(c).

6 Final HATCH with:
   a) user-defined, angle of 45 and spacing of 4
   b) select objects method and pick fourth circle AND text items
   c) preview, etc. – Figure 18.7A(d).

7 Associative hatching works with text items:
   Figure 18.7A(e): moving and scaling the two items of text from Figure 18.7A(a)
   Figure 18.7A(f): rotating the text item of Figure 18.7A(b)
   Figure 18.7A(g): mirroring the text item, then editing Figure 18.7A(d).

8 Save if required – the exercise is complete and will not be used again.

Hatch origin

1 Hatch patterns are usually added to selected areas without the need to specify an origin point.

2 There are occasions when specifying a hatch origin is useful.

3 Refer to Figure 18.7(B) and:
   a) on layer OUT, draw a rectangular shape with snap off
   b) copy this shape to another part of the screen.

4 With layer SECT current, activate the HATCH command with:
   a) predefined BRICK pattern
   b) angle of 0 and scale of 2
   c) pick points method and pick any point in the first shape – Figure 18.7B(a).

5 Repeat the HATCH command with the same parameters as step 4 and:
   a) activate Hatch origin: Specified origin active
   b) pick on Click to set new origin:

   prompt   Specify origin point
   respond **Endpoint icon and pick point indicated in Figure 18.7B(b)**
   then   pick points-preview, etc.

6 It is user preference whether to use the hatch origin option.

Hatch trim and explode

1 AutoCAD 2006 allows hatching to be trimmed and exploded.

2 *a*) Refer to Figure 18.7(C) and draw a square.
   b) Add user-defined hatching, angle and spacing to suit, but with Retain boundaries not active.
   c) Copy the hatched square to another part of the screen.

3 Fillet one corner and chamfer another corner of the square – no associative effect – Figure 18.7C(b).

4 Trim the hatching to the fillet arc and the chamfer line – Figure 18.7C(c).

5 Menu bar with **Modify-Explode** and:
   prompt   Select objects
   respond **pick the hatching**.

6 The individual lines of the exploded hatch pattern are now 'ordinary lines' – Figure 18.7C(d).

Separate hatching

1 When hatch is being added to more than one area in a single operation, the user can create separate hatching.

2 The effect is displayed in Figure 18.7(D) with:
   a) separate not active and if the hatching is modified, both areas will be altered – Figure 18.7C(a)
   b) separate active and both areas can be modified if required – Figure 18.7C(b).

Gradient fill

1 A gradient fill is a solid hatch fill that gives the blended colour effect of a surface with light on it.

2 Gradient fills can be used to suggest a solid form in a two-dimensional drawings.

3 Gradient fill is applied in the same way as hatching.

4 It is associative and can be modified.

5 Open your A3PAPER standard sheet and draw a 50-sided square then copy it to eight other places.

6 Menu bar with **Draw-Hatch** and:
   prompt   Hatch and Gradient dialogue box
   respond   **pick the Gradient tab**
   prompt   Gradient tab active
   with   the following option effects:
               a) Colour
                    1. one colour with shade to tint control
                    2. two colour effect
               b) Orientation
                    1. centred pattern
                    2. ser-defined angled pattern
               c) Fill options: linear, spherical and radial
   respond   *a*) One colour active
               b) Full tint, i.e. slider bar to right
               c) Fill option: pick middle top box
               d) Centred active
               e) Angle: scroll and pick 45
               f) Boundaries: pick Add: Pick points
               g) pick an internal point in first square
               h) right click-preview-right click
   and   selected square is gradient filled as Figure 18.7E(a).

7 Figure 18.7(E) displays some other gradient fill effects:
   (b): one colour, full shade
   (c): two colour effect, index colours selected at random
   (d): two colour effect, colour book colours selected at random.

8 Try some other gradient fill effects for yourself.

The hatching exercises are now all complete.

Assignments

1 Some AutoCAD users may not use the hatching in their draughting work, but they should still be familiar with the process.

2 The pick points method makes hatching fairly easy and for this reason I have included five interesting exercises for you to attempt.

3 These should test all your existing CAD draughting skills.

The activities are:

1 *Activity 20: Three different types of components to be drawn and hatched*
   a) A small engineering component. The hatching is user-defined with your own angle and spacing values and you have to add the dimensions.
   b) A pie chart. The text is to have a height of 8 and the hatch names and variables are given.
   c) A model airport runway system which has to be dimensioned.

2 *Activity 21: Cover plate*
   a) A relatively simple drawing to complete.
   b) The MIRROR command is useful and the hatching should no trouble. Add text and the dimensions.
   c) A gradient fill effect has been added with the Draw order set to 'Send behind boundary'.

3 *Activity 22: Protected bearing housing*
   Four views to complete, two with hatching. A fairly easy drawing.

4 *Activity 23: Steam expansion box*
   a) This activity is easier to complete than it would appear.
   b) Create the outline from lines then fillet the corners.
   c) The complete component uses many commands, e.g. offset, fillet, mirror, etc.
   d) The hatching should not be mirrored – why?

5 *Activity 24: Gasket cover*
   a) An interesting exercise to complete.
   b) Draw the 'left view' which consists only of circles use layers correctly.
   c) The right view can be completed using offset, trim and mirror.
   d) Do not dimension, but discretion for sizes not given.

Chapter 19

Drawing assistance and information

1   Up until now, all objects have been created:
    a) by picking points on the screen
    b) entering co-ordinate values
    c) referencing existing objects, e.g. midpoint, endpoint, etc.

2   There are other methods which enable objects to be positioned on the screen, these being:
    a) point filters
    b) construction lines
    c) ray lines.

3   Drawings contain information which may be useful to the user, including:
    a) co-ordinate data
    b) distances between points
    c) area of shapes, etc.

    In this chapter we will investigate several of the above concepts.

Point filters

1   These allows objects to be positioned by referencing the X and Y co-ordinate values of existing objects.

2   Open the A3PAPER standard sheet and refer to Figure 19.1.

3   *a*) draw a 50-sided square, lower left corner at 20,220
    b) ensure that grips are not active, i.e. off
    c) multiple copy this square to three other positions.

4   A circle of diameter 30 has to be created at the 'centre' of each square and this will be achieved by four different methods:
    a) *Co-ordinates*
       Activate the circle command with centre: 45,245; radius: 15.
    b) *Object snap midpoint*
       1. draw in a diagonal of the square
       2. select the CIRCLE icon from the Draw toolbar and:
          a) centre point: **Snap to Midpoint** of diagonal
          b) radius: enter 15
    c) *Object snap from*
       1. select the CIRCLE icon from the Draw toolbar and:
       2. centre: pick the **Snap from icon**
          a) base point: **Endpoint icon** and pick left end of line AB
          b) offset: enter @25,25
          c) radius: enter 15

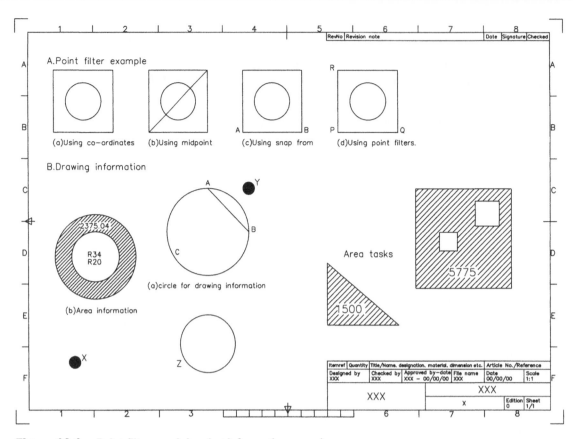

Figure 19.1 Point filters and drawing information exercises.

d) Point filters

Activate the circle command and:

prompt   Specify center point
enter   **.X <R>**
prompt   of
respond   **Midpoint icon and pick line PQ**
prompt   (need YZ)
respond   **Midpoint icon and pick line PR**
prompt   Specify radius and enter **15 <R>**.

Point identification

1   This command displays the co-ordinates of a selected point.

2   *a)* Draw a circle, centre at 150,140 with a radius of 35 – Figure 19.1B(a)
    b) Draw the line AB using quadrant object snaps
    c) Activate the Inquiry toolbar.

3   Select the **LOCATE POINT icon** from the Inquiry toolbar and:
    prompt   Specify point
    respond   **Snap to Midpoint icon and pick line AB**.

4   The command line area displays:
    X = 167.50   Y = 157.50   Z = 0.00.

5   Menu bar with **Tools-Inquiry-ID Point** and:
    prompt   Specify point
    respond   **Snap to Center icon and pick the circle**.

6   Command line display is X = 150.00   Y = 140.00   Z = 0.00.

7   The command can be activated with **ID <R>** at the command line.

Distance

1   Returns information about a line between two selected points including the distance and the angle to the horizontal.

2   Select the **DISTANCE icon** from the Inquiry toolbar and:
    prompt   Specify first point
    respond   **snap to endpoint and pick point A**
    prompt   Specify second point
    respond   **snap to endpoint and pick point B**.

3   The command prompt area will display:
    Distance = 49.50, Angle in XY plane = 315.0, Angle from XY plane = 0.0
    Delta X = 35.00, Delta Y = −35.00, Delta Z = 0.00.

4   Menu bar with **Tools-Inquiry-Distance** and:
    prompt   Specify first point
    respond   **snap to centre of circle**
    prompt   Specify second point
    respond   **pick midpoint of line AB**.

5   The display at the command prompt is:
    Distance = 24.75, Angle in XY plane = 45.0, Angle from XY plane = 0.0
    Delta X = 17.5, Delta Y = 17.5, Delta Z = 0.00.

6   *a*) Using the DISTANCE command, select point B as the first point and point A as the second point
    b) Compare the information with the step 3 information.

7   Entering **DIST <R>** at the command line will activate the command.

List

1   A command which gives useful information about a selected object.

2   Select the **LIST icon** from the Inquiry toolbar and:
    prompt   Select objects
    respond   **pick the circle then right-click**
    prompt   AutoCAD Text Window with information about the circle
    respond   **F2 to flip back to drawing screen**.

3   Menu bar with **Tools-Inquiry-List** and:
    prompt   Select objects
    respond   **pick line AB then right-click**
    prompt   AutoCAD text window
    either   *a*) F2 to flip back to drawing screen
    or   *b*) cancel icon from text window title bar.

4   Figure 19.2 is a screen dump of the AutoCAD Text window display for the two selected objects.

5   **LIST <R>** is the command line entry.

```
        CIRCLE     Layer: "OUT"
                   Space: Model space
            Handle = 2515
      center point, X=    150.00  Y=    140.00  Z=      0.00
      radius      35.00
circumference    219.91
        area    3848.45



        LINE      Layer: "OUT"
                   Space: Model space
          Handle = 2516
     from point, X=    150.00  Y=    175.00  Z=      0.00
       to point, X=    185.00  Y=    140.00  Z=      0.00
Length =    49.50,   Angle in XY Plane =   315.0
        Delta X =     35.00, Delta Y =     -35.00, Delta Z =      0.00
```

Figure 19.2   Information from the LIST command.

Area

1   This command will return the area and perimeter for selected shapes or polyline shapes.

2   It has the facility to allow composite shapes to be selected.

3   Select the **AREA icon** from the Inquiry toolbar and:
   prompt   Specify first corner point or [Object/Add/Subtract]
   enter   **O <R>** – the object option
   prompt   Select objects
   respond **pick the circle**.

4   The command line will display:
   Area = 3848.45, Circumference = 219.91.

5   *Question*: Are these values the same as from the LIST command?

6   *a*) Draw two concentric circles with radii 34 and 20 – Figure 19.1B(b).
   b) Menu bar with **Tools-Inquiry-Area** and:
   prompt   Specify first corner point or [Object/Add/Subtract]
   enter   **A <R>** – the add option
   prompt   Specify first corner point or [Object/Add/Subtract]
   enter   **O <R>** – the object option
   prompt   (ADD mode) Select objects
   respond **pick the larger circle**
   prompt   1.Area = 3631.68, Circumference = 213.63
              2.Total area = 3631.68
              3.(ADD mode) Select objects
   respond **<RETURN>** to end add selection sequence
   prompt   Specify first corner point or [Object/Subtract]
   enter   **S <R>** – the subtract option
   prompt   Specify first corner point or [Object/Add]
   enter   **O <R>** – the object option
   prompt   (SUBTRACT mode) Select objects
   respond **pick the smaller circle**

prompt   1.Area = 1256.64, Circumference = 125.66
          2.Total area = 2375.04
          3.(SUBTRACT mode) Select objects

respond   **<R><R>** to end the area sequence.

7   *Question*: Is the area figure correct for the annulus shape?

Time

1   This command displays information in the AutoCAD Text Window about the current drawing, e.g.:
    a) when it was originally created
    b) when it was last updated
    c) the length of time spent working on it.

2   The command can be activated:
    a) from the menu bar with **Tools-Inquiry-Time**
    b) by entering **TIME <R>** at the command line.

3   The command has options of Display, On, Off and Reset.

4   A useful command or a nightmare?

Status

1   This command gives:
    a) additional information about the current drawing
    b) disk space information.

2   Select the sequence **Tools-Inquiry-Status** to 'see' the status display in the AutoCAD text window.

Calculator

1   AutoCAD 2006 has a built-in calculator which can be used:
    a) to evaluate mathematical expressions
    b) to assist on the calculation of co-ordinate point data.

2   The mathematical operations obey the usual order of preference with brackets, powers, etc.

3   At the command line enter **CAL <R>** and:
    prompt   >> Expression:
    enter   **12.6*(8.2 + 5.1) <R>**
    prompt   **167.58** – is it correct?

4   Enter **CAL <R>** and:
    prompt   >>Expression:
    enter   **(5*(7 − 3))^2 <R>**
    prompt   400.

Quick calculator

1   The user can activate and display a Quick calculator with the menu bar sequence **Tools-QuickCalc**.

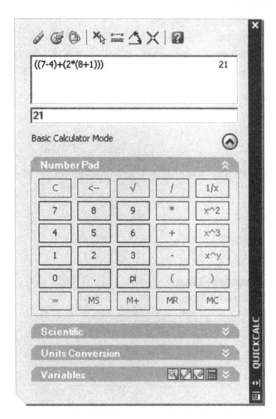

Figure 19.3   The Quick calculator.

2   Figure 19.3 displays the Quick calculator which has:
   a) a standard number pad
   b) options for:
      1. scientific usage
      2. units conversion
      3. variables
   c) auto-hide facilities.

3   *Question*: What is answer to $((7 - 4) + (2*(8 + 1)))$ – a key question?

Transparent calculator

1   A transparent command is one which can be used 'while in another command'.

2   It is activated from the command line by entering the ' symbol.

3   The calculator command has this transparent ability.

4   Activate the **DONUT** command with diameters of 0 and 10, then:
prompt	Specify center of donut
enter	**'CAL <R>** – the transparent calculator command
prompt	>>>>Expression:
enter	**CEN/4 <R>**
prompt	>>>>Select entity for CEN snap
respond	**pick circle C**
prompt	37.5, 35, 0
and	donut at position X

prompt	`Specify center of donut`
enter	**'CAL <R>**
prompt	`>>>>Expression:`
enter	**(2*MID)-(CEN) <R>**
prompt	`>>>>Select entity for` MID snap and **pick line AB**
prompt	`>>>>Select entity for` CEN snap and **pick circle C**
prompt	`185,175,0`
and	donut at position Y
prompt	`Specify center of donut` and right-click to end sequence.

5 Activate the circle command and:

prompt	`Specify center point for circle` and enter **150,50 <R>**
prompt	`Specify radius of circle`
enter	**'CAL <R>**
prompt	`>>>>Expression:`
enter	**rad/2 <R>**
prompt	`>>>>Select circle, arc or polyline segment for RAD function`
respond	**pick circle C**
and	circle at position Z.

6 *a*) Check the donut centre points with the ID command
 b) They should be 37.5,35 and 185,175
 c) These values were given at the command prompt line as the donuts were being positioned.

7 *Question*: How were the donut centre co-ordinate values calculated during step 4?

Task

1 Refer to Figure 19.1 and create the following (anywhere on the screen, but use SNAP ON to help):
 a) right-angled triangle with vertical side of 50 and horizontal side of 60
 b) square of side 80 and inside this square two other squares of side 15 and 20.

2 Find the shaded areas using the AREA command.

The drawing information exercise is now complete and can be saved if required.

Construction lines

1 Construction lines are:
 a) lines that extend to infinity in both directions from a selected point on the screen
 b) they can be referenced to assist in the creation of other objects.

2 Open the A3PAPER standard sheet, layer OUT current and display toolbars Draw, Modify and Object Snap.

3 Refer to Figure 19.4(A) and with layer OUT current, draw:
 a) a 100-sided square, lower left corner at 50,50
 b) a circle, centred on 250,190 with radius 50.

4 *a*) Make a new layer (Format-Layer) named CONLINE, colour to suit, DASHED line-type and current
 b) Set the LTSCALE value to your requirements.

5 Menu bar with **Draw-Construction Line** and:

prompt	`Specify a point or [Hor/Ver/Ang/Bisect/Offset]`
enter	**50,50 <R>**

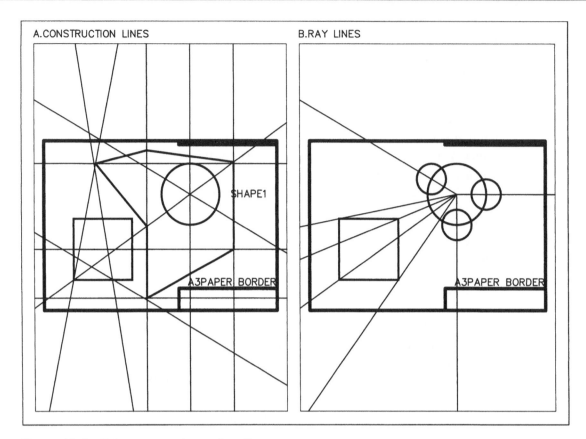

A.CONSTRUCTION LINES B.RAY LINES

SHAPE1

A3PAPER BORDER   A3PAPER BORDER

Figure 19.4 Using construction and ray lines.

and	line 'attached' to cursor through the entered point and 'rotates' as the mouse is moved
prompt	Specify through point
enter	**80,200 <R>**
prompt	Specify through point
respond	**Center icon and pick the circle**
prompt	Specify through point and **right-click**.

6   At the command line enter **XLINE <R>** and:

prompt	Specify a point or...
enter	**H <R>** – the horizontal option
prompt	Specify through point
enter	**100,20 <R>**
prompt	Specify through point
respond	**Midpoint icon and pick a vertical line of square**
prompt	Specify through point
respond	**Quadrant icon and pick top of circle**
prompt	Specify through point and right-click.

7   Select the **CONSTRUCTION LINE icon** from the Draw toolbar and:

prompt	Specify a point or [Hor/Ver/Ang/Bisect/Offset]
enter	**V <R>** – the vertical option
prompt	Specify through point
respond	**Center icon and pick the circle**
prompt	Specify through point and right-click.

8 Activate the construction line command and:
  prompt Specify a point or [Hor/Ver/Ang/Bisect/Offset]
  enter **O <R> –** the offset option
  prompt Specify offset distance or [Through] and enter **75 <R>**
  prompt Select a line object
  respond **pick the vertical line through the circle centre**
  prompt Select side to offset
  respond **offset to the right**
  prompt Select a line object
  respond **offset the same line to the left**
  prompt Select a line object and right-click.

9 Construction line command and at prompt:
  enter **A <R> –** the angle option
  prompt Enter angle of xline (0.0) or [Reference]
  enter **−30 <R>**
  prompt Specify through point
  respond **Center icon and pick the circle**
  prompt Specify through point
  enter **135,40 <R>** then right-click.

10 Construction line command for last time and at prompt:
  enter **B <R> –** the bisect option
  prompt Specify angle vertex point
  respond **Midpoint icon and pick top line of square**
  prompt Specify angle start point
  respond **pick lower left vertex of square**
  prompt Specify angle end point
  respond **Midpoint icon and pick square right vertical line**
  prompt Specify angle end point and right-click.

11 *a*) Construction lines can be:
    1. copied, moved, trimmed, etc.
    2. object snap referenced to create other objects.
  b) SHAPE1 in Figure 19.4(A) has been 'drawn' by referencing the existing construction lines.

12 *Note*: I would recommend that:
  a) construction lines are created on their own layer
  b) this layer is frozen to avoid 'screen clutter' when not in use
  c) the layer is given a colour and linetype not normally used.

13 *Task*: Try the following:
  a) At the command line enter **LIMITS <R>** and:
    prompt Specify lower left corner and enter **0,0 <R>**
    prompt Specify upper right corner and enter **10000,10000 <R>**.
  b) From menu bar select **View-Zoom-All** and:
    1. drawing appears very small at bottom of screen
    2. the construction lines 'radiate outwards' the screen edges.
  c) Using **LIMITS <R>** enter the following values:
    lower left corner: −10000, −10000
    upper right corner: 0,0.
  d) View-Zoom-All to 'see' the construction lines.
  e) Return limits to 0,0 and 420,297 then View-Zoom-All to restore the original drawing screen.

14 The construction line exercise is now complete.

15 The drawing can be saved if required, but we will not use it again.

Rays

1 Rays extend to infinity in one direction from the selected start point.

2 Make a new current layer called RAYLINE, colour to suit and with a DIVIDE linetype.

3 Freeze layer CONLINE and erase any objects to leave the original square and circle.

4 Refer to Figure 19.4(B), menu bar with **Draw-Ray** and:

   prompt   Specify start point
   respond   **Center icon and pick the circle**
   prompt   Specify through point and enter **@100<150 <R>**
   prompt   Specify through point and enter **@100<0 <R>**
   prompt   Specify through point and enter **@100<−90 <R>**
   prompt   Specify through point and right-click.

5 With layer OUT current, draw three 25-radius circles at the intersection of the ray lines and circle.

6 Make layer RAYLINE current and at the command line enter **RAY <R>** and:

   prompt   Specify start point
   respond   **pick the circle centre point**
   prompt   Specify through point
   respond   **Intersection icon and pick the four vertices of the square** then **right-click.**

7 Use the LIMITS command and enter:
   a) lower left: −10000, −10000
   b) upper right: 0,0.

8 Menu bar with **View-Zoom-All** and note position of 'our drawing'.

9 Return limits to 0,0 and 420,297 then **Zoom-Previous** to restore the original drawing screen.

10 *Note*: It is recommended that ray lines:
   a) are created on 'their own layer' with a colour and linetype to suit
   b) this layer should be frozen when not in use.

   This chapter is now complete.

Chapter 20

Polylines

1 A polyline is a single object which can consist of line and arc segments.

2 It can be drawn with varying widths.

3 It has it's own editing facility and can be activated by icon, from the menu bar or by keyboard entry.

4 A polyline is a very useful and powerful object yet it is probably one of the most under-used draw commands.

Persevere with the demonstration which follows, as it is quite long and requires several keyboard entries.

Using polylines

1 Open your A3PAPER standard file with layer OUT current and refer to Figure 20.1(A).

2 Display the Draw, Modify, Object Snap and Modify II toolbars.

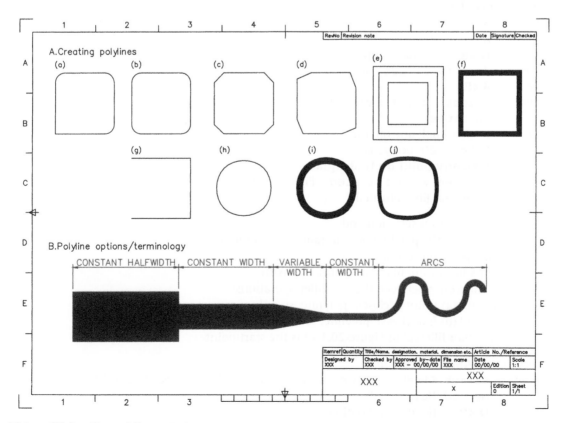

Figure 20.1 The polyline exercises.

3 Select the **POLYLINE icon** from the Draw toolbar and:

prompt	Specify start point
enter	**15,220 <R>**
prompt	Current line-width is 0.00
then	Specify next point or [Arc//Halfwidth/Length/Undo/Width]
enter	**@50,0 <R>**
prompt	Specify next point or [Arc/Close/Halfwidth/Length/Undo/ Width]
enter	**@0,50 <R>**
prompt	Specify next point or ... and enter **@−50,0 <R>**
prompt	Specify next point or ... and enter **@0,−50 <R>**
prompt	Specify next point or . . .
respond	**right-click-enter** to end command.

4 From menu bar select **Draw-Polyline** and:

prompt	Specify start point and enter **80,220 <R>**	absolute entry
prompt	Specify next point and enter **@50<0 <R>**	relative polar entry
prompt	Specify next point and enter **140,270 <R>**	absolute entry
prompt	Specify next point and enter **@−50,0 <R>**	relative absolute entry
prompt	Specify next point and enter **C <R>**	close option.

5 Select the **COPY icon** and:

prompt	Select objects
respond	*a*) pick any point on SECOND square (all four lines are highlighted with a single pick) *b*) right-click
prompt	Specify base point
and	**multiple copy the square to eight other parts of the screen.**

6 Select the **FILLET icon** and:

prompt	Select first object or [Undo/Polyline/Radius/Trim/ Multiple]
enter	**R <R>** – the radius option
prompt	Specify fillet radius
enter	**8 <R>**
prompt	Select first object or [Undo/Polyline/Radius/Trim/ Multiple]
enter	**P <R>** – polyline option
prompt	Select 2D polyline
respond	**pick any point on first square**
prompt	3 lines were filleted – Figure 20.1A(a)
and	command line returned.

7 *a*) Repeat the fillet icon selection and:
   1. enter P <R> – the polyline option, radius still set to 8
   2. pick any point on second square
   3. prompt 4 lines were filleted – Figure 20.1A(b)
   b) Note the difference between the two fillet operations:
   1. Figure 20.1(a) is not a 'closed' polyline, so only three corners filleted.
   2. Figure 20.1(b) is a 'closed' polyline, so all four corners filleted.
   c) The corner 'not filleted' in Figure 20.1(a) is the start point.

8 Select the **CHAMFER icon** and:

prompt	Select first line or [Undo/Polyline/Distance/Angle/Trim/ method/Multiple]
enter	**D <R>** – the distance option
prompt	Specify first chamfer distance and enter **8 <R>**

prompt	Specify second chamfer distance and enter **8 <R>**
prompt	Select first line or [Undo/Polyline/Distance/Angle/Trim/ method/Multiple]
enter	**P <R>** – the polyline option
prompt	Select 2D polyline
respond	**pick any point on third square**
prompt	4 lines were chamfered – Figure 20.1A(c)
and	command line returned.

9 *Tasks*
  a) 1. set chamfer distances to 12 and 5
     2. chamfer the fourth square remembering to enter P <R> to activate the polyline option
     3. result is Figure 20.1A(d)
     4. note the orientation of the 12 and 5 chamfer distances
  b) 1. set an offset distance to 5 and offset the fifth square 'outwards'
     2. set an offset distance to 8 and offset the fifth square 'inwards'
     3. the complete square is offset with a single pick – Figure 20.1A(e).

10 Select the **EDIT POLYLINE icon** from the Modify II toolbar and:

prompt	Select polyline or [Multiple]
respond	**pick the sixth square**
prompt	Enter an option [Open/Join/Width/Edit vertex/Fit/Spline/ Decurve/Ltype gen/Undo]
enter	**W <R>** – the width option
prompt	Specify new width for all segments
enter	**4 <R>**
prompt	Enter an option [Open/Join/Width/Edit vertex/Fit/Spline/ Decurve/Ltype gen/Undo]
respond	**right-click-Enter** to end command – Figure 20.1A(f).

11 Menu bar with **Modify-Object-Polyline** and:

prompt	Select polyline or [Multiple]
respond	**pick the next square**
prompt	Enter an option [Open/Join/Width/Edit vertex/Fit/Spline/ Decurve/Ltype gen/Undo]
enter	**O <R>** – the open option
prompt	Enter an option [Close/Join/Width/Edit vertex/Fit/ Spline/Decurve/Ltype gen/Undo]
enter	**right-click-Enter**
and	square displayed with 'last segment' removed: Figure 20.1A(g).

12 At the command line enter **PEDIT <R>** and:

prompt	Select polyline and **pick the next square**
prompt	options and enter **S <R>** – the spline option
prompt	options and enter **X <R>** – exit command
and	square displayed as a splined curve, in this case a circle – Figure 20.1A(h).

13 Activate the polyline edit command, pick the next square then:
  a) enter W <R> then 5 <R>
  b) enter S <R> then X <R> – Figure 20.1A(i).

14 *Task*
  a) set a fillet radius to 9
  b) fillet the next square – remember P
  c) use the polyline edit command with options: width 3, spline, exit – Figure 20.1(j).

15 *Notes*
   a) if a polyline is drawn closed, then the edit polyline option is: Open
   b) if the polyline was not closed, then the option is: Close.

Polyline options

1 The polyline command has options displayed at the prompt line when the start point has been selected.

2 Options are activated by entering the capital letter corresponding to the option.

3 The options are:
   a) Arc   draws an arc segment
   b) Close   closes a polyline shape to the start point
   c) Halfwidth   draws a variable width polyline from halfwidths
   d) Length   length of line segment entered
   e) Undo   undoes the last option entered
   f) Width   draws a variable width polyline.

4 *a*) The options are displayed in Figure 20.1(B) exercise.
   b) Using the options, create a polyline of your own.

5 *Task*: Before leaving this exercise:
   a) MOVE the complete polyline shape with a single pick from its start point by @25,25
   b) with **FILL <R>** at the command line, toggle fill off
   c) REGEN the screen
   d) turn FILL on then REGEN the screen
   e) this exercise is now complete. Save if required.

Line and arc segments

1 A continuous polyline object can be created from a series of line and arc segments of varying width.

2 We will now use several polyline options to create a shape.

3 This shape will be used later in the chapter.

4 Open A3PAPER, layer OUT current, toolbars to suit.

5 Refer to Figure 20.2, select the polyline icon and:

prompt	*enter*	*fig-ref*	*comment*
start point	60,70	point 1	start point
next pt/options	L		length option
length of line	45	point 2	
next pt/options	W		width option
starting width	0		
ending width	10		
next pt/options	@120,0	point 3	
next pt/options	A		arc option
arc end/options	@50,50	point 4	
arc end/options	L		back to line option
next pt/options	W		width option
starting width	10		
ending width	0		
next pt/options	@0,100	point 5	
next pt/options	230,240	point 6	
next pt/options	W		width option

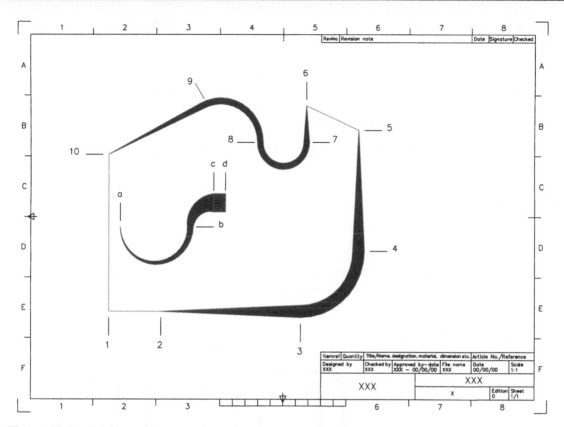

Figure 20.2 Polyline and arc segments.

prompt	enter	fig-ref	comment
starting width	0		
ending width	5		
next pt/options	@30<−90	point 7	
next pt/options	A		arc option
arc end/options	@−40,0	point 8	
arc end/options	@−50,30	point 9	
arc end/options	L		back to line option
next pt/options	W		width option
starting width	5		
ending width	0		
next pt/options	50,180	point 10	
next pt/options	C	point 1	close option and back to start

6   *a*) If your entries are correct the polyshape will be the same as that displayed in Figure 20.2.

   b) Polyline mistakes can be rectified as each segment is being constructed with the U (undo) option.

   c) Both the line and arc segments have their own command line option entries.

7   Repeat the polyline icon selection and:

prompt	enter	fig-ref	comment
start point	70,140	point a	
next pt/options	A		arc option
arc end/options	W		width option
starting width	0		
ending width	5		
arc end/options	@60<0	point b	

arc end/options	W		width option
starting width	5		
ending width	15		
arc end/options	@20,20	point c	
arc end/options	L		back to line option
next pt/options	@10,0	point d	
next pt/options	right-click and Enter		

8   Save the drawing layout as C:\MYCAD\POLYEX for later in the chapter.

Polyline tasks

1   Polyline shapes can be used with the modify commands.

2   To demonstrate their use, refer to Figure 20.3 and:
    either   *a)* open the A3PAPER template file
    or   *b)* erase all objects from the drawing area.

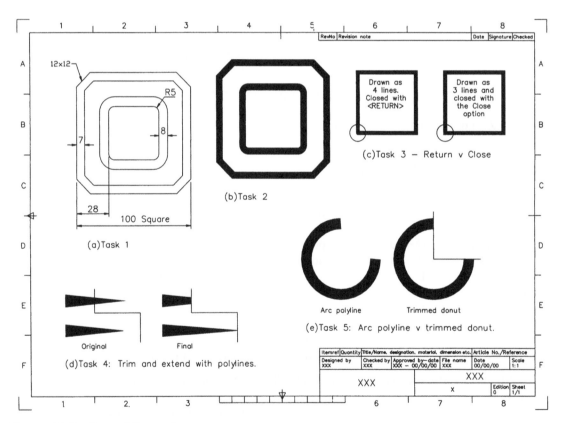

Figure 20.3   The polyline tasks.

3   *Task 1: Figure20.3(a)*
    a) Draw a 100 closed polyline square.
    b) Use the sizes given to complete the component – it is easier than you may think.
    c) Commands are OFFSET, CHAMFER, FILLET.

4   *Task 2: Figure 20.3(b)*
    a) Draw a 100-sided closed polyline square with constant width 7.
    b) Use the fillet, chamfer and offset sizes from the Task 1 to complete the component.

5 *Task 3: Figure 20.3(c)*
   a) Draw a 50-sided polyline square of width 5 ended closed with a <RETURN>.
   b) Draw a 50-sided polyline square of width 5 ended closed with the close option.
   c) Note the difference at the polyline start point.

6 *Task 4: Figure 20.3(d)*
   a) Polylines can be trimmed and extended.
   b) Try these operations with an 'arrowhead' type polyline with starting width 10 and ending width 0.

7 *Task 5: Figure 20.3(e)*
   a) Draw a polyline arc with following entries:

prompt	*enter*	*comment*
start point	pick to suit	
next pt/options	A	arc option
arc end/options	W	width option
starting width	10	
ending width	10	
arc end/options	CE	centre point option
center point	@0,−30	centre point relative to the start point
options	A	angle option
arc end/included angle	270	angle value
arc end/options	<R>	

   b) Draw a donut with an ID of 50 and an OD of 70 picking a centre point to suit.
   c) Draw two lines and trim the donut to these lines.
   d) Decide which method is easier.

8 This exercise is now complete. There is no need to save these tasks.

Modifying polylines

1 Polylines have their own editing facility.

2 This gives the user access to several extra options in addition to the existing modify commands.

Editing line and arc segments

1 Open the earlier saved exercise POLYEX and refer to Figure 20.4.

2 Note that Figure 20.4 displays a scaled version of the polyline exercises.

3 Select the **EDIT POLYLINE icon** from the Modify II toolbar and:
   prompt   Select polyline
   respond **pick any point on outer polyline shape** – Figure 20.4(a)
   prompt   Enter an option [Open/Join/Width/Edit vertex/Fit/Spline/
             Decurve/Ltype gen/Undo]
   respond **enter the following in respond to the prompts**

prompt	*enter*	*fig-ref*	*comments*
Open/Join . . .	W		constant width
new width	3	(b)	width value
options	D	(c)	decurve
options	S	(d)	spline
options	F	(e)	fit option
options	D	(f)	decurve again
options	O	(g)	open from start point

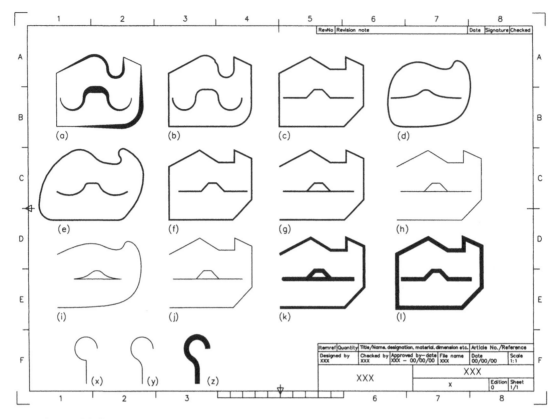

Figure 20.4 Editing polyline and arc segments.

prompt	enter	fig-ref	comments
options	W		constant width
new width	0	(h)	width value
options	S	(i)	spline
options	D	(j)	decurve
options	W		constant width
new width	5	(k)	width value
options	c	(l)	close shape
options	ESC		end command

The join option

1   This is a very useful option as it allows several individual polylines to be 'joined' into a single polyline object. Refer to Figure 20.4 again.

2   Use the MIRROR command to mirror the polyarc shape about a vertical line through the right end of the object.

3   At the command line enter **PEDIT <R>** and:
    prompt   Select polyline
    respond   **pick any point on original polyarc**
    prompt   Enter an option [Close/Join/Width/Edit . . .
    enter   **J <R>**
    prompt   Select objects
    respond   **pick the two polyarcs then right-click**
    prompt   3 segments added to polyline
    then   Enter an option [Close/Join/Width
    enter   **X <R>** to end command.

4   The line and arc segments will now be one polyline object.

5   Menu bar with **Modify-Object-Polyline** and:
      a) pick the mirrored joined polyshape
      b) enter the same options as step 3 of the editing line and arc segments exercise
      c) EXCEPTION: enter C(lose) instead of O(pen) and (O)pen instead of (C)lose.

6   Note the prompt when the Edit Polyline command is activated:
      a) Open/Join/Width, etc. if the selected polyshape is 'closed'
      b) Close/Join/Width, etc. if the selected polyshape is 'opened'.

7   The join option of the polyline edit command is very useful as it allows 'non-polyline objects' to be converted into polylines.

8   Refer to Figure 20.4 and:
      a) draw a vertical line and a three point arc, any size as Figure 20.4(x)
      b) add a fillet between these two objects – Figure 20.4(y)
      c) activate the polyline edit command and:

prompt	`Select polyline`
respond	**pick the line**
prompt	`Object selected is not a polyline`
	`Do you want to turn it into one <Y>`
enter	**Y <R>**
prompt	`Enter an option`
enter	**J <R>** – the join option
prompt	`Select objects`
respond	**window the line, fillet and arc then right-click**
prompt	`Enter an option`
enter	**W <R> then 3 <R>** – Figure 20.4(z)
prompt	`Enter an option`
respond	**X <R>** to end the command.

9   This exercise is now complete and can be saved if required.

Edit vertex option

1   The options available with the Edit Polyline command usually 'redraws' the selected polyshape after each entry, e.g. if S is entered at the options prompt, the polyshape will be redrawn as a splined curve.

2   The Edit vertex option is slightly different from this.

3   When E is entered as an option, the user has another set of options. Refer to Figure 20.5 and:
      a) erase all objects from the screen or open A3PAPER
      b) draw an 80-sided closed square polyshape and multiple copy it to three other areas of the screen.

4   *Variable widths*
      Menu bar with **Modify-Object-Polyline** and:

prompt	`Select polyline or [Multiple]`
respond	**pick the first square**
prompt	`Enter an option [Open/Join/Width/Edit vertex/Fit/Spline/`
	`Decurve/Ltype gen/Undo]`
enter	**W <R> and then 5 <R>**
prompt	`Enter an option [Open/Join/Width/Edit vertex/Fit/Spline/`
	`Decurve/Ltype gen/Undo]`
enter	**E <R>** – the edit vertex option

prompt	Enter a vertex editing option [Next/Previous/Break/ Insert/. . .
and	an X is placed at the start vertex, e.g. lower left
enter	**W <R>**
prompt	Specify starting width for next segment and enter **5 <R>**
prompt	Specify ending width for next segment<5> and enter **0 <R>**
prompt	Enter a vertex editing option [Next/Previous/Break . . .
enter	**N <R>** and X moves to next vertex (lower right for me)
prompt	Enter a vertex editing option [Next/Previous/Break . . .
enter	**W <R>**
prompt	Specify starting width for next segment and enter **0 <R>**
prompt	Specify ending width for next segment and enter **5 <R>**
prompt	Enter a vertex editing option [Next/Previous/Break . . .
enter	**X <R>** – to exit the edit vertex option
then	**X <R>** – to exit the edit polyline command – Figure 20.5(a).

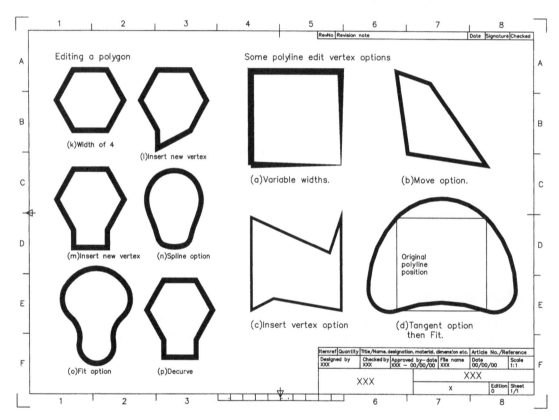

Figure 20.5 The polyline edit vertex option and editing a polygon.

5 *Moving a vertex*

Select the **Edit Polyline icon** and:

a) pick the second square

b) enter a constant width of 2

c) enter **E <R>** – the edit vertex option:

prompt	Enter a vertex editing option [Next/Previous/Break
enter	**N <R>** until X at lower left vertex – probably is?
then	**M <R>** – the move option
prompt	Specify new location for marked vertex
enter	**@10,10 <R>**
prompt	Enter a vertex editing option [Next/Previous/Break . . .

enter	**N <R>** until X at diagonally opposite vertex
then	**M <R>**
prompt	`Specify new location for marked vertex` and enter **@−50, −10 <R>**
prompt	`Enter a vertex editing option [Next/Previous/Break`
enter	**X <R> then X <R>** – Figure 20.5(b).

6 *Inserting a new vertex*

At the command line enter **PEDIT <R>** and:

a) pick the third square

b) enter a constant width of 1

c) pick the edit vertex option

d) enter **N <R>** until X at lower left vertex then:

prompt	`Enter a vertex editing option [Next/Previous/Break . . .`
enter	**I <R>** – the insert new vertex option
prompt	`Specify location of new vertex`
enter	**@20,10 <R>**
prompt	`Enter a vertex editing option [Next/Previous/Break . . .`
prompt	**N <R>** until X at opposite corner then enter **I <R>**
prompt	`Specify location of new vertex`
enter	**@−10,−30 <R>**
then	**X <R> and X <R>** – Figure 20.5(c).

7 *Tangent-Fit Options*

Activate the edit polyline command:

a) select the fourth square

b) set a constant width of 2

c) select the edit vertex option with the X at lower left vertex and:

prompt	`Enter a vertex editing option [Next/Previous/Break . . .`
enter	**T <R>** – the tangent option
prompt	`Specify direction of vertex tangent`
enter	**20 <R>**
and	**note arrowed line direction**
prompt	`Enter a vertex editing option [Next/Previous/Break . . .`
enter	1. **N <R>** until X at lower right vertex
	2. **T <R>**
prompt	`Specify direction of vertex tangent`
enter	**−20 <R>** – note arrowed line direction
prompt	`Enter a vertex editing option [Next/Previous/Break . . .`
enter	**X <R>** – to end the edit vertex options
prompt	`Enter an option [Open/Join . . .`
enter	**F <R>** – the fit option
then	**X <R>** – to end command and give Figure 20.5(d)
and	the final fit shape 'passes through' the vertices of the original polyline square.

Editing a polygon

1 A polygon is a polyline and can, therefore, be edited with the Edit Polyline command.

2 Draw a 6-sided polygon inscribed in a 40-radius circle.

3 Activate the Edit Polyline command, pick the polygon then enter the following option sequence:

enter	*fig-ref*	*comment*
W then 4	(k)	constant width
E		edit vertex option

N until X at lower left vertex

I then @0,−20	(l)	new vertex inserted
I then @40,0	(m)	new vertex inserted
X		end edit vertex option
S	(n)	spline curve
F	(o)	fit option
D	(p)	decurve option
X		end command

Assignments

Three activities involving the use of polylines:

1   *Activity 25: Shapes*

   a) Some basic polyline shapes created from line and arc segments.

   b) All relevant sizes are given for you and use discretion as appropriate. Snap on helps.

2   *Activity 26: Hooks and eyes*

   a) Three shapes which have to be drawn as polylines.

   b) You can create them as line and arcs then use the JOIN polyline edit option.

3   *Activity 27: Printed circuit board*

   a) A practical use for polylines in a drawing.

   b) It is harder to complete than you may think, especially with the dimensions being given as ordinate.

   c) Use your discretion when positioning the various 'lines' and add the dimensions.

Chapter 21

Additional draw and modify commands

1   In this chapter we will investigate some additional draw and modify commands.

2   These commands will give the user some useful draughting tools.

Getting started

1   Open your A3PAPER standard sheet with later OUT current.

2   Activate the Draw, Modify, Dimension and Object Snap toolbars.

3   Refer to Figure 21.1 and:
    a) set the point style and size indicated with **Format-Point Style** from the menu bar
    b) draw the following objects:
        1. LINE   from any suitable start point, to @85<15
        2. CIRCLE   with any suitable centre point and a radius of 30

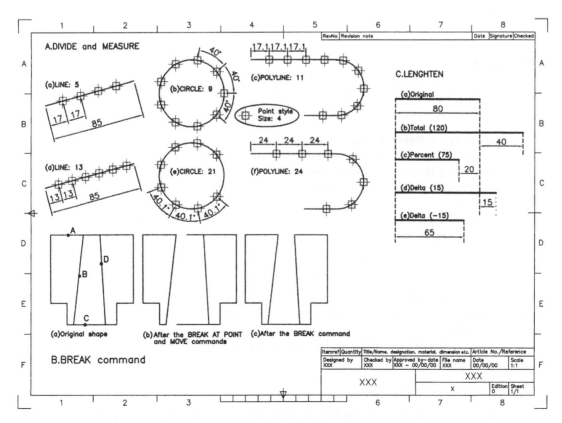

Figure 21.1   Using the DIVIDE, MEASURE, BREAK and LENGTHEN commands.

3. POLYLINE with:

    a) line segment from any suitable start point to @80,0

    b) arc segment to @0,−50

    c) line segment to @ −30,0

c) copy the three objects below the originals.

Divide

1   A selected object is 'divided' into an equal number of segments, the user specifying this number.

2   The current point style is 'placed' at the division points.

3   Menu bar with **Draw-Point-Divide** and:

    prompt   Select object to divide

    respond   **pick the line**

    prompt   Enter the number of segments or [Block]

    enter   **5 <R>**

    and   the line will be divided into five equal parts, and a point is placed at the end of each segment length as Figure 21.1A(a).

4   At the command line enter **DIVIDE <R>** and:

    prompt   Select object to divide

    respond   **pick the circle**

    prompt   Enter the number of segments or [Block]

    enter   **9 <R>**

    and   the circle will be divided into nine equal arc lengths and nine points placed on the circle circumference as Figure 21.1A(b).

5   Divide the POLYLINE into 11 segments – Figure 21.1A(c).

Measure

1   A selected object is 'divided' into a number of user-specified equal lengths.

2   The current point style is placed at each measured length.

3   Menu bar with **Draw-Point-Measure** and:

    prompt   Select object to measure

    respond   **pick the line** – the copied one of course!

    prompt   Specify length of segment or [Block]

    enter   **13 <R>**

    and   the line is divided into measured lengths of 13 units from the line start point – Figure 21.1A(d).

4   At command line enter **MEASURE <R>** and:

    prompt   Select object to measure

    respond   **pick the circle**

    prompt   Specify length of segment or [Block]

    enter   **21 <R>**

    and   the circle circumference will display points every 21 units from the start point – Figure 21.1A(e).

5   *Question*: Where is this start point?

6   Use the MEASURE command on the POLYLINE with a segment length of 24 – Figure 21.1A(f).

Task

1 Using the Dimension Style Manager dialogue box, select:
   a) the Modify option for the style A3DIM
   b) the Primary Units tab and set Linear and Angular Dimensions Precision to 0.0.

2 Add the linear and angular dimensions displayed. The Snap to Node is used for points.

3 This completes the exercise – no need to save.

Break

1 This command allows a selected object to be broken:
   a) at a specified point
   b) between two specified points with an erase effect.

2 Refer to Figure 21.1(B):
   a) draw the original shape as Figure 21.1B(a) from eight lines
   b) the actual shape size is unimportant, so use snap on
   c) ensure that the top and bottom horizontal lines have been created as single objects
   d) copy the complete shape to another part of the screen.

3 We want to modify the shape:
   a) by splitting it into two parts
   b) by erasing part of it.

4 Select the **BREAK AT POINT icon** from the Modify toolbar and:
   prompt   Select object
   respond **pick line A**
   prompt   Specify first break point
   respond **Snap to Endpoint of 'top' of line B**
   and   command line returned.

5 Select the **BREAK AT POINT icon** and:
   prompt   Select object
   respond **pick line C**
   prompt   Specify first break point
   respond **Snap to Endpoint of 'lower' end of line B**
   and   command line returned.

6 Now move the lines to the left of line B (six lines in total) to a suitable point – Figure 21.1B(b).

7 Select the **BREAK icon** from the Modify toolbar and:
   prompt   Select object
   respond **pick line A of the copied shape**
   prompt   Specify second point or [First point]
   enter   **F <R>** – the first point option
   prompt   Specify first break point
   respond **Snap to Endpoint of 'top' of line B**
   prompt   Specify second break point
   respond **Snap to Endpoint of 'top' of line D**.

8 The segment of line A between the two snapped endpoints will be erased – Figure 21.1B(c).

9 The exercise is now complete and need not be save.

10 *a*) When using BREAK, **an entry of @ at the second break point prompt** ensures that the first and second break points are the same.
   b) The TRIM command could have been used to obtain Figure 21.1B(c).

Lengthen

1   *a*) This command will alter the length of objects (including arcs).
     b) It cannot be used with CLOSED objects.

2   *a*) Continue with Figure 21.1.
     b) Draw a horizontal line of length 80 – Figure 21.1C(a).
     c) Multiple copy it four times below the original.

3   Menu bar with **Modify-Lengthen** and:
     prompt   Select an object or [DElta/Percent/Total/DYnamic]
     enter   **T <R>** – the total option
     prompt   Specify total length or [Angle]<1.00>
     enter   **120 <R>**
     prompt   Select an object to change or [Undo]
     respond   **pick the second line at the right end**
     then   right click-Enter – Figure 21.1A(b).

4   Menu bar with **Modify-Lengthen** and:
     prompt   Select an object or [DElta/Percent/Total/DYnamic]
     enter   **P <R>** – the percent option
     prompt   Enter percentage length<100.00>
     enter   **75 <R>**
     prompt   Select an object to change or [Undo]
     respond   **pick the next line**
     then   right click-Enter – Figure 21.1C(c).

5   At the command line enter **LENGTHEN <R>** and:
     enter   **DE <R>** – the delta option
     enter   **15 <R>** – the delta length
     respond   pick the next line – Figure 21.1C(d).

6   *Task*
     a) Use the DELTA option with a −15 value and select the last copied line – Figure 21.1C(e).
     b) Linear dimension the five lines to check the various options have been performed correctly.

7   The drawing can now be saved if required.

Align

1   A very powerful command which combines the move and rotate commands into one operation.

2   *a*) Refer to Figure 21.2.
     b) Draw a right-angled triangle and a rectangle (own sizes) – Figure 21.2A(a).
     c) Copy the two objects to three other parts of the screen (snap on will help).

3   We want to align:
     a) side 23 of the triangle onto side XY of the rectangle
     b) side XY of the rectangle onto side 23 of the triangle.

4   Menu bar with **Modify-3D Operation-Align** and:
     prompt   Select objects
     respond   **pick the three lines of triangle A then right-click**
     prompt   Specify first source point
     respond   **snap to endpoint and pick point 3**
     prompt   Specify first destination point

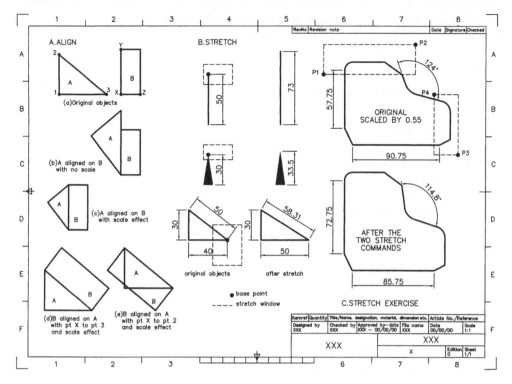

Figure 21.2 Using the ALIGN and STRETCH commands.

respond	**snap to endpoint and pick point X**
and	a line is drawn between points 3 and X
prompt	Specify second source point
respond	**snap to endpoint and pick point 2**
prompt	Specify second destination point
respond	**snap to endpoint and pick point Y**
and	a line is drawn between points 2 and Y
prompt	Specify third source point or <continue>
respond	**right-click as no more selections needed**
prompt	Scale objects based on alignment points? [Yes/No]<N>
enter	**N <R>**.

5 The triangle is moved and rotated onto the rectangle with sides 23 and XY in alignment – Figure 21.2A(b).

6 Repeat the Align selection and:
   a) pick the three lines of the next triangle then right-click
   b) pick the same source and destination points as step 3
   c) enter Y <R> at the Scale prompt
   d) the triangle will be aligned onto the rectangle, and side 23 scaled to side XY – Figure 21.2A(c).

7 At the command line enter **ALIGN <R>** and:

prompt	Select objects
respond	**pick the next rectangle then right-click**
prompt	Specify first source point
respond	**Snap to Midpoint icon and pick line XY**
prompt	Specify first destination point
respond	**Snap to Midpoint icon and pick line 23**
prompt	Specify second source point and **pick point X**

prompt	Specify second destination point and **pick point** 3
prompt	Specify third source point and **right-click**
prompt	Scale objects based on alignment points
enter	**Y <R>**.

8 The selected rectangle will be aligned onto the triangle as Figure 21.2A(d) and scaled to suit.

9 Repeat the align command and:
   a) pick the last rectangle then right-click
   b) midpoint of side XY as first source point
   c) midpoint of side 23 as first destination point
   d) point X as second source point
   e) point 2 as second destination point
   f) right-click at third source point prompt
   g) Y to scale option – Figure 21.2A(e).

10 Note that the orientation of the aligned object is dependent on the order of selection of the source and destination points.

Stretch

1 This command does what it says – it 'stretches' objects.

2 If hatching and dimensions have been added to the object being stretched, they will both be affected by the command – remember that hatch and dimensions are associative.

3 Select a clear area of the drawing screen, refer to Figure 21.2(B) and draw to your own sizes:
   a) a vertical dimensioned line
   b) a dimensioned variable width polyline
   c) a dimensioned triangle.

4 Select the **STRETCH icon** from the Modify toolbar and:

prompt	Select object to stretch by crossing-window or crossing-polygon
then	Select objects
enter	**C <R>** – the crossing option
prompt	Specify first corner
respond	**window the top of vertical line and dimension**
prompt	2 found
then	Select objects
respond	**right-click**
prompt	Specify base point or [Displacement]
respond	**pick top end of line (endpoint)**
prompt	Specify second point or <use first point as displacement>
enter	**@0,23 <R>**.

5 The line and dimension will be stretched by the entered value.

6 Menu bar with **Modify-Stretch** and:

prompt	Select objects
enter	**C <R>** – crossing option
prompt	First corner
respond	**window the top of polyline and dimension then right-click**
prompt	Specify base point and **pick top end of polyline**
prompt	Specify second point and enter **@0,3.5 <R>**.

7 The polyline and dimension are stretched by the entered value.

8 At the command line enter **STRETCH <R>** and:
   a) enter C <R> for the crossing option
   b) window the vertex of triangle indicated then right-click
   c) pick indicated vertex as the base point
   d) enter @10,0 as the displacement.

9 The triangle is stretched as are the dimensions.

Stretch example

1 Open your WORKDRG and refer to Figure 21.2(C) and:
   a) erase the centre lines
   b) scale the component by a factor of 0.55
   c) linear dimension the two lines and angle displayed.

2 Activate the STRETCH command and:
   a) enter C <R> – crossing option
   b) first corner: pick a point P1
   c) opposite corner: pick a point P2 then right-click
   d) base point: pick any suitable point
   e) second point: enter @0,15 <R>.

4 Repeat the STRETCH command and:
   a) activate the crossing option
   b) pick a point P3 for the first corner
   c) pick a point P4 for the opposite corner
   d) pick a suitable base point
   e) enter @−5,0 as the second point.

5 The component and dimensions will be stretched relative to the entered values.

Chapter 22

The array command

1   Array is a command which allows multiple copying of objects in either:
     a) a rectangular pattern
     b) a circular (polar) pattern.

2   It is one of the most powerful and useful of the commands available, yet is one of the easiest to use.

Getting started

1   Open the A3PAPER file with layer OUT current and the toolbars Draw, Modify and Object snap.

2   *a*) Refer to Figure 22.1 and draw the I-BEAM shape using the given reference sizes.
     b) Do NOT ADD the dimensions.

3   Multiple copy the I-BEAM shape from the point indicated to the points:
     A(10,210), B(275,265), C(175,160), D(260,50) and E(30,160).
     Note that the donuts in Figure 22.1 are for reference only.

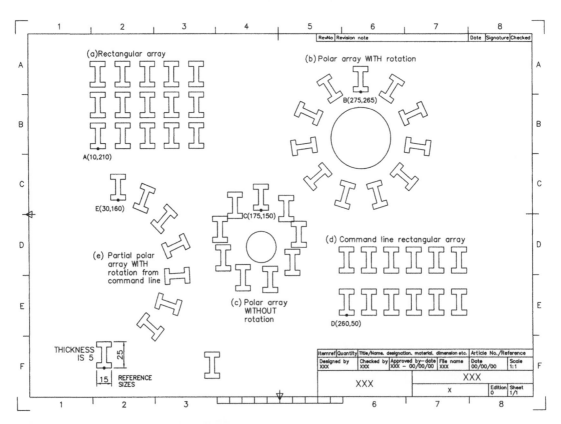

Figure 22.1   Using the ARRAY command.

4 Draw two circles:
   a) centre at 275,220 with radius 30
   b) centre at 175,115 with radius 15.

5 Move the original I-BEAM shape to a 'safe place' on the screen.

Rectangular array

1 Select the **ARRAY icon** from the Draw toolbar and:
   prompt   Array dialogue box
   with   *a*) options: Rectangular Array or Polar Array
             b) information for creating the array pattern
             c) a preview display
   respond *a*) ensure rectangular array active
             b) alter Rows: 3 and Columns: 5
             c) alter Row offset: 30
             d) alter Column offset: 25
             e) ensure Angle of array: 0
             f) pick Select objects
   prompt   Select objects at the command line
   respond **window the shape at A then right-click**
   prompt   Array dialogue box
   with   12 objects selected and a preview display – Figure 22.2
   prompt   **pick OK**.

2 The shape at A will be copied 14 times into a three-row and five-column matrix
   pattern as Figure 22.2(a).

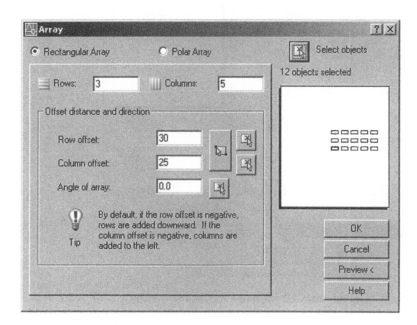

Figure 22.2   The dialogue box for the rectangular array.

Polar array with rotation

1 Menu bar with **Modify-Array** and:
   prompt   Array dialogue box
   respond *a*) Polar Array active
             b) pick Pick Center Point icon

prompt	Select center point of array at the command line
respond	**snap to Center icon and pick larger circle**
prompt	Array dialogue box
with	Selected centre point co-ordinates
respond	*a*) Method: Total number of items & Angle to fill
	b) Total number of items: 11
	c) Angle to Fill: 360
	d) Rotate items as copied: active, i.e. tick
	e) pick Select objects
prompt	Select objects at the command line
respond	**window the shape at B then right-click**
prompt	Array dialogue box as Figure 22.3(a)
respond	**pick Preview<**
and	*a*) preview of the entered polar array values
	b) Array message – Figure 22.3(b)
respond	**pick Accept if the array is correct, or Modify to return** to the dialogue box for alterations.

2   *a*) The shape at B is copied in a circular pattern about the selected centre point.
    b) The objects are 'rotated' about the centre point as they are copied – Figure 22.1(b).

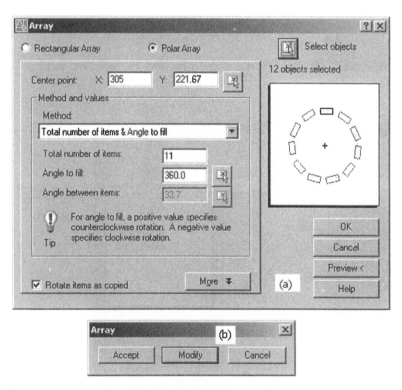

Figure 22.3   The dialogue box and AutoCAD message for the Polar array.

Polar array without rotation

1   At the command line enter **ARRAY <R>** and use the Array dialogue box with the following data:
    a) Polar Array
    b) Center point: pick and snap to center icon of smaller circle
    c) Method: Total number of items & Angle to fill
    d) Total number of items: 9

e) Angle to fill: 360
f) Rotate items as copied: Not active, i.e. no tick
g) Select objects: window shape at C the right-click
h) preview then accept.

2 *a*) The original shape is copied about the selected centre point.
   b) The selected objects are not 'rotated' as copied – Figure 22.1(c).

Rectangular array from command line

1 The previous three examples have used the Array dialogue box to enter the details required for the various arrays.

2 The command can also be activated from the command line, so enter **–ARRAY <R>** and:

prompt	`Select objects`
respond	**Window the shape at D then right-click**
prompt	`Enter the type of array [Rectangular or Polar]`
enter	**R <R>** – the rectangular option
prompt	`Enter the number of rows(---)<1>` and enter **2 <R>**
prompt	`Enter the number of columns(|||)<1>` and enter **6 <R>**
prompt	`Enter the distance between rows or specify unit cell (---)`
enter	**40 <R>**
prompt	`Specify the distance between columns(|||)`
enter	**22.5 <R>**.

3 The shape at D will be copied into a 2 × 6 column matrix pattern as Figure 22.1(d).

Polar array with partial fill angle

1 Activate the array command and:
   a) objects: window the shape at D
   b) type of array: P
   c) center point: enter X: 35 and Y: 100
   d) number of items: 7
   e) angle to fill: −130
   f) rotate objects: Y
   g) the result is Figure 22.1(e).

2 Your drawing should now resemble Figure 22.1 and can be saved if required.

Array options

1 Both the rectangular and polar array commands have variations to those displayed in Figure 22.1.

2 These variations include:
   a) rectangular: allows the angle of the array to be altered
   b) polar: allows for different method items and fill angle.

3 *a*) erase the array exercises (save first?) to leave the original (moved to a safe place) shape
   b) refer to Figure 22.4.

4 Multiple copy the original shape from the point indicated to the points:
   P(15,170), Q(120,170), R(75,110), S(190,110), X(260,305), Y(260,215) and Z(260,65).

5 *a*) Rotate the shapes at P and Q by 20 and the shapes at R and S by −10.
   b) Scale the shapes at X, Y and Z by 0.5.
   c) In each case pick the indicated donut point as the base.

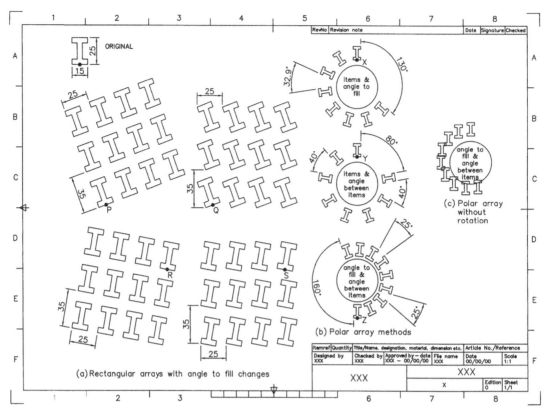

Figure 22.4 The ARRAY options.

6 Draw three circles, radius 20 with centres at the points: 260,280; 260,190; 260,100.

7 Using the array dialogue box four times, activate the rectangular pattern and set:

	1st	*2nd*	*3rd*	*4th*
a) objects	shape P	shape Q	shape R	shape S
b) rows, columns	3,4	3,4	3,4	3,4
c) row offset	35	35	−35	−35
d) column offset	25	25	−25	−25
e) angle of array	20	0	−10	0

8 The result of the four rectangular operations is Figure 22.4(a).

9 Select the ARRAY command and with the polar array pattern active, enter the following three sets of data:

	1st	*2nd*	*3rd*
a) objects	shape X	shape Y	shape Z
b) method	items & angle to fill	items & angle between	angle to fill & angle between
c) center point	circle X	centre Y	centre Z
d) items	8	8	–
e) angle to fill	230	–	220
f) angle between	–	40	25
g) rotate items	yes	yes	yes

10 The results of these three operations is Figure 22.4(b).

11 *Task*
  a) Linear/align dimension as shown and note the 35 and 25 values for the rectangular array.

b) Angular dimension the polar arrays and note the fill angle and angle between the objects values.

c) Comment on the 160 degree fill angle. This operation had an angle to fill of 220, and therefore this angle should be 140?

d) Try the polar array operations with no rotate items as copied – Figure 22.4(c) displays one operation.

12 This exercise is now complete and can be saved if required.

Assignments

Several activities to complete and all involve using the array command as well as hatching, adding text, etc. I have tried to make these activities varied and interesting so I hope you enjoy attempting them. As with all activities:

a) start with your A3PAPER file – drawing or template?

b) use layers correctly for outlines, text, hatching, etc.

c) when complete, save as C:\MYCAD\:ACT?? or similar

d) use your discretion for any sizes which have been omitted.

1 *Activity 28: Car wheel design*
   a) Six designs of varying difficulty
   b) You can create as given or use your imagination
   c) No sizes are given as none are needed.

2 *Activity 29: Ratchet and saw tip blade*
   a) Two typical examples of a use for the array command.
   b) The ratchet tooth shape is fairly straightforward and the saw blade tooth is more difficult than you would think. Or is it?
   c) Draw each tooth in a clear area of the screen, then copy them to the appropriate circle points. The trim command is extensively used.

3 *Activity 30: Fish design*
   a) Refer to Figure 22.5 and create the fish design using the sizes given.
   b) Save this drawing as **C:\MYCAD\FISH** for future work.

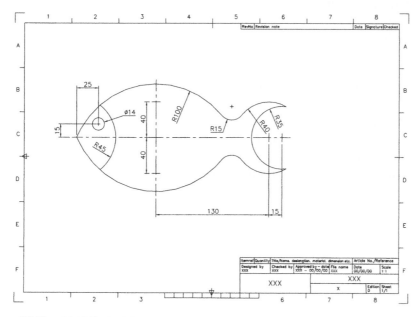

Figure 22.5   C:\MYCAD\FISH information for Activity 30.

c) When the 'fish' has been completed and saved, create the following array patterns:
  1. A 5 × 3 angular rectangular array at 10 degrees with a 0.2 scale. The distances between the rows and columns are at your discretion.
  2. Three polar arrays with:
     a) scale 0.25 with radius of 90 and 9 items
     b) scale 0.2 with radius of 60 and 7 items
     c) scale 0.15 with radius of 35 and 7 items.

4 *Activity 31: The light bulb*
   a) An interesting activity from previous books.
   b) Creating the R10 radius arc is quite tricky.
   c) The basic bulb is copied and scaled three times. The polar array centre is at your discretion.
   d) The position of the smaller polar bulb is relative to the larger polar array, but how is it positioned? This problem is for you to decide.
   e) The arc aligned text is interesting – how to create, align, etc.
   f) No hints are given on how to draw the basic bulb.

5 *Activity 32: Bracket and Gauge*
   a) Two engineering type applications for the array command.
   b) The bracket is fairly easy, the hexagonal rectangular array positioned to suit yourself.
   c) The gauge drawing requires some thought with the polar arrays. I drew a vertical line and arrayed 26 items twice, the fill angles being +150 and −150. Think about this!! The other gauge fill angles are +140 and −140. What about the longer line in the gauge? The pointer position is at your discretion. Another method is to array lines and trim to circles?

6 *Activity 33: Pinion gear wheel*
   a) Another typical engineering application of the array command.
   b) The design details are given and the basic tooth shape is not too difficult.
   c) Copy and rotate the second gear wheel – but at what angle? Think about the number of teeth.

7 *Activity 34: Propeller blades*
   a) Draw the outline of the blade using the sizes given. Trimmed circles are useful.
   b) When the blade has been drawn, use the COPY, SCALE and polar ARRAY commands to produce the following propeller designs:
     1. a 2-bladed at a scale of 0.75
     2. a 3-bladed at a scale of 0.65
     3. a 4-bladed at a scale of 0.5
     4. a 5-bladed at a scale of 0.35.
   c) Note that the 4- and 5-bladed designs require some 'tidying up'.

8 *Activity 35: Leaf design*
   a) Draw the leaf from the reference sizes.
   b) Mirror effect is useful for the outline then 2D solid or hatch.
   c) The arrays are with a one-fourth size object.
   d) The angular array requires some thought for the angle and the row/column offsets.

Chapter 23

Changing object properties

1   All drawn objects have properties, e.g. linetype, colour, layer, position, lineweight, etc.

2   Text has properties such as height, style, width factor, obliquing angle, etc.

3   Object properties can be changed using the:
    a) command line CHANGE command
    b) Properties palette.

Getting started

1   Open A3PAPER with layer OUT current, toolbars to suit and refer to Figure 23.1.

2   Draw a 40-unit square and copy it to six other parts of the screen.

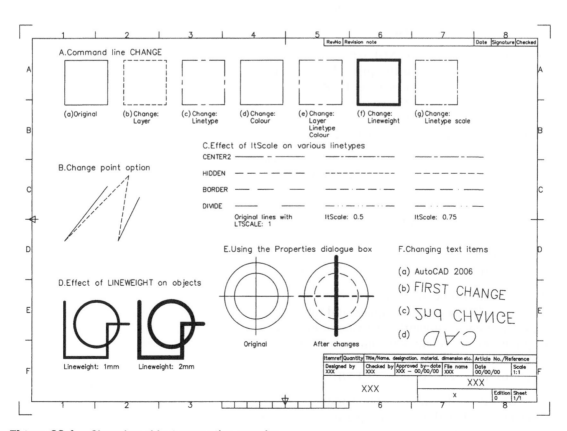

Figure 23.1   Changing object properties exercise.

3   *a*) At the command line enter **LTSCALE <R>** and:
    prompt   Enter new linetype scale factor
    enter   1 **<R>**
   b) At the command line enter **LWDISPLAY <R>** and:
    prompt   Enter new value for LWDISPLAY
    enter   **ON <R>**

Command line CHANGE

1   At the command line enter **CHANGE <R>** and:
    prompt   Select objects
    respond **window the second square then right-click**
    prompt   Specify change point or [Properties]
    enter   **P <R>** – the properties option
    prompt   Enter property to change[Color/Elev/LAyer/LType/ltScale/
                LWeight/Thickness/PLotstyle]
    enter   **LA <R>** – the layer option
    prompt   Enter new layer name
    enter   **HID <R>**
    prompt   Enter property to change, i.e. any more property changes
    respond **right click-Enter**.

2   The square will be displayed as brown hidden lines – Figure 23.1A(b)

3   Repeat the CHANGE command line entry and:
   a) objects: window the third square then right-click
   b) change point or properties: enter P <R> for properties
   c) options: enter LT <R> for linetype
   d) new linetype: enter CENTER2 <R> for center linetypes
   e) options: right-click/Enter to end command.

4   The square will be displayed with red center lines – Figure 23.1A(c).

5   Use the command line CHANGE with:
   a) objects: window the fourth square then right-click
   b) change point or properties: enter P <R>
   c) options: enter C <R> – the colour option
   d) new colour: enter BLUE <R>
   e) options: press <R> – Figure 23.1A(d).

6   Use the CHANGE command with the fifth square and:
   a) enter P <R> then LA <R> – the layer option
   b) new layer: DIMS <R>
   c) options: enter LT <R>
   d) new linetype: CENTER2 <R>
   e) options: enter: C <R>
   f) new colour: enter GREEN <R>
   g) options: right-click/Enter – Figure 23.1A(e).

7   With the command entry CHANGE:
   a) objects: window the sixth square and right-click
   b) change point or properties: activate properties
   c) options: enter LW <R> – the lineweight option
   d) new lineweight: enter 0.5 <R>
   e) options: press <R> – Figure 23.1A(f).

8   CHANGE <R> at the command line and:
   a) objects: window the seventh square and right-click
   b) activate properties

c) options: enter LT <R> then CENTER2 <R> i.e. center linetype

d) options: enter S <R> – the linetype scale option

e) new linetype scale: enter 1 <R>

f) options: press <R> – Figure 23.1A(g).

9 Compare the center linetype appearance of Figures 23.1A(c) and(g).

10 Menu bar with **Format-Layer** and:

a) make layer 0 current

b) freeze layer OUT then OK

c) only the hidden line square and a green center line square displayed? – Figures 23.1A(b) and (e)

d) thaw layer OUT and make it current.

11 *Notes*

a) This exercise with the CHANGE command has resulted in:

Figure 23.1(A)	Square appearance	Layer
a	red continuous lines	OUT
b	brown hidden lines	HID
c	red center lines	OUT
d	blue continuous lines	OUT
e	green center lines	DIMS

b) I would suggest to you that only Figure 23.1A(a) and (b) are 'ideally' correct, i.e. the correct colour and linetype for the layer being used.

c) The other squares demonstrate that it is possible to have different colours and linetypes on named layers.

d) This can become confusing and is not recommended.

The change point option

1 The above exercise has only used the Properties option of the command line CHANGE command, but there is another option – change point.

2 Draw two lines:

a) start point: 40,150   next point: 80,200

b) start point: 90,150   next point: 110,190

3 Activate the CHANGE command and:

prompt   Select objects

respond   **pick the two lines then right-click**

prompt   Specify change point or [Properties]

respond   **pick about the point 100,210 on the screen**.

4 The two lines are redrawn to this point – Figure 23.1(B).

5 This could be a very useful drawing aid?

LTSCALE and ltScale

1 One of the options available with the CHANGE command is ltScale.

2 This is a system variable and allows individual objects to have their line type appearance changed.

3 The final appearance of the object depends on the value entered and the value assigned to LTSCALE.

4 The LTSCALE system variable is **GLOBAL**, i.e. if it is altered, all objects having center lines, hidden lines, etc. will alter in appearance and this may not be to the user's requirements. Hence the use of the ltScale option of CHANGE.

5 If LTSCALE is globally set to 0.6, and the value for ltScale is entered as 0.5, the selected objects will be displayed with an effective value of 0.3.

6 If LTSCALE is 0.6 and 2 is entered for ltScale, the effective value for selected objects is 1.2.

7 This effect is shown in Figure 23.1(C) using the Center2, Hidden, Border and Divide linetypes.

Lineweight

1 LINEWEIGHT allows selected objects to be displayed at varying width.

2 Although this may seem similar to a polyline-width object, it is not.

3 Objects can be drawn directly with varying lineweight or with the LWeight option of the CHANGE command.

4 It is necessary to toggle the LWDISPLAY system variable to ON before the lineweight effect is displayed.

5 This was achieved at the start of this chapter.

6 Right-click on **LWT in the Status bar pick Settings** and:
   prompt   Lineweight Settings dialogue box
   respond   *a*) ensure Millimeters active
               b) scroll at Lineweights and pick 1.00 mm
               c) dialogue box as Figure 23.2
               d) note the default value then pick OK.

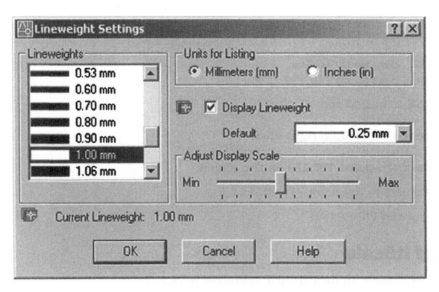

Figure 23.2 The Lineweight Settings dialogue box.

7 Now draw some lines and circles.

8 Finally set the lineweight back to the default value – 0.25?

9 Figure 23.1(D) displays some objects at the two set lineweights.

Pickfirst

1   When objects require to be modified, the normal procedure is to:
    a) activate the command
    b) then select the objects.

2   This procedure can be reversed with the PICKFIRST system variable.

3   Draw a line and circle anywhere on the screen, then:
    a) menu bar with **Modify-Erase**
    b) select the two objects and right-click
    c) and the objects are erased – but you knew this already!

4   Draw another line and circle.

5   At the command line enter **PICKFIRST <R>** and:
    prompt   Enter new value for PICKFIRST
    enter   1 <R>
    and   a pickbox will be displayed on the cursor crosshairs
    note   do not confuse the pickbox with the grips box – they are totally different.

6   Pick the line and circle with the pickbox, then:
    a) menu bar with **Modify-Erase**
    b) the objects are erased.

7   Thus PICKFIRST allows the user to alter the selection process and:

    PICKFIRST: 0 – activate the command then select the objects

    PICKFIRST: 1 – select the objects then activate the command.

Changing properties using the Properties palette

1   With layer OUT current, draw:
    a) two concentric circles
    b) two lines to represent center lines.

2   Ensure PICKFIRST is set to 1.

3   With the pickbox, select the horizontal line then the **PROPERTIES icon** from the Standard toolbar and:
    prompt   Properties palette
    with   two sets of information about the object selected:
                a) General: colour, layer, linetype, etc.
                b) Geometry: start and end point of line, etc.

4   With the palette active:
    a) pick Layer line and scroll arrow appears
    b) scroll and pick layer CL
    c) alter linetype scale value to 0.5 – Figure 23.3(a)
    d) cancel the dialogue box – top left X box
    e) press ESC.

5   The selected line will be displayed as a green center line.

6   Pick the vertical line then the Properties icon and from the Properties palette alter:
    a) colour: Magenta
    b) linetype: Hidden2
    c) lineweight: 1.2
    d) cancel palette then ESC.

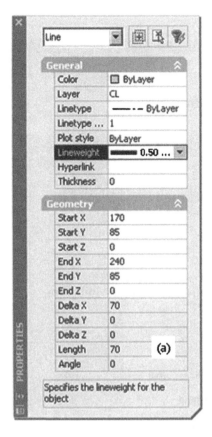

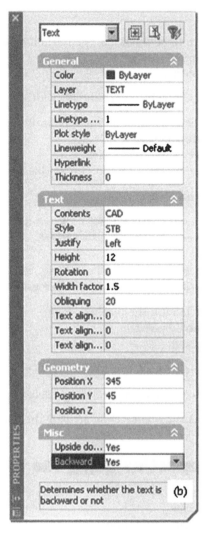

Figure 23.3 The Properties palette for changing (a) a selected line and (b) a selected item of text.

7   Finally pick the smaller circle then the Properties icon and alter:
    a) layer: HID
    b) linetype scale: 2
    c) cancel and ESC.

8   Figure 23.1(E) displays the before and after effects of using the Pickfirst-Properties palette method for changing the properties of selected objects.

9   The Properties palette gives useful information about selected objects:
    a) lines: start and end point, delta values, length, angle
    b) circles: centre point, radius, diameter, area, circumference
    c) these values can be altered from the palette, and the object will be modified accordingly
    d) you can try this for yourself.

Changing text

1   Text objects have several properties:
    a) general:layer, colour, etc.
    b) specific:style, height, width factor, etc.

2   Text properties can be altered with the command line CHANGE or using the Properties palette.

3   Create the following two new text styles:

Name	STA	STB
font	romant	italict
height	0	10
width	1	1
oblique	0	5
backwards	N	N
upside-down	N	Y
vertical	N	N

4   With layer TEXT current and STA the current style, refer to Figure 23.1(F) and:
    a) enter the text item AutoCAD 2006 at height 5 and rotation 0 at a suitable part of the screen
    b) Figure 23.1F(a).

5   Multiple copy this item of text to three other parts of the screen.

6   At the command line enter **CHANGE <R>** and:

prompt	Select objects
respond	**pick the second text item then right-click**
prompt	Specify change point or [Properties]
respond	**<RETURN>**
prompt	Specify new text insertion point <no change>
respond	**<RETURN>** – no change for text start point
prompt	Enter new text style <STA>
respond	**<RETURN>** – no change to text style
prompt	Specify new height and enter **8 <R>**
prompt	Specify new rotation angle and enter **-5 <R>**
prompt	Enter new text <AutoCAD 2006>
enter	**FIRST CHANGE <R>** – Figure 23.1F(b).

7   With **CHANGE <R>** at the command line, pick the third item of text and:
    a) change point: enter <R>
    b) new text insertion point: enter <R>
    c) new text style: enter STB <R>
    d) new rotation angle: enter −2
    e) new text: enter 2nd CHANGE – Figure 23.1F(c)
    f) *Question*: Why no height prompt?

8   With PICKFIRST on (i.e. set to 1) pick the fourth item of text then the Properties icon and:

prompt	Properties palette
with	four sections for the selected text item: General; Text; Geometry and Misc
respond	alter the following:
	1. contents: CAD
	2. text style: STB
	3. height: 12
	4. width factor: 1.5
	5. obliquing: 20
	6. Backwards: Yes
and	palette as Figure 23.3(b)
then	cancel the palette then ESC
and	changes displayed in Figure 23.1F(d).

9   This exercise is now complete and can be saved if required.

Combining the ARRAY and CHANGE commands

1 Combining the array command with the properties command can give interesting results.

2 Open your standard A3PAPER sheet and refer to Figure 23.4.

2 Draw the two arc segments as trimmed circles using the information given in Figure 23.4(a).

3 Draw the polyline and 0 text item using the reference data.

4 With the ARRAY command, polar array (twice) the polyline and text item using an arc centre as the array centre point:
   a) for 4 items, angle to fill +30 degrees with rotation
   b) for 7 items, angle to fill −60 degrees with rotation – Figure 23.4(b).

5 Using (a) the command line CHANGE or (b) the Properties icon, pick each text item and alter:
   a) the text values to 10, 20, 30, etc. to 90
   b) the text height to 8.

6 The final result should be as Figure 23.4(c).

7 Save if required as the exercise is complete.

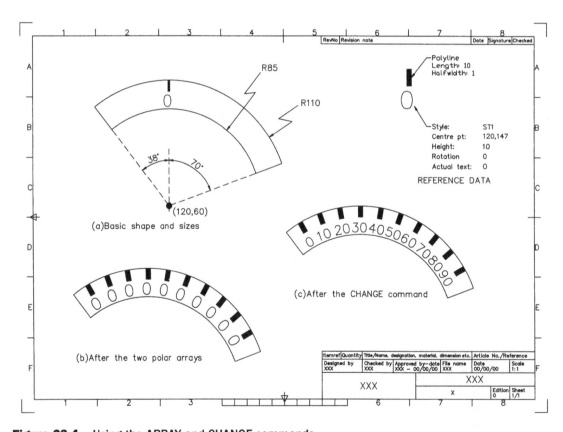

Figure 23.4   Using the ARRAY and CHANGE commands.

Notes

1 Figure 23.1 demonstrated that it was possible to change the properties of objects independent of the current layer.

2 This means that if layer OUT (red, continuous) is current, objects can be created on this layer as green center lines, blue hidden lines, etc.

3 This is a practice **I WOULD NOT RECOMMEND** until you are proficient at using the AutoCAD draughting package.

4 If green center lines have to be created, use the correct layer, or make a new layer if required.

5 Try not to 'mix' different types of linetype and different colours on the one layer.

6 Remember that this is only a recommendation – the choice is always left to the user.

Assignments

Three activities are included which involve using the array and change commands.

1 *Activity 36: Array patterns with text styles*
   a) Open C:\MYCAD\STYLEX from Chapter 13 to display a blank screen but with 12 created and saved text styles (ST1-ST12).
   b) Create a 30-unit square at A(10,10) then read the note (f) before proceeding.
   c) Multiple copy the square from point A to the points B(85,220), C(200,70), D(300,55) and E(20,255).
   d) Array the squares using the following information:
      A: rectangular with two rows and five columns, both distances 35
      B: polar for 10 items, full circle with rotation, the centre point of the array being 100,170
      C: rectangular with five rows and two columns and 35 distances
      D: angular rectangular (angle of array −10) with five rows and two columns, the distances both 35
      Remember to rotate first!!
      E: rectangular for one row and 10 columns, the distance 35.
   e) Add/or change the text items using the style names listed.
   f) *Notes*:
      1. Text can be added after the squares are arrayed, or before the first square is multiple copied.
      2. After array: the text items are added using single line text, the style, rotation angle and start point being entered by the user. This is fairly straightforward with the exception of the polar and angular arrays, i.e. what are the rotation angles?
      3. Before the multiple copy: two text items are added to the original square and it is then multiple copied and arrayed. The added text items can then be altered with the change properties command.
      4. It is the user's preference as to which method is used.
      5. I added the text to the original square, copied, arrayed then changed properties.

2 *Activity 37: Telephone dials*
   a) The 'old-fashioned' type has circles arrayed for a fill angle of ?? Is the text arrayed or just added to the drawing as text? You have to decide!
   b) The modern type has a polyline 'button' with middled text. The array distances are at your discretion as is the outline shape.

3 *Activity 38: Flow gauge and dartboard*
   a) Flow gauge: A nice simple drawing to complete, but it takes some time! The text is added during the array operation.
   b) Dartboard: Draw the circles then array the 'spokes'. The filled sections are trimmed donuts. The text is middled, height 10 and ROMANT. Array then change properties?

Chapter 24

Text tables

1 A table is an object containing data in rows and columns, the user:
   a) creating and positioning the empty table
   b) adding the content to the table cells.

2 All tables can:
   a) contain three types of cell – title, header and data
   b) be created downwards or upwards
   c) be customised with colour and text styles to user requirements.

3 Figure 24.1 displays:
   a) the basic table terminology
   b) a DOWN table with a title cell, 3 header cells and 6 data cells. All the text entries display the same text style
   c) an UP table with a title cell, no header cells and 12 data cells. This table has been created using different text styles.

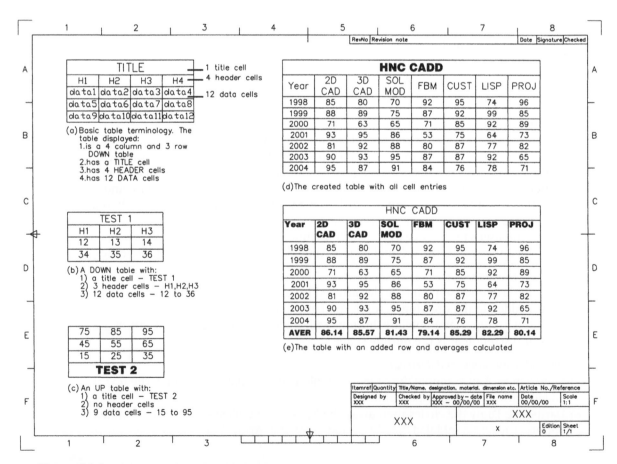

Figure 24.1 Table terminology and exercise.

Creating a table

1 Open your A3PAPER template file

2 *a*) We want to create a table to display a set of results for a CAD course from a well known college.
   b) There are several topics offered in this course, including 2D draughting, 3D draughting, solid modelling, feature based modelling, customisation, LISP programming and a project.
   c) It is the 'pass rates' for these topics over a 7-year period which have to be presented in both tabular and graphical form.

3 Make layer TEXT current and select from the menu bar **Draw-Table** and:

prompt	Insert Table dialogue with three sections:
	a) table style settings
	b) insertion behaviour
	c) column & row settings.
respond	**pick . . .** at the Table Style name
prompt	Table Style dialogue box with Standard style listed
with	*a*) list of current table styles
	b) preview of the current table.
respond	**pick New**
prompt	Create New Table Style dialogue box
respond	*a*) New Style Name: TB1
	b) Start With: Standard – Figure 24.2
	c) pick Continue.

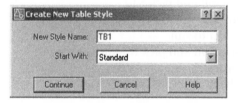

Figure 24.2   The Create New Table Style dialogue box.

prompt	New Table Style: TB1 dialogue box
with	three tab selections – Data (probable active), Column Heads, Title
respond	**select the appropriate named tab and alter as follows:**

    a) Title tab active
        1. Cell properties
            a) Include Title row active, i.e. tick
            b) Text style: scroll and pick ST3
            c) Text height: 6
            d) Colours: to suit
            e) Alignment: scroll and pick Middle Centre.
        2. Border properties
            a) Borders: none active
            b) Grid lineweight and colour: to suit.
        3. General
            a) Table direction: down
        4. Cell margins
            a) Both horizontal and vertical set to 1.5 – Figure 24.3

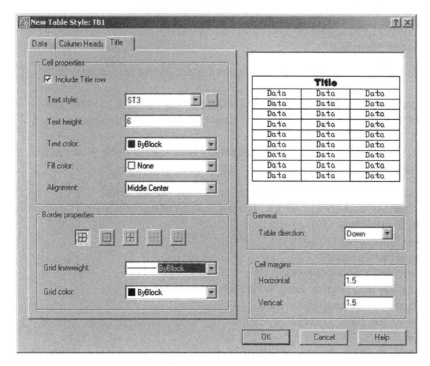

Figure 24.3   The New Table Style dialogue box for TB1 – Title tab.

        b) Column Heads tab active
                1. Cell properties
                    a) Include Header row: active
                    b) Text style: scroll and pick ST2
                    c) Text height: 5
                    d) Colours: select to suit
                    e) Alignment: scroll and pick Middle Centre.
                2. Border properties
                    a) Leave as given.
                3. General
                    a) Table direction: down.
                4. Cell margins: both horizontal and vertical set to 1.5
        c) Data tab active
                1. Cell properties
                    a) Text style: scroll and pick ST1
                    b) Text height: 4
                    c) Colours: investigate and decide yourself
                    d) Alignment: scroll and pick Middle Centre.
                2. Border properties
                    a) Leave as given.
                3. General
                    a) Table direction: down.
                4. Cell margins: both horizontal and vertical set to 1.5.

then	**pick OK**
prompt	`Table Style dialogue box`
with	*a*) TB1 added to the Styles list and highlighted
	b) a display of the new TB1 table.
respond	*a*) pick Set Current
	b) pick Close.
prompt	`Insert Table dialogue box`
with	Table Style name: TB1

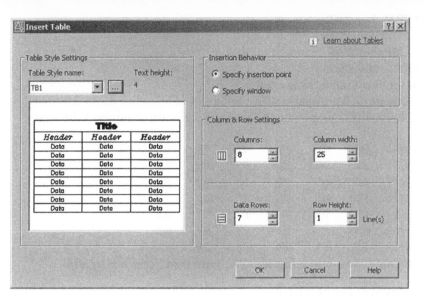

Figure 24.4   The Insert Table dialogue box for table style TB1.

respond	*a*) Insertion Behaviour: Specify insertion point active
	b) Column & Row Settings.
	1. Columns: 8 with column width 25
	2. Data Rows: 6 with Row Height 1 – Figure 24.4.
prompt	`Specify insertion point`
respond	**pick to suit**
and	*a*) blank table TB1 inserted
	b) Text Formatting bar displayed with:
	1. ST3 as the text style
	2. Arial Black as the font
	3. 6 as the height.
	c) the title bar cell highlighted with flashing cursor.
respond	*a*) enter HNC CADD
	b) press the TAB key.
prompt	*a*) cursor 'jumps' to next cell – the first header cell
	b) cell is highlighted
	c) Text Formatting bar displays: ST2, italicc, 5 – Figure 24.5.
respond	*a*) enter YEAR
	b) press the TAB key and cursor jumps to next header cell
	c) enter 2D CAD then TAB
	d) enter 3D CAD then TAB
	e) enter SOL MOD then TAB

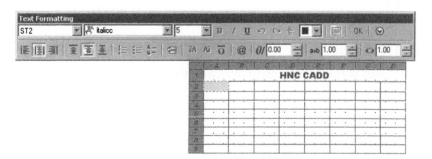

Figure 24.5   The Text Formatting bar for the first header cell.

         f) enter FBM then TAB

         g) enter CUST then TAB

         h) enter LISP then TAB

         i) enter PROJ then TAB.

prompt   *a*) cursor jumps to the first data cell

         b) Text Formatting bar displays: ST1, romans, 4

respond   enter the following data, pressing TAB after each entry to move to the next data cell

1998	85	80	70	92	95	74	96
1999	88	89	75	87	92	99	85
2000	71	63	65	71	85	92	89
2001	93	95	86	53	75	64	73
2002	81	92	88	80	87	77	82
2003	90	93	95	87	87	92	65
2004	95	87	91	84	76	78	71

then   when last data (71) is entered, **pick OK from text formatting bar.**

4   The table will be displayed as Figure 24.1(d)

5   Data entered in a table can be edited with:

    a) PICKFIRST set to 1: a double-left-click in the appropriate cell will display the text formatting bar and the user can edit as required

    b) PICKFIRST set to 0: the command line entry **TABLEDIT** will allow the user to select the cell to be edited.

6   Tables can be used for many different purposes, e.g. parts list, stock control, scheduling, etc. and the fact that the tables can be customised gives the user a very powerful draughting aid.

Inserting a new row and calculating averages

1   The created table requires an additional row to display the average of each topic over the recorded 7-year period.

2   Ensure PICKFIRST set to 1.

3   Left-click the 2004 cell to highlight the cell then right-click and:

    prompt   Shortcut menu displayed

    respond   *a*) pick Insert Rows

             b) pick Below

             c) press the ESC key

    and   a complete row of blank cells is added to the table.

4   Double left-click in the leftmost cell of the new row and:

    prompt   Text Formatting bar

    respond   *a*) enter AVER

             b) scroll and select the ST3 text style

             c) pick OK

    and   AVER added to the cell.

5   Left-click the cell to the right of the AVER cell, then right-click and:

    prompt   Shortcut menu displayed

    respond   *a*) pick Insert Formula

             b) pick Average.

    prompt   Select first corner of table cell range

    respond   **pick any part of the 85 value 2D CAD cell**

    prompt   Select second corner of table cell range

respond	**drag the window and pick any part of the 95 value 2D CAD cell**
prompt	`Text Formatting bar`
respond	*a*) scroll and select the ST3 text style
	b) note formula given as **=Average(B3:B9)**
	c) pick OK form Text Formatting bar
and	86.14 entered in the cell, which is the 2D CAD average for the 7-year period.

6 The averages for the other topics can be calculated:

  a) as step 5

  b) using cut-and-paste which is the method I used. Try this by:

    1. left-click in the 86.14 cell

    2. right-click to display the shortcut menu

    3. pick Copy

    4. move to the cell to the right of 86.14 and left-click

    5. right-click to display the shortcut menu and pick Paste

    6. repeat steps 4 and 5 in the other cells.

7 The table with the additional row and the averages is displayed in Figure 24.1(e) and still has the table style name TB1.

Exporting table data

1 While inserting a customised table into a drawing is useful to the user, the data contained in the table can be exported from the drawing and imported into other application software packages (e.g. Excel).

2 This export process is very straightforward.

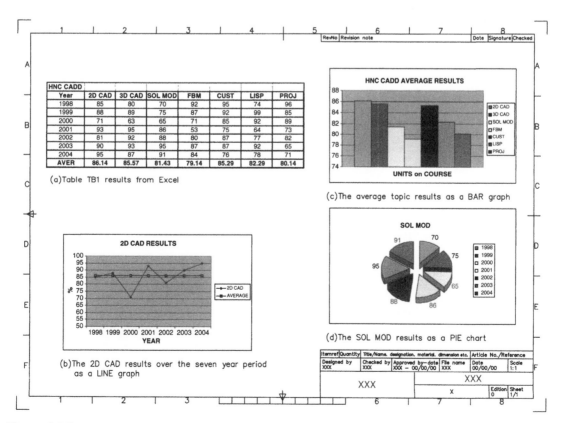

Figure 24.6 The results from TB1 using Excel.

3   At the command line enter **TABLEEXPORT <R>** and:
    prompt   Select a table
    respond   **pick any point on table TB1**
    prompt   Export Data dialogue box
    respond   *a*) Save in: your named folder, i.e. MYCAD
                  b) File name: TB1
                  c) File type: Comma Delimited (*.csv)
                  d) pick Save.
    and   command prompt returned.

4   From the Windows taskbar select **Start-Programs-Excel** (or your own selection)
    to display a blank spreadsheet.

5   Menu bar with **File-Open** and:
    prompt   Open dialogue box
    respond   *a*) Look in: scroll and pick your MYCAD folder
                  b) File types: scroll and pick Text Files
                  c) pick TB1 then Open.

6   The saved table will be displayed in the spreadsheet but without the CAD text styles.

7   The table data can be used as any normal Excel data (e.g. charts, averages, etc.).

8   Figure 24.6 displays the TB1 data from Excel 'copy-and-pasted' back into AutoCAD.
    This data is:
    a) original table TB1 in Excel format
    b) 2D CAD results over the 7-year period as a LINE graph with the average
    c) average results for the seven topics as a BAR graph
    d) SOL MOD results as a 3D PIE CHART.

9   The information obtained from exporting a table is at user discretion.

    This completes the table exercise and chapter.

Tolerances

1   In a previous chapter we investigated how dimensions could be customised to user requirements by setting and saving a dimension style.

2   The dimension style A3DIM was created and saved with our A3PAPER standard sheet.

3   In this chapter we will:
   a) create and use several new dimension styles
   b) add tolerance dimensions
   c) investigate geometric tolerance
   d) investigate dimension variables.

4   The process for creating new dimension styles involves altering the values for specific variables which control how the dimensions are displayed on the screen.

Tolerance dimensions

1   Dimensions can be displayed with tolerances and limits.

2   The formats available are symmetrical, deviation, limits and basic.

3   Figure 25.1(a) displays these formats along with the standard default dimension type, i.e. no tolerances displayed.

To use tolerance dimensions correctly, a dimension style should be created for each 'type' which is to be used in the drawing. As usual we will investigate the topic with an exercise so:

1   Open your A3PAPER standard sheet and refer to Figure 25.1.

2   Draw a line, circle and angled lines and copy them to four other areas of the screen.

3   Dimension one set of objects with your standard A3DIM dimension style setting – Figure 25.1(p).

Creating the tolerance dimension styles

1   *a*) Using the Dimension Style Manager dialogue box, create four new named dimension styles, all starting with A3DIM.
   b) The four new style names are DIMTOL1, DIMTOL2, DIMTOL3 and DIMTOL4.

2   With the Dimension Style Manager dialogue box:
   a) pick each new dimension style
   b) pick Modify then select the:
      1. Primary Units tab and alter both the Linear and Angular precision to 0.000.
      2. Tolerances tab and alter as follows:

	DIMTOL1	*DIMTOL2*	*DIMTOL3*	*DIMTOL4*
Method	Symmetrical	Deviation	Limits	Basic
Precision	0.000	0.000	0.000	—

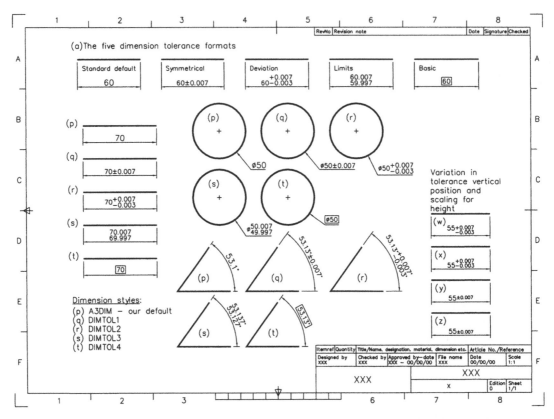

Figure 25.1 Dimension tolerance exercise.

	DIMTOL1	DIMTOL2	DIMTOL3	DIMTOL4
Upper value	0.007	0.007	0.007	—
Lower value	—	0.003	0.003	—
Scaling for height	1	1	1	—
Vertical position	Middle	Middle	Middle	Middle

3 a) Making each new dimension style current, dimension a set of three objects
   b) The effect is displayed in Figure 25.1 with:
      (q): style DIMTOL1
      (r): style DIMTOL2
      (s): style DIMTOL3
      (t): style DIMTOL4.

4 *Task:* Investigate the vertical position and the scaling for height options available in the Tolerances tab dialogue box. Figure 25.1 displays the following:

figure	method	vertical pos	scaling for height
(w)	deviation	top	0.8
(x)	deviation	bottom	0.8
(y)	symmetrical	top	0.7
(z)	symmetrical	bottom	0.7.

Comparing dimension styles

1 It is possible to compare dimension styles with each other to note changes.

2 To demonstrate this comparison, select from the menu bar **Dimension-Style** and:
   prompt   Dimension Style Manager dialogue box
   respond   a) ensure A3DIM is the current style
                 b) pick DIMTOL1 then Compare.

prompt	Compare Dimension Styles dialogue box
with	*a*) Compare: DIMTOL1
	b) With: A3DIM
and	comparison between the different dimension variables of DIMTOL1 and A3DIM – Figure 25.2
respond	note the differences then Close both dialogue boxes.

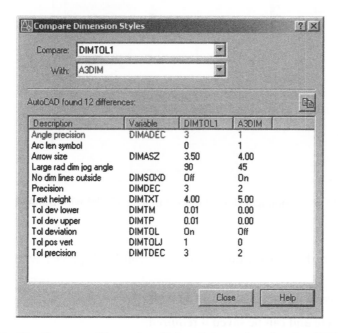

Figure 25.2 The Compare Dimension Styles dialogue box for the A3DIM and DIMTOL1 styles.

Dimension variables

1 The comparison between the two dimension styles displays some of the AutoCAD dimension variables.

2 All dimension variables (dimvars) can be altered by keyboard.

3 Make A3DIM the current dimension style and note the dimensions which used this style – the Figure 25.1(p) dimensions in Figure 25.1.

4 At the command line enter **DIMASZ <R>** and:
prompt   Enter new value for DIMASZ <4.00>
enter   **12 <R>**.

5 Pick the **Dimension Style icon** from the Dimension toolbar and:
prompt   Dimension Style Manager dialogue box
with   *a*) Current Style: A3DIM
           b) Styles: A3DIM
                           └──────<style overrides>
           c) Description: A3DIM + Arrow size = 12.00 – Figure 25.3.
respond *a*) move pointer to <style overrides>
           b) right-click the mouse.
prompt   Shortcut menu displayed.
respond *a*) pick Save to current style
           b) close the Dimension Style Manager dialogue box.

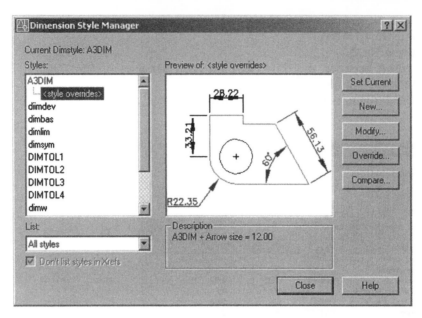

Figure 25.3 The Dimension Style Manager dialogue box for style A3DIM with DIMASZ set to 12.

6 The dimensions which used the A3DIM dimension style will be displayed with an arrow size of 12.

7 Now change the DIMASZ variable for A3DIM back to 4.

8 This exercise is now complete and can be saved if required.

Geometric tolerancing

To demonstrate geometric tolerancing:

1 Open your A3PAPER standard sheet.

2 Menu bar with **Dimension-Tolerance** and:
   prompt   Geometric Tolerance dialogue box
   with   Feature control frames
   respond   **move pointer to box under Sym and left-click**
   prompt   Symbol dialogue box as Figure 25.4.
   respond   *a*) press ESC to cancel the Symbol dialogue box
               b) study the Geometric Tolerance dialogue box
               c) pick Cancel at present.

Feature control frames

1 Geometric tolerances define the variations which are permitted to the form or profile of a component, its orientation and location as well as the allowable runout from the exact geometry of a feature.

2 Geometric tolerances are added by the user in feature control frames, these being displayed in the Geometric Tolerance dialogue box.

3 *a*) a feature control frame consists of two/three sections or compartments, displayed in Figure 25.5(a)
   b) these are:
      1. **The geometric characteristic symbol**
         a) This is a symbol for the tolerance which is to be applied

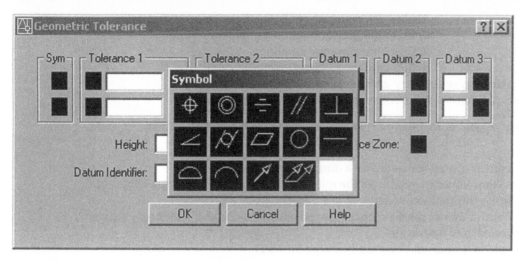

Figure 25.4 The Symbol and Geometric Tolerance dialogue boxes.

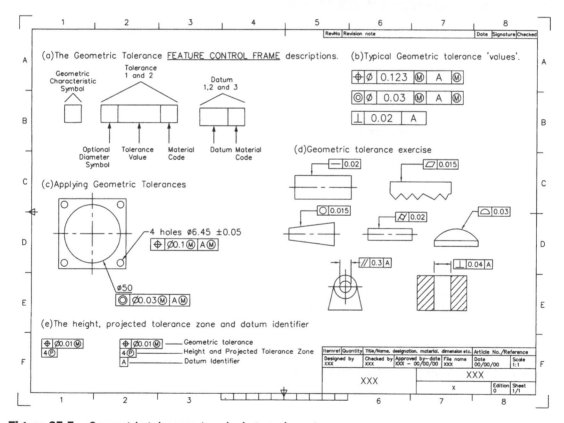

Figure 25.5 Geometric tolerance terminology and usage.

b) These geometric symbols can represent location, orientation, form, profile and runout

c) The symbols include symmetry, flatness, straightness, angularity, concentricity, etc.

2. **The tolerance section (1 and 2)**

   a) This creates the tolerance value in the Feature Control Frame

   b) The tolerance indicates the amount by which the geometric tolerance characteristic can deviate from a perfect form. It consist of three boxes:

   1. first box: inserts an optional diameter symbol
   2. second box: the actual tolerance value

3. third box: displays the material code which can be:
                                   a) M: at maximum material condition
                                   b) L: at least material condition
                                   c) S: regardless of feature size.

3. **The primary, secondary and tertiary datum information**
    a) This can consist of the reference latter and the material code.

4   Typical types of geometric tolerance 'values' are displayed in Figure 25.5(b).

5   *Notes*
    a) Feature control frames are added to a dimension by the user.
    b) A feature control frame contains all the tolerance information for a single dimension
    c) The tolerance information must be entered by the user.
    d) Feature control frames can be copied, moved, erased, stretched, scaled and rotated once 'inserted' into a drawing.
    e) Feature control frames can be modified with DDEDIT.

Geometric tolerance example

1   Refer to Figure 25.5(c) and create a square with circles as shown – own sizes.

2   Diameter dimension the larger circle using the A3DIM dimension style with DIMS layer current.

3   Select the **TOLERANCE icon** from the Dimension toolbar and:
    prompt   Geometric Tolerance dialogue box
    respond   **pick top box under Sym**
    prompt   Symbol dialogue box
    respond   **pick Concentricity symbol** (second top left)
    prompt   Geometric Tolerance dialogue box
    with   Concentricity symbol displayed
    respond   *a*) at Tolerance 1, pick the left box
                     b) at Tolerance 1, enter 0.03
                     c) at Tolerance 1, pick the right (MC) box and:
                           prompt   Material Condition dialogue box
                           respond   **pick M**
                     d) at Datum 1, enter A
                     e) at Datum 1, pick the right (MC) box and pick M
                     f) Geometric Tolerance dialogue box as Figure 25.6
    then   pick OK
    prompt   Enter tolerance location
    respond   **pick 'under' the diameter dimension.**

4   Menu bar with **Dimension-Tolerance** and:
    prompt   Geometric Tolerance dialogue box
    respond   **pick top Sym box**
    prompt   Symbol dialogue box
    respond   **pick Position symbol** (top left)
    prompt   Geometric Tolerance dialogue box
    respond   *a*) at Tolerance 1 pick left box (for diameter symbol)
                     b) at Tolerance 1 enter value of 0.1
                     c) at Tolerance 1 pick right (MC) box and pick M
                     d) at Datum 1 enter datum A
                     e) at Datum 1 pick right (MC) box and pick M
                     f) pick OK
    prompt   Enter tolerance location
    respond   **pick to suit** – refer to Figure 25.5(c).

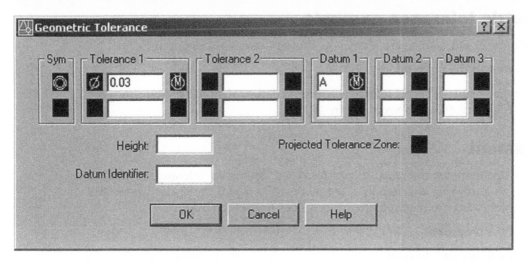

Figure 25.6   The Geometric Tolerance example dialogue box.

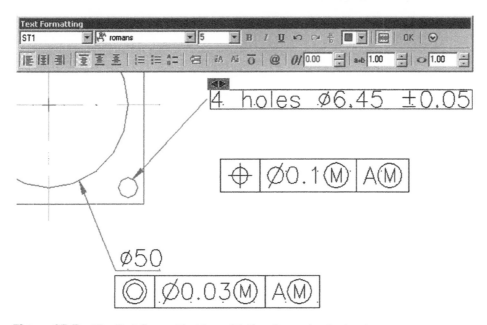

Figure 25.7   The Text Formatting bar with the dimension text entry.

5   Select the **QUICK LEADER icon** from the Dimension toolbar and:
    prompt   Specify first leader point
    respond   **pick a smaller circle**
    prompt   Specify next point
    respond   **pick a point to suit then right-click**
    prompt   Specify text width
    enter   **0 <R>**
    prompt   Enter first line of annotation text
    respond   **right-click**
    prompt   Text Formatting Editor dialogue box
    enter   **4 holes %%C6.45 %%P0.05** – Figure 25.7
    then   pick OK.

Geometric tolerance example

1   Refer to Figure 25.5(d) and create 'shapes' as shown – size is not important.

2   Add the geometric tolerance information.

3   Investigate the height, projected tolerance zone and datum identifier options from the Geometric Tolerance dialogue box as Figure 25.5(e).

Assignment

A single assignment for tolerance dimensions and geometric tolerancing. Adding dimensions to a drawing is generally a tedious process and if dimension styles and geometric tolerancing are required, the process does not get any easier.

Activity 39: *Three components*

1   The components are easy to create.

2   Several dimension styles using the symmetrical, limits, deviation and basic tolerance methods are required.

3   You have to enter the upper and lower values.

Chapter 26

Using various units and paper sizes

1 Three of the most commonly asked questions are:
   a) can I draw with imperial units?
   b) can you set a scale at the start of a drawing?
   c) how do you work with very large/small components?

2 The answer to these three questions is yes:
   a) you can draw in feet and inches
   b) you can set a scale at the start of a drawing
   c) you can draw large/small components on an A3-sized sheet.

3 The most difficult concept for many CAD users is that all work should be completed full size.

4 In this chapter we will investigate three different types of exercise:
   a) a component drawn and dimensioned in inches
   b) a large-scale drawing
   c) a small-scale drawing.

5 Each exercise will involve modifying the existing A3PAPER standard sheet.

6 Hopefully all users will appreciate what it means by drawing everything full size.

Drawing using Imperial units (inches)

1 To draw with Imperial units we require a new standard sheet which could be created:
   a) from scratch
   b) using the Wizard facility
   c) altering the existing A3PAPER standard sheet which is set to metric sizes.

2 We will alter our existing standard sheet.

3 The reason is that A3PAPER has already been 'customised' to our requirements (i.e. it has layers, a text style and a dimension style). This may save us some time?

4 The conversion from a metric standard sheet to one which will allow us to draw in inches is relatively straightforward. We have to alter several parameters (e.g. units, limits, drawing aids and the dimension style variables).

5 Open your A3PAPER standard sheet.

6 Units: Menu bar with **Format-Units** and:
   prompt Drawing Units dialogue box
   respond set the following:
       a) *Length*
           Type: Engineering with Precision: 0'–0.00".

    b) *Angle*
     Type: Decimal Degrees with Precision: 0.00.
    c) *Insertion scale*
     Units to scale inserted content: Inches
  then pick OK.

7 Limits: Menu bar with **Format-Drawing Limits** and:
  prompt `Specify lower left corner`
  enter **0,0 <R>**
  prompt `Specify upper right corner`
  enter **16″,12″ <R>** [note shift 2 (″) for the inches symbol].

8 Drawing aids:
  a) menu bar with **View-Zoom-All**
  b) set grid spacing to 0.5
  c) set snap spacing to 0.25
  d) set global LTSCALE value to 1.

9 Dimension Style.
  Menu bar with **Dimension Style** and:
  prompt `Dimension Style Manager dialogue box`
  respond **create a new dimension style** with:
    a) New Style Name: STDIMP
    b) Start with: A3DIM
    c) Tab alterations as follows:
     1. Lines
      Baseline spacing: 0.75″
      Extend beyond dim line: 0.175″
      Offset from origin: 0.175″.
     2. Symbols and Arrows
      Arrowheads: Closed Filled
      Arrow size: 0.175″
      Centre Marks for Circles: None.
     3. Text
      Style: ST1 with height: 0.18″
      Offset from dim line: 0.08″
      Text alignment: with dimension line.
     4. Primary Units
      Unit Format: Engineering
      Precision: 0′–0.00″
      Degrees: Decimal
      Precision: 0.00
      Zero suppression: both ON (i.e. tick)
      Scale factor: 1.

10 Make STDIMP the current dimension style.

11 *a*) make layer OUT current
  b) save the drawing as **C:\MYCAD\STDIMP** – it may be useful.

12 Refer to Figure 26.1 and draw the component as shown. Add all text and dimensions. You should have no problems in completing the drawing or adding the dimensions. What should be the value for text height? Use the grid spacing as a guide.

13 When complete, save the drawing – plot?

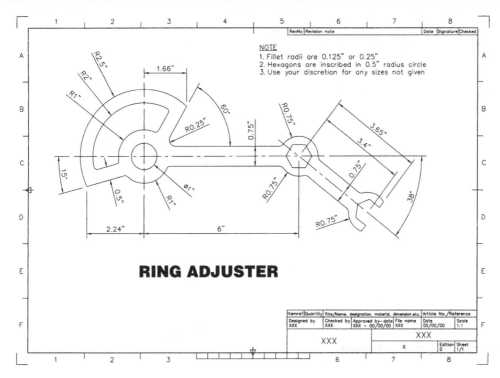

Figure 26.1 Using inches to draw a component.

Large-scale drawing

1 This exercise will require the A3PAPER standard sheet to be modified to allow a very large drawing to be drawn (full size) on A3paper.

2 *a*) Open A3PAPER standard sheet and make layer 0 current
   b) Left-click the Model from the layout tabs
   c) Zoom-all.

3 At the command line enter **MVSETUP <R>** and:

   prompt   Enable paper space? (No/Yes)
   enter   **N <R>**
   prompt   Enter   units   type   [Scientific/Decimal/Engineering/
              Architectural/Metric]
   enter   **M <R>** – for metric units
   prompt   AutoCAD Text Window with Metric scales
              Enter the scale factor
   enter   **50 <R>** i.e. scale of 1:50
   prompt   Enter the paper width and enter **420 <R>**
   prompt   Enter the paper height and enter **297 <R>**.

4 A black A3 border will be displayed. This is our drawing area.

5 With the menu bar **Tools-Inquiry-ID Point**, obtain the co-ordinates of the lower left and upper right. corners of the black border. These should be:
   a) lower left: X = 0.00   Y = 0.00   Z = 0.00
   b) upper right: X = 21000.00   Y = 14850.00   Z = 0.00.

6 *a*) Return to the A3PAPER layout with a left-click on ISO A3 Title Block from the layout tabs
   b) Zoom-all
   c) Set the grid to 400 and the snap to 300.

7 With layer OUT current:
   a) refer to Figure 26.2 and complete the factory plan layout using the sizes as given

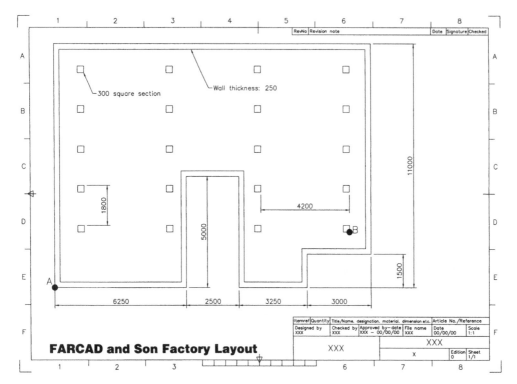

Figure 26.2 A large scale drawing.

b) point A has co-ordinates (800,3200)
c) the wall thickness is 250
d) erase the black border created during the MVSETUP command.

8 *a*) Create the 300 square section column at the point B(14800,5700)
   b) Rectangular array the 300 square (5 rows and 4 columns) using the offset information given.

9 With the **Dimension Style icon** create a new dimension style and alter as follows:
   a) New Style Name: STDLRG
   b) start with: A3DIM
   c) Fit tab: Use overall scale: alter to 50
   d) make STDLRG the current dimension style
   e) close the Dimension Style dialogue box.

10 Add the given dimensions and appropriate text – but what will be the text height?

11 Save this drawing as **C:\MYCAD\LARGESC** as it will be used:
   a) as a block activity
   b) to demonstrate the model/paper space concept.

12 If you have access to a plotter/printer, obtain a hard copy on A3 paper – the problem is the plot figures (50 is a useful number to know).

Small-scale drawing

1 This exercise requires the A3 standard sheet to be modified to suit a very small-scale drawing, which (as always) will be drawn full size.

2 *a*) Open A3PAPER standard sheet and make layer 0 current
   b) Left-click Model from the layout tabs
   c) Zoom-all.

3   At the command line enter **MVSETUP <R>** and:
prompt   Enable paper space? (No/Yes)
enter   **N <R>**
prompt   Enter units type [Scientific/Decimal/Engineering/
           Architectural/Metric]
enter   **M <R>** – for metric units
prompt   AutoCAD Text Window with Metric scales
           Enter the scale factor
enter   **0.001 <R>** i.e. scale of 1:50
prompt   Enter the paper width and enter **420 <R>**
prompt   Enter the paper height and enter **297 <R>**.

4   A black A3 border will be displayed – the drawing area.

5   Set UNITS to:
    a) Length precision: 0.0000
    b) Angle precision: 0.00.

6   ID the lower left and upper right corners and:
    a) lower left: X = 0.0000   Y = 0.0000   Z = 0.0000
    b) upper right: X = 0.4200   Y = 0.2970   Z = 0.0000.

7   *a*) Return to the A3PAPER layout with a left-click on ISO A3 Title Block from the lay-
        out tabs
    b) Zoom-all
    c) Set the grid to 0.01 and the snap to 0.005
    d) Erase the black border.

8   With layer OUT current:
    a) refer to Figure 26.3 and draw the toggle switch (full size) using the sizes
        given

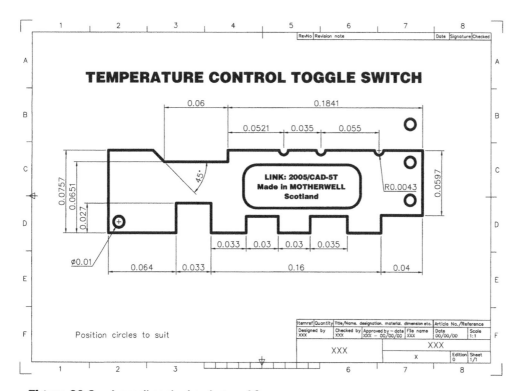

Figure 26.3   A small scale drawing on A3 paper.

    b) add all dimensions after altering the A3DIM dimension style with:
       1. Fit tab: overall scale: 0.001
       2. Primary units tab: Linear precision: 0.0000.
    c) Add text, but you have to decide on the height.

9   Save when the drawing is complete.

Assignment

Activity 40: House Plan

1   A single assignment requiring a large-scale drawing to be created on A3paper.

2   The assignment requires a house plan to be drawn at a scale of 1:50.

3   I would suggest:
    a) Open the LARGESC drawing saved earlier in the chapter as it has all the required 'settings'
    b) Erase the factory layout
    c) Complete the house plan using the metric sizes given
    d) Add the text and dimensions and:
       1. remember that the drawing is 50 times larger than the A3paper
       2. text and dimensions should be 50 times larger than normal
       3. the text height is easy to enter
       4. the dimension overall scale factor should have been 'set' to 50
    e) As an additional task, modify the dimension style primary tab to Architectural units and note the effect.

Chapter 27

Multilines, complex lines and groups

1 Layers have allowed the user to display continuous, centre and hidden linetypes.

2 AutoCAD has the facility to display multilines and complex lines, these being defined as:
   a) Multiline
      1. Parallel lines which can consist of several linear elements of differing linetype and colour
      2. They must be created by the user.
   b) Complex
      1. Lines which can be displayed containing text items and shapes
      2. They can be created by the user
      3. AutoCAD has several 'stored' complex linetypes.

Multilines

1 Multilines consist of between 1 and 16 parallel line elements.

2 The ends of multilines can be capped with lines and arcs or be left uncapped.

3 The basic multiline terminology is displayed in Figure 27.1(a).

4 To investigate how to use multilines:
   a) open your A3PAPER standard sheet with layer OUT current
   b) refer to Figure 27.1.

5 From the menu bar select **Draw-Multiline** and:
prompt	Current settings: Justification=Top, Scale=??, Style= STANDARD
then	Specify start point or [Justification/Scale/STyle]
enter	**S <R>** – the scale option
prompt	Enter mline scale<??>
enter	**10 <R>**
prompt	Specify start point and enter **20,40 <R>**
prompt	Specify next point and enter **@80,0 <R>**
prompt	Specify next point and enter **@70<110 <R>**
prompt	Specify next point and enter **@0,50 <R>**
prompt	Specify next point and **right-click-Enter**.

6 Menu bar again with **Draw-Multiline** and:
   a) set scale to 5
   b) draw a square of side 40 from the point 20,60 using the close option – Figure 27.1(b).

7 From the menu bar select **Format-Multiline Style** and:
prompt	Multiline Style dialogue box
respond	**pick New**

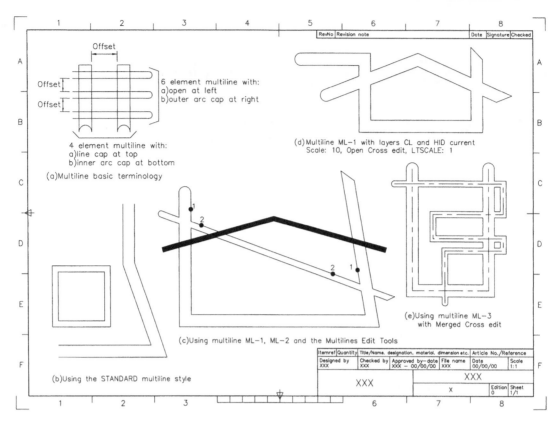

Figure 27.1 Multiline terminology and usage.

prompt	Create New Multiline Style dialogue box
respond	*a*) New Style Name: enter ML-1
	b) pick Continue
prompt	New Multiline Style: ML-1 dialogue box
respond	*a*) Description: MY FIRST ATTEMPT
	b) Caps:
	1. Start – Outer arc with Angle 90
	2. End – Line with Angle 45.
	c) Fill color: None
	d) Display joints: not active (blank)
	e) Elements: leave as given
	f) pick OK
prompt	Multiline Style dialogue box
with	ML-1 added to the Styles list
respond	*a*) pick ML-1
	b) pick Set Current – Figure 27.2
	c) pick OK.

8   At the command line enter **MLINE <R>** and:

prompt	Specify start point or [Justification/Scale/Style]
enter	**ST <R>** – the style option
prompt	Enter mline style name and enter **ML-1 <R>**
prompt	Specify start point or [Justification/Scale/Style]
enter	**S <R>** – the scale option
prompt	Enter mline scale and enter **10 <R>**
prompt	Specify start point and enter **135,165 <R>**
prompt	Specify next point and enter **@0,−100 <R>**

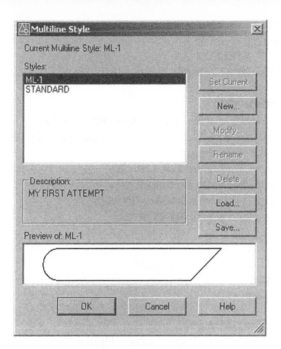

Figure 27.2 The Multiline Style dialogue box for ML-1.

prompt	Specify next point and enter **@150,0 <R>**
prompt	Specify next point and enter **@120<100 <R>**
prompt	Specify next point and **right-click-Enter.**

9 Menu bar with **Format-Multiline Style** and:

prompt	Multiline Style dialogue box
respond	**pick New**
prompt	Create New Multiline Style dialogue box
respond	*a*) New Style Name: enter ML-2
	b) Start With: ML-1
	c) pick Continue
prompt	New Multiline Style: ML-2 dialogue box
respond	*a*) Description: MY SECOND ATTEMPT
	b) Caps:
	1. Start – Line with Angle 90
	2. End – Line with Angle 90
	c) Fill colour: Blue
	d) Display joints: not active (blank)
	e) Elements: leave as given
	f) pick OK
prompt	Multiline Style dialogue box
with	ML-2 added to the Styles list
respond	*a*) pick ML-2
	b) pick Set Current
	c) pick OK.

10 Activate the MLINE command and:

prompt	Specify start point or [Justification/Scale/STyle]
enter	**ST <R>** – the style option
prompt	Enter mline style name (or ?)
enter	**? <R>** – the query option

prompt	AutoCAD Text Window with:
	Name Description
	ML-1 MY FIRST ATTEMPT
	ML-2 MY SECOND ATTEMPT
	STANDARD

respond   **cancel the text window (F2)**
prompt   Enter multiline style name and enter **ML-2 <R>**
prompt   Specify start point or [Justification/Scale/STyle
enter   **S <R> then 4 <R>** – the scale option
prompt   Specify start point and enter **110,120 <R>**
prompt   Specify next point and enter **@100<15 <R>**
prompt   Specify next point and enter **@100<−15 <R>**
prompt   Specify next point and press **<RETURN>**.

11 Repeat the MLINE command and:
    a) set a scale of 5
    b) make ML-1 the current multiline
    c) draw from 110,150 to 310,80 then <R>.

12 Menu bar with **Modify-Object-Multiline** and:
    prompt   Multilines Edit Tools dialogue box – Figure 27.3
    respond   **pick Open Cross then OK** (middle row left)
    prompt   Select first multiline and **pick multiline 1**
    prompt   Select second multiline and **pick multiline 2**
    prompt   Select first multiline
    then   **pick as required** until all the 'crossed' multilines are opened – Figure 27.1(c).

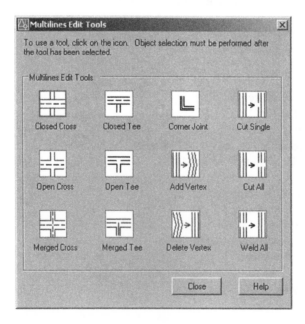

Figure 27.3 The Multilines Edit Tools dialogue box.

13 *Tasks*
    a) The ML-1 and ML-2 multilines have been created with layer OUT current
    b) Make layers CL and HID current then draw multilines the ML-1 style, using your own scale values
    c) Dimension the multilines in Figure 27.1(b) to determine which element has been drawn at the correct 'length'.

Creating a 3 element multiline

1 The two multilines (ML-1 and ML-2) have both been created with two continuous linetype elements.

2 We will now investigate how to create a three element multiline with varying linetype and colour.

3 From the menu bar select **Format-Multiline Style** and:
prompt   Multiline Styles dialogue box
respond   **pick New**
prompt   Create New Multiline Style dialogue box
respond   *a*) New Style Name: enter ML-3
           b) Start With: ML-1
           c) pick Continue
prompt   New Multiline Style: ML-3 dialogue box
respond   *a*) Description: MY THIRD ATTEMPT
           b) Caps:
               1. Start – Outer arc with Angle 90
               2. End – Outer arc with Angle 90.
           c) Fill color: None
           d) Display joints: not active (blank)
           e) Elements:
               1. pick the 0.5 offset line
               2. pick Add
               3. a 0.0 offset line added between the 0.5 and −0.5 lines
               4. Colour: scroll and pick Blue
               5. pick Linetype
               6. pick Center from the list then OK
               7. dialogue box as Figure 27.4
               8. pick OK
prompt   Multline Style dialogue box with ML-3 added to Styles list
respond   pick OK.

4 Now use the created ML-3 multiline style to draw some line segments – Figure 27.1(e).

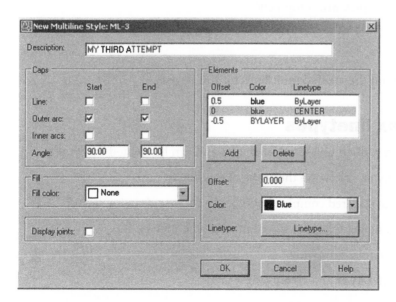

Figure 27.4 The dialogue box for creating the new ML-3 multiline style.

Saving multilines

1 When multilines have been created, they are only available for use in the current drawing.

2 If these multilines are required in other drawings they must be saved to a named folder.

3 Menu bar with **Format-Multiline Style** and:

prompt   Multiline Styles dialogue box
respond   **pick ML-1 then Save**
prompt   Save Multiline Style dialogue box
respond   *a*) Save in: scroll and pick C:\MYCAD or your named folder
            b) File name: alter to MYMULTIS.MLN
            c) pick Save
prompt   Multiline Styles dialogue box
respond   *a*) pick ML-2
            b) save to C:\MYCAD\MYMULTIS.MLN file
            c) save the ML-3 multiline style to the new MLN file
prompt   Multiline Styles dialogue box
respond   **pick OK**.

4 Close the multiline exercise drawing after saving.

5 Open your A3PAPER drawing and menu bar with **Format-Multiline Styles** and:

prompt   Multiline Styles dialogue box
and   only STANDARD listed
respond   **pick Load**
prompt   Load Multiline Styles dialogue box
respond   **pick File**
prompt   Load Multiline Style form File dialogue box
respond   *a*) Look in: scroll and select C:\MYCAD or your named folder
            b) pick MYMULTIS
            c) pick Open
prompt   Load Multiline Styles dialogue box
respond   pick ML-1 then
prompt   Multiline Styles dialogue box with ML-1 listed
respond   *a*) select Load, pick ML-2 then OK
            b) select Load, pick ML-3 then OK
prompt   Multiline Styles dialogue box with all 3 created multilines available
prompt   pick OK.

6 Now save A3PAPER with the created multilines for future use.

Using Complex linetypes

1 Complex linetypes can contain text items or shapes and can be created by the user.

2 Creating complex linetypes is beyond the scope of this book.

3 We will use the AutoCAD stored complex linetypes to create a drawing layout.

4 AutoCAD has seven complex linetypes stored in a named linetype file.

5 Open your A3PAPER standard drawing.

6 Menu bar with **Format-Layer** and create a new layer:
   a) Name: L1
   b) Colour: to suit

c) Linetype: pick 'Continuous' in L1 layer line and:

prompt	Select Linetype dialogue box
respond	**pick Load**
prompt	Load or Reload Linetypes dialogue box
with	acadiso.lin file displayed
respond	**scroll and pick GAS_LINE then OK**
prompt	Select Linetype dialogue box
with	GAS_LINE displayed
respond	**pick GAS_LINE then OK**
prompt	Layer Properties Manager dialogue box
with	layer L1 displayed with GAS_LINE linetype
respond	**pick OK**.

7   Using step 2 as a guide, create another six layers and load the appropriate linetype using the following information:

Name	*Colour*	*Linetype*
L2	to suit	BATTING
L3	to suit	FENCELINE1
L4	to suit	FENCELINE2
L5	to suit	HOT_WATER_SUPPLY
L6	to suit	TRACKS
L7	to suit	ZIGZAG

8   *a*) With each layer current, refer to Figure 27.5 and create a small housing estate layout using your imagination and design ability

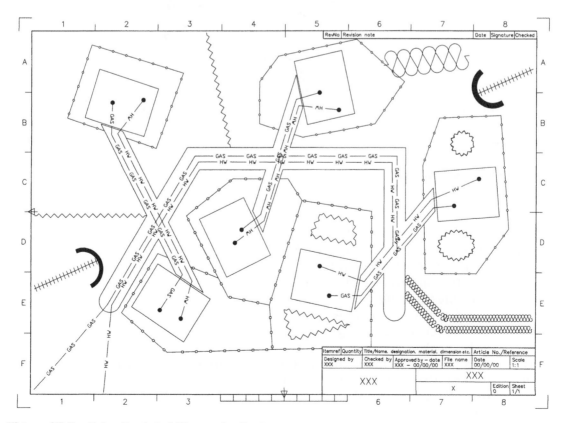

Figure 27.5   Using the AutoCAD complex linetypes.

b) Using the CHANGE PROPERTIES command (command line CHANGE or icon) use the ltScale option to optimise the appearance of the new added linetypes.

9   The exercise is now complete and can be saved if required.

Groups

1   A group is a named collection of objects saved by the user within a drawing.

2   Open your standard sheet, layer OUT current, refer to Figure 27.6 and:
   a) draw the reference shape using your discretion for any sizes not given
   b) ensure the lowest vertex is at the point 210,120.

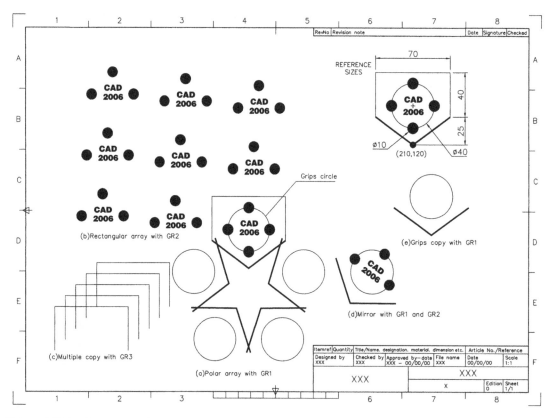

Figure 27.6   Group exercise.

3   At the command line enter **GROUP <R>** and:

prompt	Object Grouping dialogue box
respond	*a*) at Group Name enter: GR1
	b) at Description enter: FIRST GROUP
	c) ensure Selectable is ON, i.e. tick in box
	d) pick Create Group: New<
prompt	Select objects for grouping
then	Select objects
respond	**pick the two inclined lines and the large circle then right-click**
prompt	Object Grouping dialogue box
with	*Group Name Selectable*
	GR1 Yes
respond	**pick OK**.

4 At the command line enter **GROUP <R>** and:
   prompt   `Object Grouping dialogue box`
   respond   *a*) at Group Name enter: GR2
                b) at Description enter: SECOND GROUP
                c) Selectable ON
                d) pick Create Group: New<
   prompt   `Select objects`
   respond   **pick the items of text and the top three filled circles then right-click**
   prompt   `Object Grouping dialogue box with GR2 listed`
   respond   **pick OK**.

5 Repeat the GROUP command and:
   a) Name: GR3
   b) Description: THIRD GROUP
   c) Create Group: New<
   d) Objects: pick the two vertical and the horizontal lines
   e) pick OK.

6 At the command line enter **–ARRAY <R>** and:
   prompt   `Select objects`
   enter   **GROUP <R>**
   prompt   `Enter group name` and enter **GR1 <R>**
   prompt   `3 found`, then `Select objects`
   respond   **right-click**
   prompt   `Enter type of array` and enter P <R>
   prompt   `Specify center point of array` and enter **210,100 <R>**
   prompt   `Enter the number of items` and enter **5 <R>**
   prompt   `Specify the angle to fill` and enter **360 <R>**
   prompt   `Rotate arrayed objects` and enter **Y <R>**.

7 The named group (GR1) will be displayed as Figure 27.6(a).

8 Select the **ARRAY icon** and:
   prompt   `Array dialogue box`
   respond   **pick Select objects**
   prompt   `Select objects at command line`
   enter   **GROUP <R> then GR2 <R><R>** – why two returns?
   prompt   `Array dialogue box`
   respond   alter as follows:
                a) Rectangular array
                b) Rows: 3 and Columns: 3
                c) Offsets: Row 55, Column –70 and Angle –5
                d) preview then Accept – Figure 27.6(b).

9 With the COPY command:
   a) objects and enter GROUP <R>
   b) group name and enter GR3 <R><R>
   c) base point and enter 175,145 <R> – why these co-ordinates?
   d) second point and enter the following five co-ordinate pairs:
       @–150,–100; @–140,–90; @–130,–80; @–120,–70; @–110,–60.

10 Select the MIRROR icon and:
   a) select objects and enter GROUP <R>
   b) group name and enter GR1 <R>
   c) 3 found then select objects and enter GROUP <R>

d) group name and enter GR2 <R>

e) 5 found, 8 total then select objects and right-click

f) first point of mirror line – Center icon and pick bottom right circle

g) second point – Center icon and pick top right circle

h) delete source objects and enter N <R> – Figure 27.6(d).

11 Ensure GRIPS are on (GRIPS 1 at command line) and pick the circle indicated in Figure 27.6.

prompt	`grip boxes at circle and inclined lines`, i.e. GR1
respond	**make the circle centre grip the base grip**
prompt	STRETCH and enter <R>
prompt	MOVE and enter C <R>
prompt	`Specify move point` and enter @175<10 <R> – Figure 27.6(e)
then	exit and cancel the grip command.

12 The exercise is complete and can be saved if required.

13 Do not quit the drawing.

14 *Tasks*

a) Erase the original text item – complete group (GR2) erased?

b) Undo this effect with the UNDO icon

c) Select the explode icon and select any group object and:

prompt	`0 found, 1 group`
	`3 were not able to be exploded` (or similar message)

d) At the command line enter GROUP <R> and from the dialogue box:

1. pick GR2 line
2. pick Explode
3. pick OK
4. The text item can now be erased.

Assignment

Activity 41: Rail layout

1 A model railway enthusiast has in his collection a old style three-rail track, the middle rail being 'live'.

2 It is this layout that you have to create using only multilines.

3 One problem with multilines is that there is not a facility to draw arcs segments, so how did I create the curved three rail track?

4 Think about using multiline ML-1 three times and changing the layer of the middle multiline.

5 No more help.

Chapter 28

Blocks and attributes

1 A block is part of a drawing which is 'stored away' for future recall **within the drawing in which it was created**.

2 The block may be a nut, a diode, a tree, a house or any part of a drawing.

3 Blocks are used when repetitive copying of objects is required, but they have another important feature – text can be attached to them.

4 This text addition to blocks is called **attributes**.

5 Block attribute text can be exported from an AutoCAD drawing as input to other software applications.

6 *a*) Remember the phrase '**within the drawing in which it was created'**
   b) We will refer to this statement in a later chapter.

Getting started

1 Open the A3PAPER standard sheet and refer to Figure 28.1.

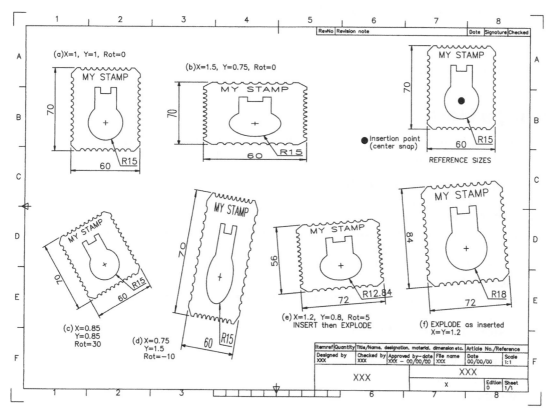

Figure 28.1 Creating and inserting a block.

2   Draw the stamp shape using the reference sizes given, or your own design, but with:
    a) the outline on layer OUT
    b) the stamp design on layer OUT but green (CHANGE command?)
    c) a text item on layer TEXT
    d) three dimensions on layer DIMS
    e) remember to use your discretion for the design and any size not given.

3   Make layer OUT current.

4   At this stage save as **C:\MYCAD\STAMP** for future recall.

Creating a block

1   Blocks can be created from both the command line and via a dialogue box.

2   This first exercise will consider the keyboard entry method.

3   At the command line enter **–BLOCK <R>** and:
prompt	Enter block name or [?]
enter	**STAMP <R>**
prompt	Specify insertion base point
respond	**Center icon and pick circle as indicated**
prompt	Select objects
respond	**window the stamp and dimensions**
prompt	'??' found then Select objects
respond	**right-click**.

4   The stamp shape may disappear from the screen. If it does (or does not), don't panic!

5   *Notes*
    a) the stamp shape has been 'stored' as a block within the current drawing
    b) this drawing (with the stamp block) has not yet been saved
    c) the -BLOCK keyboard entry will 'bypass' the dialogue box.

6   If the stamp shape did not disappear, erase it now.

Inserting a block

1   Created blocks can be inserted into the current drawing by either direct keyboard entry or via a dialogue box and we will investigate both methods.

2   At the command line enter **–INSERT <R>** and:
prompt	Enter block name or [?]
enter	**STAMP <R>**
prompt	Specify insertion point or [Basepoint/Scale/X/Y/Z/Rotate/ Pscale/PX/PY/PZ/Protate]
and	'ghost' image of block attached to cursor
enter	**60,220 <R>**
prompt	Enter X scale factor, specify opposite corner or [Corner/ XYZ]
enter	**1 <R>** – an X scale factor of 1
prompt	Enter Y scale factor <use X scale factor>
enter	**1 <R>** – a Y scale factor of 1
prompt	Specify rotation angle<0>
enter	**0 <R>**.

3   The stamp block is inserted full-size as Figure 28.1(a).

4   Repeat the **–INSERT** command and:
    prompt   Block name and enter **STAMP <R>** (STAMP should be default name?)
    prompt   Insertion point and enter **190,220 <R>**
    prompt   X scale factor and enter **1.5 <R>**
    prompt   Y scale factor and enter **0.75 <R>**
    prompt   Rotation angle and enter **0 <R>**
    and   Stamp block inserted as Figure 28.1(b).

5   From the Draw toolbar select the **INSERT BLOCK icon** and:
    prompt   Insert dialogue box
    with   STAMP as the Name
                (from the command line INSERT)
    respond   *a*) ensure that the three Specify On-screen prompts are active, i.e. tick in
                    box for Insertion point, Scale, Rotation
                b) ensure that Explode is **NOT** active (i.e. no tick)
                c) ensure that Uniform Scale **IS** active (i.e. tick)
                d) pick OK
    prompt   Specify insertion point and enter **60,100 <R>**
    prompt   Specify scale factor and enter **0.85 <R>**
    prompt   Specify rotation angle and enter **30 <R>**
    and   Stamp block inserted as Figure 28.1(c).

6   Menu bar with **Insert-Block** and:
    prompt   Insert dialogue box
    with   STAMP as block name
    respond   *a*) deactivate the three On-screen prompts (no tick)
                b) alter Insertion Point to X: 160, Y: 90, Z: 0
                c) deactivate the Uniform Scale (no tick)
                d) alter Scale to X: 0.75, Y: 1.5, Z: 1
                e) alter Rotation to Angle: −10
                f) ensure Explode not active (no tick)
                g) dialogue box as Figure 28.2
                h) pick OK – Figure 28.1(d).

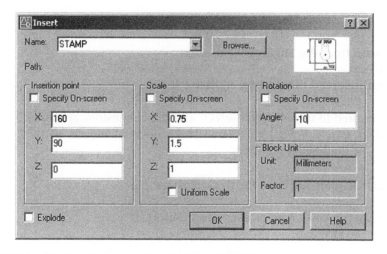

Figure 28.2   The Insert dialogue box for block STAMP.

7   *Notes*
    a) 1. An inserted block is a single object
        2. Select the erase icon and pick any point on one of the inserted blocks then right-
           click.

3. The complete block is erased

4. Undo the erase effect.

b) 1. Blocks are inserted into a drawing with layers 'as used'.

2. Freeze the DIMS layer and turn off the TEXT layer

3. The inserted blocks will be displayed without dimensions or text

4. Now thaw the DIMS layer and turn on the TEXT layer.

c) 1. Blocks can be inserted at varying X and Y scale factors and at any angle of rotation

2. The default scale is X = Y = 1 (i.e. the block is inserted full size)

3. The default rotation angle is 0.

d) Dimensions which are attached to inserted blocks are not altered if the scale factors are changed

e) A named block can be redefined and will be discussed later.

Exploding a block

1 The fact that a block is a single object may not always be suitable to the user, i.e. it may be necessary to copy a certain part of the block.

2 AutoCAD uses the EXPLODE command to 'convert' an inserted block back to its individual objects.

3 The explode option can be used:
a) after a block has been inserted
b) during the insertion process.

4 At the command line enter **–INSERT <R>** and:

prompt	Block name and enter **STAMP**
prompt	Insertion point and enter **265,100**
prompt	X scale factor and enter **1.2**
prompt	Y scale factor and enter **0.8**
prompt	Rotation angle and enter **5**.

5 Note the dimensions of this inserted block.

6 Select the **EXPLODE icon** from the Modify toolbar and:

prompt	Select objects
respond	**pick the last inserted block then right-click**.

7 The exploded block is restored to its individual objects and the dimensions are scaled to the factors entered in step 4 – Figure 28.1(e). The individual objects of this exploded block can now be modified if required.

8 Select the **INSERT BLOCK icon** and from the Draw dialogue box:
a) Name: STAMP
b) Deactivate the three Specify On-screen prompts (i.e. no tick)
c) Insertion Point X: 375; Y: 115; Z: 0
d) Scale X: 1.2; Y: 0.8; Z: 1
e) Rotation angle: 5
f) Explode: ON, i.e. tick and *note the scale factor values*
g) pick OK.

9 The block is exploded as it is inserted at a scale of X = Y = 1.2 and the dimensions display this scale effect as Figure 28.1(f).

10 Compare the dimensions of Figure 28.1(e) with those of Figure 28.1(f) and note the dimension 'orientation' of the two exploded blocks.

11 *Notes*
   a) a block exploded after insertion will retain the original X and Y inserted scale factors
   b) a block exploded as it is inserted has X = Y scale factors.

Block exercise

1 Open the A3PAPER standard sheet with layer OUT current.

2 Refer to Figure 28.3 and draw the SEAT shape as given. Do not add the dimension or donut.

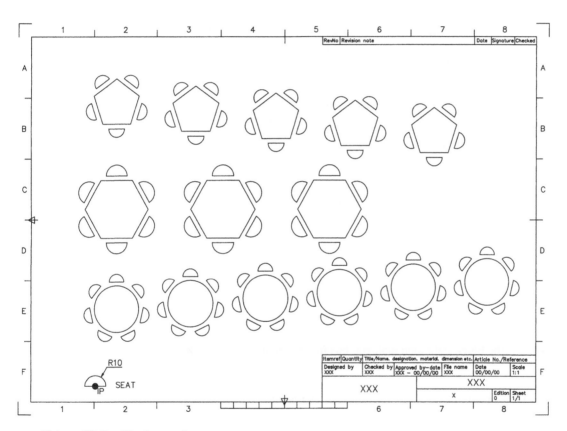

Figure 28.3 Block exercise.

3 Menu bar with **Draw-Block-Make** and:
   prompt   Block Definition dialogue box
   respond   *a*) Name: enter SEAT
              b) Base point: select Pick point
   prompt   Specify insertion base point at command line
   respond   **midpoint icon and pick line indicated by donut in Figure 28.3**
   prompt   Block Definition dialogue box
   with   Base point: X and Y co-ordinates of selected point
   respond   **Objects: pick Select objects**
   prompt   Select objects at command line
   respond   **window the seat then right-click**
   prompt   Block Definition dialogue box
   with   preview of selected objects

respond   *a*) Objects: ensure Delete is active (black dot)
              b) Settings:
                    1. Block unit: Millimetres
                    2. Scale uniformly: active
                    3. Allow exploding: active
              c) Description: enter SEAT BLOCK
              d) Open in Block Editor: not active
and   dialogue box as Figure 28.4
respond   **pick OK**.

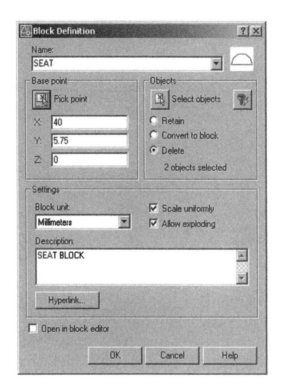

Figure 28.4 The Block Definition dialogue box for the created SEAT block.

4   The seat shape should disappear (but entering OOPS <R> will restore the shape).

5   *Notes*
     The Objects part of the Block dialogue box allows the user one of three options, these being:
     a) Retain   will keep the selected objects as individual objects on the screen after block creation
     b) Convert to block   will display the selected objects on the screen as a block after block creation
     c) Delete   will remove the selected objects from the screen after block creation (our selection)
     d) It is user preference as to which of these options is used
     e) I generally select retain.

6   Draw the following three objects (the centre points being important):
     a) pentagon, centred on 80,150, inscribed in a 30 radius circle
     b) hexagon, centred on 210,150 and inscribed in a 30 radius circle
     c) circle, centre 340,150 and radius 30.

7 Select the INSERT BLOCK icon from the Draw toolbar and use the Insert dialogue box to insert the block SEAT three times using the following information:

Insertion point	Scale	Rotation	Explode
80,120	X = Y = 1	180	not active
210,120	X = Y = 1	180	not active
340,185	X = Y = 1	0	not active

8 Using the ARRAY dialogue box, array the three inserted seat blocks with the following information:

	First array	Second array	Third array
objects	pick left SEAT block	pick middle SEAT block	pick right seat block
type	Polar	Polar	Polar
centre	80,150	210,150	340,150
method	items and angle	items and angle	items and angle
items	5	6	7
angle	360	360	360
rotate	Y	Y	Y

9 Select the **MAKE BLOCK icon** from the Draw toolbar and:

prompt Block Definition dialogue box
respond a) enter Name: TABLE1
  b) enter Base point as X: 80; Y: 150; Z: 0
  c) at Objects: pick Select objects
prompt Select objects at command line
respond **window the 5 seat arrangement then right-click**
prompt Block Definition dialogue box with icon preview display
respond a) Objects: ensure Delete is active (black dot)
  b) Settings: 1. Block unit: Millimetres
          2. Scale uniformly: active
          3. Allow exploding: active.
  c) Description: enter 5 SEATER TABLE
  d) Open in Block Editor: not active
respond **pick OK**
and the 5-seater table and chairs will disappear.

10 Repeat the Make Block icon selection and create blocks of the 6- and 7-seat table layouts with the following data:

	Second block	Third block
name	TABLE2	TABLE3
base point	210,150	340,150
select objects	window 6 seats and table	window 7 seats and table
objects	delete active	delete active
units	millimetres	millimetres
scale	uniformly	uniformly
exploding	not active	not active
block editor	not active	not active
description	6 SEATER TABLE	7 SEATER TABLE

11 Now erase any objects from the screen.

12 With layer OUT current, menu bar with **Insert-Block** and using the Insert dialogue box, insert each table arrangement using the following data:

Name	Insertion point	Scale	Rotation	Explode
TABLE1	60,260	0.75	0	not active
TABLE2	60,165	1	0	not active
TABLE3	60,75	0.7	0	not active

13 Menu bar with **Modify-Array** then use the Array dialogue box with the inserted tables and:

Objects	Type	Rows	Columns	RowOffset	ColumnOffset	Angle of Array
TABLE1	Rect	1	5	0	75	−5
TABLE2	Rect	1	3	0	100	0
TABLE3	Rect	1	6	0	70	4

14 The final layout should be as Figure 28.3 and can be saved with a suitable name.

Notes on blocks

Several points are worth discussing about blocks:

1 *The insertion point*
 a) When a block is being inserted using the Insert dialogue box, the user has the option to decide whether the parameters are entered via the dialogue box or at the command line.
 b) The Specify On-screen selection allows this option.

2 *Exploding a block*
 a) I would recommend that blocks are inserted before they are exploded.
 b) This maintains the original X and Y scale factors.
 c) Remember that blocks do not need to be exploded.

3 *Command line*
 a) Entering −BLOCK and −INSERT will 'byepass' the dialogue boxes.
 b) Parameters can then be entered from the keyboard.

4 *The ? option*
 a) Both the command line −BLOCK and −INSERT commands have a query (?) option which will list all the blocks in the current drawing.
 b) At the command line enter **−BLOCK <R>** and:

prompt	Enter block name (or ?) and enter **? <R>**
prompt	Enter block(s) to list<*> and enter *** <R>**
prompt	AutoCAD Text Window
with	Defined blocks

 "ISO A3 title block"
 "SEAT"
 "TABLE1"
 "TABLE2"
 "TABLE3"

User Blocks	External Blocks	Dependent Blocks	Unnamed Blocks
5	0	0	1

5 *Making a block*
 a) Blocks can be created by command line entry or by a dialogue box.
 b) The following is interesting:
  1. command line   the original shape disappears from the screen
  2. dialogue box   options are available to retain the shape.

6 *Nested blocks*
 a) These are 'blocks within blocks'.
 b) With the menu bar selection **Tools-Inquiry-List** then pick any 7-seat table arrangement and:

prompt	AutoCAD text Window
with	**details about the selected object**.

c) This information tells the user that the selected object is a block with name TABLE3, created on layer OUT as Figure 28.5(a).

```
                     BLOCK REFERENCE  Layer: "OUT"
                               Space: Model space
                         Handle = 2406
           Block Name:  "TABLE3"
                    at point, X=     60.00  Y=     75.00  Z=     0.00
        X scale factor:        0.70
        Y scale factor:        0.70
        rotation angle:        0.0          (a)LIST with the 7 SEATER TABLE
        Z scale factor:        0.70
              InsUnits: Millimeters
        Unit conversion:       1.00
        Scale uniformly: Yes
   (a)  Allow exploding: Yes


                     BLOCK REFERENCE  Layer: "OUT"
                               Space: Model space
                         Handle = 2423
           Block Name:  "SEAT"
                    at point, X=     60.00  Y=     99.50  Z=     0.00
        X scale factor:        0.70
        Y scale factor:        0.70
        rotation angle:        0.0            (b)LIST with the SEAT
        Z scale factor:        0.70
              InsUnits: Millimeters
        Unit conversion:       1.00
        Scale uniformly: Yes
   (b)  Allow exploding: Yes
```

Figure 28.5 Using the LIST command with selected items.

d) With the EXPLODE icon, pick the previously selected 7-seat table arrangement then right-click.

e) With the LIST command, pick any seat of the exploded block and the text window will display details about the object, i.e. it is a block with name SEAT, created on layer OUT – Figure 28.5(b).

f) This is an example of a nested block, i.e. block SEAT is contained within block TABLE2.

g) If you exploded one of the seats of the exploded table, what would you expect to be displayed with the LIST command?

Using blocks

1 Open your A3PAPER standard sheet with layer OUT current.

2 Refer to Figure 28.6 and draw the original I beam shape as Figure 28.6(x) then:
   a) make a block of the I beam with block name: BEAM
   b) insertion point: at the I beam 'centre'
   c) objects: window the shape – no dimensions
   d) retain option active
   e) explode option active
   f) scale uniformly and block editor options both **NOT** active.

3 Draw an inclined line, a circle, an arc and a polyline shape of line and arc segments – discretion for sizes, but use Figure 28.6 as a guide for the layout.

4 Menu bar with **Draw-Point-Measure** and:
   prompt   Select object to measure
   respond   **pick the line**
   prompt   Specify length of segment or [Block]
   enter   **B <R>** – the block option

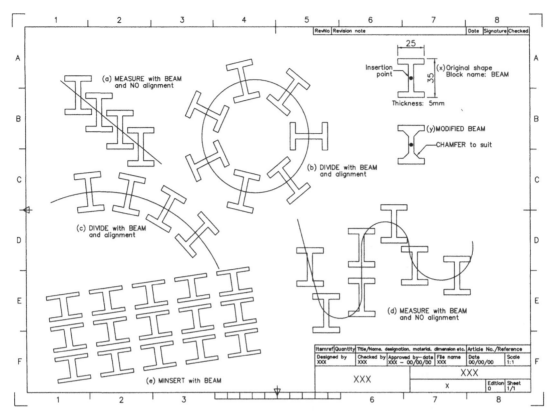

Figure 28.6   Using blocks.

prompt	Enter name of block to insert
enter	**BEAM <R>**
prompt	Align block with object
enter	**N <R>**
prompt	Specify length of segment
enter	**25 <R>** – Figure 28.6(a).

5   Menu bar with **Draw-Point-DIVIDE** and:

prompt	Select object to divide
respond	**pick the circle**
prompt	Enter the number of segments or [Block] and enter **B <R>**
prompt	Enter name of block to insert and enter **BEAM <R>**
prompt	Align block with object and enter **Y <R>**
prompt	Enter the number of segments and enter **7 <R>** – Figure 28.6(b).

6   Use the *DIVIDE* and *MEASURE* commands and:
   a) divide the arc using block BEAM, with alignment and five segments – Figure 28.6(c)
   b) measure the polyline using block BEAM, no alignment with a segment length of 45 –
       Figure 28.6(d).

7   The *MINSERT* or *multiple insert command*
    At the command line enter **MINSERT <R>** and:

prompt	Enter block name
enter	**BEAM <R>**
prompt	Specify insertion point or [Basepoint/Scale/Rotate/ Pscale/Protate]

respond	**pick a suitable point on screen**
prompt	Enter X scale factor and enter **1.25 <R>**
prompt	Enter Y scale factor and enter **0.65 <R>**
prompt	Specify rotation angle and enter **8 <R>**
prompt	Enter number of rows and enter **3 <R>**
prompt	Enter number of columns and enter **5 <R>**
prompt	Enter distance between rows and enter **28 <R>**
prompt	Specify distance between columns and enter **41 <R>**.

8 The block BEAM is arrayed in a 4 × 6 rectangular matrix – Figure 28.6(e).

9 Using the **EXPLODE icon**, pick any of the minsert blocks and:
   prompt   1 was minserted
   i.e. **you cannot explode minserted blocks**.

10 *Redefining a block*
   a) Refer to Figure 28.6 and add a chamfer effect to the original shape as Figure 28.6(y)
   b) At the command line enter **–BLOCK <R>** and:

prompt	Enter block name and enter **BEAM <R>**
prompt	Block "BEAM" already exist. Redefine it? [Yes/No]<N>
enter	**Y <R>**
prompt	Specify insertion base point
respond	**pick the same point as previously**
prompt	Select objects
respond	**window the circle and lines then right-click**.

   c) Interesting result? – all the beams will be displayed with a chamfer effect.

11 *Note*: the scale uniformly option:
   a) when active, insert is with the same X and Y values
   b) when not active, the user can enter different X and Y values.

12 This exercise is now complete – save if required.

Layer 0 and blocks

1 Blocks have been inserted with the objects drawn on their 'correct' layers.

2 Layer 0 is the AutoCAD default layer and can be used for block creation with interesting results.

3 Open your A3PAPER template and draw anywhere on the screen:
   a) a 50 unit square on layer 0 – black
   b) a 20-radius circle inside the square on layer OUT – red
   c) two centre lines on layer CL – green.

4 With the command line entry –BLOCK:
   a) make a block of the complete shape with block name TRY
   b) use the circle centre as the insertion point
   c) block disappears?

5 Make layer OUT current and with the command line –INSERT:
   a) insert block TRY at 55,210, full size with no rotation
   b) the square is inserted with red continuous lines.

6 Make layer CL current and:
   a) –INSERT block TRY at 135,210, full size and with no rotation
   b) the square has green centre lines.

7 With layer HID current.
   a) –INSERT the block at 215,210, full size with no rotation
   b) the square will be displayed with coloured hidden lines.

8 Make layer 0 current and:
   a) –INSERT the block at 295,210, full size with no rotation
   b) the square is black.

9 With layer 0 still current, freeze layer OUT and:
   a) no red circles
   b) no red square from first insertion (when layer OUT was current).

10 Thaw layer OUT and, still with layer 0 current, insert block TRY using the Insert dialogue box with:
   a) explode option active, i.e. tick in box
   b) insertion point: 175,115
   c) full size with 0 rotation
   d) square is black – it is on layer 0.

11 *a*) explode the first three inserted blocks and the square should be black
   b) it has been 'transferred' to layer 0 with the explode command.

12 Finally make layer OUT current, freeze layer 0 and:
   a) no black squares
   b) no objects from fourth insertion – why?
   c) the screen displays four red circles with green centre lines.

13 This completes this block exercises – no save necessary.

Attributes

1 An attribute is an item of text attached to a block or a wblock (Chapter 29).

2 Attributes allow the user to add repetitive type text to frequently used blocks.

3 The blocks/text could be:
   a) weld symbols containing appropriate information
   b) electrical components with values
   c) windows and doors containing codes, number off, material, etc.

4 Attributes used as text items are useful to the CAD user, but their main advantage is that attribute data can be extracted from a drawing and stored in an attribute extraction file.

5 This data could then be used as input to other computer packages (e.g. databases, spreadsheets, etc.) for creating a Bill-of-Material or an Inventory.

6 Attribute data can be edited but this is beyond the scope of the book.

7 We will investigate.
   a) attaching attributes to a block
   b) inserting an attribute block into a drawing.

Started the attribute exercise

1 The attribute example for demonstration is a fisherman's trophy and will use a previously created and saved drawing.

2 The fish 'symbol' will represent the block for adding the attributes.

3 The added attributes will give information about the type of fish, the year it was caught and the river.

4 Open the **C:\MYCAD\FISH** drawing created during Chapter 22 and refer to Figure 28.7

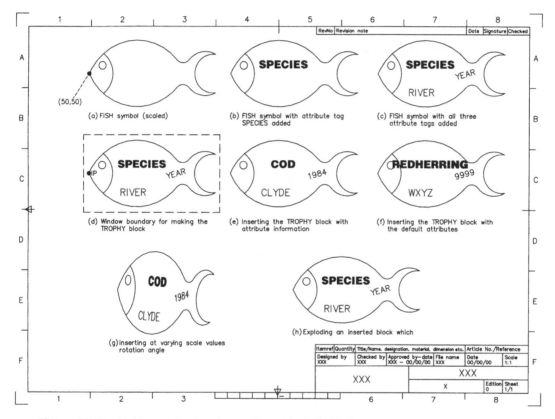

Figure 28.7 Making and using the attribute block TROPHY.

5 Erase any centre lines, text and dimensions.

6 *a*) Move the complete shape from 'the nose' to the point 50,50.
   b) Scale the shape about the point (50,50) by 0.5 – Figure 28.7(a)
   c) *Note*: the 50,50 point is essential for positioning the text items during the attribute exercise.

Defining the attributes

1 Before a block containing attributes can be inserted into a drawing, the attributes must be defined.

2 Make layer TEXT current then menu bar with **Draw-Block-Define Attributes** and:
   prompt   Attribute Definition dialogue box
   with   options for Mode, Attribute, Insertion Point and Text

respond   with the following:

    a) At Mode: leave all four options un-selected – no tick

    b) At Attribute enter:
        1. Tag: SPECIES
        2. Prompt: What displayed?
        3. Value: REDHERRING.

    c) At Insertion point, cancel the on-screen prompt and enter:
        1. X: 100   2. Y: 55   3. Z: 0.

    d) At Text options alter:
        1. Justification: scroll and pick Center
        2. Text Style: scroll and pick ST3
        3. Height: 7
        4. Rotation: 0.

    e) Align below previous attribute definition not active

    f) Lock position in block: active – Figure 28.8

    g) pick OK.

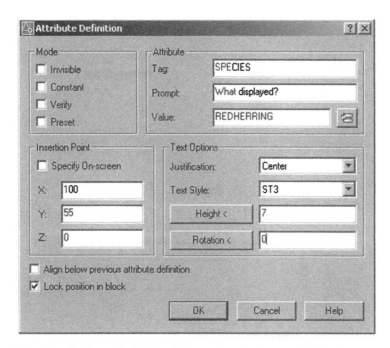

Figure 28.8   The Attribute Definition dialogue box for the tag: SPECIES.

3   The attribute tag SPECIES will be displayed in the fish symbol as Figure 28.7(b).

4   Activate the Attribute Definition command two more times and enter the following attribute information in the Attribute Definition dialogue box using the same procedure as step 2.

Item	First entry	Second entry
Attribute modes	blank	blank
Attribute tag	RIVER	YEAR
Attribute prompt	Where caught?	What year?
Default value	WXYZ	9999
Insertion Pt X	80	125
Insertion Pt Y	30	45
Insertion Pt Z	0	0

Justification	Left	Left
Text style	ST2	ST1
Height	6	5
Rotation	0	15

5   When all the attribute information has been entered, the fish symbol will display the three tags – Figure 28.7(c).

6   *Notes*
   a) When attributes are used for the first time, the words Tag, Prompt and Value can cause confusion and the following descriptions may help to overcome this confusion:
      1. tag   is the actual attribute 'label' which is attached to the drawing at the specified text start point. This tag item can have any text style, height and rotation.
      2. prompt   is an aid to the user when the attribute data is being entered with the inserted block.
      3. value   is an artificial name/number for the attribute being entered. It can have any alpha-numeric value.
   b) The Insertion point in the Attribute Definition dialogue box refers to the attribute text tag and not to the block.

7   In our first attribute definition sequence, we created the following attribute information:
   a) Tag: SPECIES
   b) Prompt: What displayed?
   c) Default value: REDHERRING
   d) Text insertion point for SPECIES, centred on 100,55 with style ST3, height 7 and 0 rotation angle.

Creating the attribute block

1   Menu bar with **Draw-Block-Make** and:
   prompt   Block Definition dialogue box
   respond   enter/activate the following:
            a) Name: TROPHY
            b) Base point: X: 50   Y: 50   Z: 0
            c) Objects: 1. Select and window the symbol and attributes as Figure 28.7(d) then right-click
                       2. Delete active.
            d) Settings: 1. Block unit: Millimeters
                        2. Scale uniformly: not active
                        3. Allow exploding: active.
            e) Open in block editor: not active
            f) Description: TROPHY block with three attributes
   and   dialogue box as Figure 28.9
   then   pick OK.

2   The symbol and attributes have been made into a block and should disappear from the screen as we activated this option from the dialogue box.

Testing the created block with attributes

1   Now that the block with attributes has been created, we want to 'test' the attribute information it contains.

2   This requires the block to be inserted into the drawing, and this will be achieved with both command line and dialogue box entries.

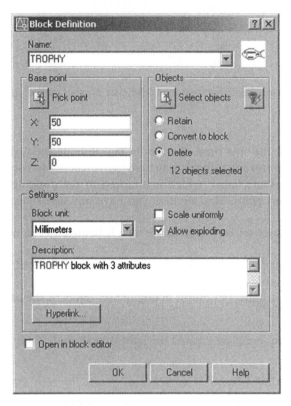

Figure 28.9 The Block Definition dialogue box for the TROPHY block.

3 Make layer OUT current and at the command line enter **ATTDIA <R>** and:
prompt   Enter new value for ATTDIA <?>
enter   **0 <R>**.

4 ATTDIA is a system variable, and when set to 0 will only allow attribute values to be entered from the keyboard.

5 At the command line enter **–INSERT <R>** and:
prompt   Enter block name and enter **TROPHY <R>**
prompt   Specify insertion point and **pick any point to suit**
prompt   Enter X scale and enter **1 <R>**
prompt   Enter Y scale and enter **1 <R>**
prompt   Specify rotation angle and enter **0 <R>**
prompt   What displayed? <REDHERRING> and enter **COD <R>**
prompt   Where caught?<WXYZ> and enter **CLYDE <R>**
prompt   What year? <9999> and enter **1984 <R>**.

6 The fish trophy symbol will be displayed with the attribute information as Figure 28.7(e).

7 *Note*s
   a) the prompt and defaults values are displayed as entered in the Attribute Definition dialogue box.
   b) the order of the last three prompt lines (i.e. displayed, caught and year) may not be in the same order as listed in step 5. Don't worry if they are not the same.

8   Now insert the trophy block twice more with **–INSERT from the command line** using:

   a) any suitable point, full size with 0 rotation and accept the default values, i.e. right-click or <R> at each attribute prompt line – Figure 28.7(f).

   b) another point on the screen with the X-scale factor as 0.75, the Y scale factor as 1.25, the rotation angle −5. Use the same attribute entries as step 4 (i.e. COD, CLYDE and 1984). The result should be as Figure 28.7(g).

9   Explode any inserted block which contains attribute information and the tags will be displayed – Figure 28.7(h).

10   We are now ready to insert the 'real' attribute data.

Attribute information

1   The fisherman's trophy cabinet contains five prime examples of what he has caught over the past few years.

2   Each catch is represented in the trophy cabinet by the block symbol containing the appropriate attribute information.

3   The attribute data to be displayed is:

Species	River	Year
SALMON	SPEY	1992
TROUT	TAY	1994
PIKE	DART	2000
CARP	NENE	2002
EELS	DERWENT	2004

4   Attribute information can be added to an inserted block:

   a) from the keyboard – as previous example

   b) via a dialogue box which will now be discussed.

5   Erase all objects from the screen and make layer OUT current.

6   At the command line enter **ATTDIA <R>** and:

   prompt   Enter new value for ATTDIA<0>

   enter   **1 <R>**

7   Menu bar with **Insert-Block** and:

   prompt   Insert dialogue box

   with   Block name: TROPHY – from previous insertion

   respond   *a*) ensure all on-screen prompts not active (i.e. no tick)

              b) insertion point at X: 40; Y: 230; Z: 0

              c) explode not active

              d) scale X: 1.5; Y: 1.5; Z: 1

              e) rotation angle 0

              f) pick OK.

   prompt   Edit Attributes dialogue box

   with   Prompts and default values from the attribute definition sequence as Figure 28.10(a)

   respond   *a*) alter What displayed to SALMON

              b) alter Where caught to SPEY

              c) alter What year to 1992

              d) dialogue box as Figure 28.10(b)

              e) pick OK.

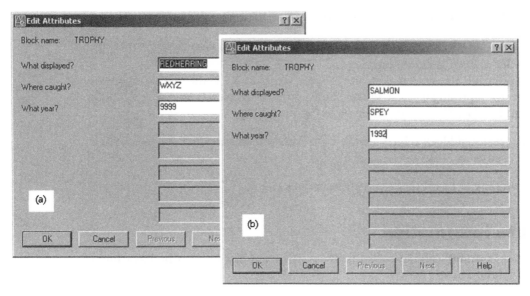

Figure 28.10   The Edit Attributes dialogue box for TROPHY.

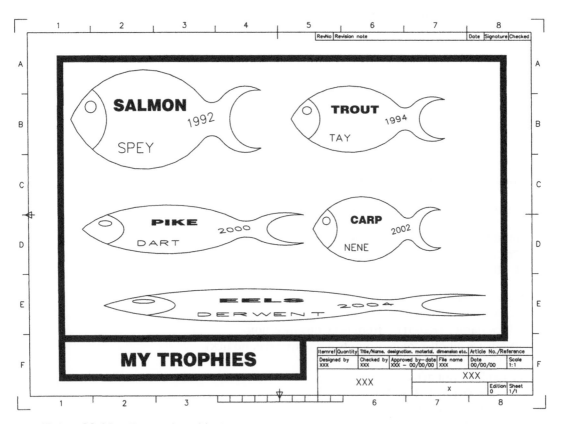

Figure 28.11   The trophy cabinet.

8   The trophy block will be inserted with the attribute information displayed.

9   Using step 7 as a guide with the attribute data of species, river and year, refer to Figure 28.11 and insert the TROPHY block to complete the cabinet – use your imagination with the scales, then complete the cabinet and save?

10   This completes our brief 'taster' into attributes.

Assignments

Three activities for this chapter, two with blocks only and one with blocks and attributes.

1 *Activity 42: In Line Cam and Roller Follower*
   a) Draw the CAM and FOL shapes using the reference sizes given (use your discretion for omitted sizes).
   b) Make a block of each shape with the block names CAM and FOL. Use the given insertion points.
   c) Insert the CAM block using the information given on the activity drawing.
   d) Insert the FOL block:
      1. insertion point: 25,90
      2. X- and Y-scale: 0.5 with 0 rotation.
   e) Rectangular array the inserted FOL block:
      1. for 1 row and 10 columns
      2. column offset: 40.
   f) The followers have to be moved vertically downwards until they 'touch' the cams. This sounds easier than it may seem.
   g) Complete the tasks detailed in the activity drawing.

2 *Activity 43: Modifying an existing drawing layout.*
   a) Open the LARGSC drawing created in Chapter 26 and:
      1. erase the square columns and any dimensions
      2. modify the lower right corner
      3. position the circular outline at the midpoint of the right vertical wall
   b) Draw the BEAM shape using the sizes given at any suitable part of the screen then make a block of the shape with name BEAM, using the insertion point indicated.
   c) Insert block BEAM twice using the following:

	First	*Second*
insertion point	4000,3750	16750,5000
scale	full size	full size
rotation	45	0

   d) Produce the rectangular and polar array using the inserted blocks as displayed in the drawing and erase any wrongly positioned beams.
   e) Save the modified layout as C:\MYCAD\LARGESCMOD.

3 *Activity 44: Stamp Page layout*
   An activity making attributes using a previously saved drawing. The activity involves the stamp enthusiast from Chapter 28 and you have to design a page of the stamp album:
   a) Open the saved STAMP drawing
   b) Refer to the activity drawing which displays:
      1. the original stamp design
      2. the modified stamp
      3. the stamp with attribute tags.
   c) Add four attribute definitions using your own text style, height and layout. The attributes to be added to the stamp design are:

Tag	Prompt	Value	Mode
Country	Not applicable	NZ	Constant
Colour	Stamp colour?	123	–
Face Value	Stamp denomination?	345	–
Design	Stamp design?	ABC	–

   d) Make a block of the stamp and four attributes, the block name being STAMPDES.

e) Insert the STAMPDES block and design a page of a stamp album. You can add your own attributes or use the following:

Design	Colour	Denomination
FLAG	Red	50c
HOUSE	Blue	60c
BOAT	Green	70c
MAP	Purple	80c
PLANE	Yellow	$1
CAR	Orange	$2
WHALE	Violet	$3
KIWI	Ultramarine	$10
MONARCH	Maroon	$20

f) When the page layout is complete remember to save.

g) *Note*

In this activity the attribute for Country was set with a constant mode, and NZ entered as the value. You should be able to reason out why the attribute mode was set to constant?

Chapter 29

Wblocks and external references

1 Blocks are useful when shapes/objects are required for repetitive insertion in a drawing, but they are drawing specific, i.e. they can only be used within the drawing in which they were created (for now).

2 There are, however, blocks which can be created and accessed by all AutoCAD users, i.e. they are global.

3 These are called **WBLOCKS (write blocks)** and they are created in a similar manner to 'ordinary' blocks.

4 Wblocks are stored and recalled from a 'named folder/directory' which we will assume to C:\MYCAD.

Getting ready

1 The wblock we will create is a parts list table which can be inserted into any drawing if required, so open your A3PAPER standard sheet with layer TEXT current.

2 Refer to Figure 29.1 and menu bar with **Draw-Table** and:

prompt	Insert Table dialogue box
respond	**at Table Style Name, pick . . .**
prompt	Table Style dialogue box
respond	**pick New**
prompt	Create New Table Style dialogue box
respond	*a*) New Style Name: enter PARTS LIST
	b) pick Continue
prompt	New Table Style: PARTS LIST dialogue box
respond	activate the named tabs and enter/alter as follows:

Title tab
a) Cell Properties
   1. Include Title row active
   2. Text style: scroll and pick ST3
   3. Text height: 6
   4. Alignment: Middle Center
b) Border Properties: leave as given
c) General: Table direction: UP
d) Cell margins: leave as given.

Column heads tab
a) Cell Properties
   1. Include Header row active
   2. Text style: scroll and pick ST2
   3. Text height: 5
   4. Alignment: Middle Center

    b) Border Properties: leave as given

    c) General: Table direction: UP

    d) Cell margins: leave as given.

    Data tab

    a) Cell Properties

        1. Text style: scroll and pick ST1

        2. Text height: 4

        3. Alignment: Middle Center

    b) Border Properties: leave as given

    c) General: Table direction: UP

    d) Cell margins: leave as given.

then	**pick OK**
prompt	Table Style dialogue box
with	*a*) PARTS LIST added to the styles names
	b) preview of the PARTS LIST table
respond	**pick Set Current then OK**
prompt	Insert Table dialogue box
respond	*a*) Specify insertion point active
	b) Columns: 4 and Data Rows: 8
prompt	Specify insertion point at command line
respond	**pick a point to suit**
prompt	Text Formatting bar display
respond	**pick OK** to display a blank table as Figure 29.1(a).

3   Make layer TEXT current and set the PICKFIRST variable to 1.

4   Left-click in the title cell of the table denoted by 1 in Figure 29.1(a) and:

    prompt   Title cell is highlighted

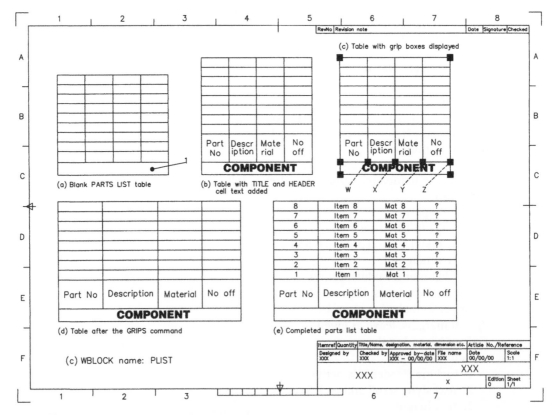

Figure 29.1   Creating the table for a wblock.

enter	**COMPONENT**
and	a) Text Formatting bar displayed with ST3 as the current style
	b) Grid markers A,B,C,D displayed for the columns
	c) Grid markers 1–10 displayed for the rows
respond	**press the TAB key**
and	a) header cell A2 becomes highlighted, i.e. active
	b) Text Formatting bar displays ST2 as the text style
enter	a) Part No then TAB to activate cell B2
	b) Description then TAB to activate cell C2
	c) Material then TAB to activate cell D2
	d) No off the TAB to activate cell A3
then	pick OK from the Text Formatting bar.

5   At this stage the table should resemble Figure 29.1(b) and requires some re-alignment of the cells.

6   Turn grips on with GRIPS 1 at the command line.

7   a) Pick any point on the table to display cold grips as Figure 29.1(c).
   b) Pick grip W to make it hot, then stretch the Part No column to the right until Parts No is a single line entry.
   c) Make grip X hot and drag to right until Description is a single line entry.
   d) With grip Y hot, drag to right until Material is a single line display.
   e) Finally make grip Z hot and drag to right.
   f) Press the ESC key to end the grips operation.

8   When the grips command has been completed, the table should resemble Figure 29.1(d).

9   You could now add other text items to the data cells. Figure 29.1(e) is an idea only as we have no idea at present what text will be displayed in a parts list table.

Making the wblock

1   At the command line enter **WBLOCK <R>** and:
prompt	Write Block dialogue box (similar to Block dialogue box)
respond	a) Source: ensure Objects active (dot)
	b) Base point: select Pick point
prompt	Specify insertion base point at the command line
respond	**Endpoint icon and pick the lower left corner of the table**
prompt	Write Block dialogue box
with	co-ordinates of selected point displayed
respond	a) Objects: delete from drawing active
	b) pick Select objects
prompt	Select objects at the command line
respond	**pick any part of the table and right-click**
prompt	Write Block dialogue box
with	a) *object selected*
respond	a) Destination
	1. File name and path: enter **C:\MYCAD\BORDER**
	2. Insert units: Millimeters – Figure 29.2
	b) pick OK
and	as the file is created, a WBLOCK preview is displayed.

2   *Notes*
   a) As the delete from the drawing option was active, the table will disappear from the current drawing.
   b) Entering OOPS at the command line will restore it, if required.

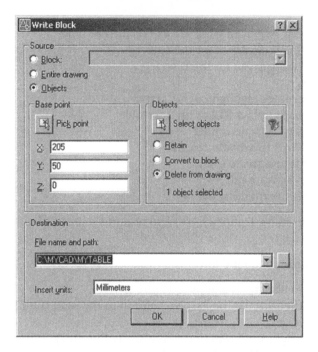

Figure 29.2 The Write Block dialogue box for the created table.

Inserting wblocks

1   *a*) Close the existing drawing and pick no to any saved changes prompt (why?)
    b) open A3PAPER.

2   Refer to Figure 29.3 and:
    a) draw the sectional pulley assembly
    b) no sizes have been given, but you should be able to complete the drawing – use the
        snap to help
    c) add the hatching and the 'balloon' effect using donut-line-circle-middled text.

3   Select the **INSERT BLOCK icon** from the Draw toolbar and:
    prompt   Insert dialogue box
    respond **pick Browse**
    prompt   Select Drawing File dialogue box
    respond *a*) Look in: navigate and pick **C:\MYCAD**
               b) Name: scroll and pick **MYTABLE**
               c) pick Open
    prompt   Insert dialogue box with MYTABLE listed
    respond *a*) ensure Insertion point, Scale and Rotation are Specify On-screen, i.e.
                   tick in three boxes
               b) ensure Explode is active
               c) pick OK
    prompt   Specify insertion point
    enter   **285,30 <R> or select any suitable point**
    prompt   Specify scale factor for XYZ axes
    enter   **1 <R>**
    prompt   Specify rotation angle
    enter   **0 <R>**
    and   the wblock MYTABLE will be displayed at the selected insertion point.

4   Now complete the drawing by editing the table for the entries displayed in Figure 29.3.

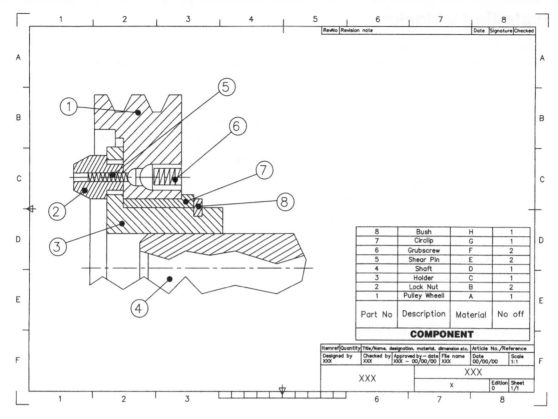

Figure 29.3 Inserting a wblock exercise.

5   When the table alterations have been completed, save the drawing (own file name), but do not exit AutoCAD.

6   *Notes*
   a) The MYTABLE wblock has been inserted into the A3PAPER drawing.
   b) This wblock was 'stored' in the C:\MYCAD folder.
   c) If the computer system is networked, then all CAD users could access the wblock.

About WBLOCKS

1   Think about the following statement. It is true.

   EVERY DRAWING IS A WBLOCK AND EVERY WBLOCK IS A DRAWING

2   When a wblock is being created, the user has three source options:
   a) *Block*:
      allows a previously created block to be 'converted' to a wblock
   b) *Entire drawing*:
      the user can enter a new file name for the existing drawing
   c) *Objects*:
      allows parts of a drawing to be saved as a wblock, i.e. as a drawing file. This is probably the most common source selection option.

3   As with blocks, the same three objects options are available:
   a) Retain
   b) Convert to Block
   c) Delete.

Exploding wblocks

1 *a*) Erase all objects from the screen
   b) With layer OUT current, draw a circle of radius 25 and a square of side 50 – do not use rectangle.

2 Make WBLOCKS of the circle and square using:
   a) circle: base point at the circle centre and file name: C:\MYCAD\CIR
   b) square: base point at any corner of the square and file name: C:\MYCAD\SQ.

3 Now close the drawing (no save) and re-open A3PAPER.

4 Menu bar with **Insert-Block** and:
   a) browse and select CIR from C:\MYCAD
   b) insert at a suitable point, full size with 0 rotation
   c) explode not active.

5 Menu bar with **Insert-Block** and:
   a) browse and pick SQ from your named folder
   b) explode active from the dialogue box
   c) insert at any suitable point with zero rotation.

6 With **LIST <R>** at the command line:
   a) pick the circle and **BLOCK REFERENCE** listed for the object selected
   b) pick any line of square and LINE listed for the selected object.

7 This exercise has demonstrated that unexploded wblocks become blocks within the current drawing.

8 This exercise is complete. Do not save.

External references

1 Wblocks contain information about objects, colour, layers, linetypes, dimension styles, etc. and all this information is inserted into the drawing with the wblock.

2 All this information may not be required by the user, and it also takes time and uses memory space.

3 Wblocks have another disadvantage:
   Drawings with several wblocks are not automatically updated if one of the original wblocks is altered.

4 External references (or xrefs) are similar to wblocks in that they are created by the user and can be inserted into a drawing.

5 They have one major advantage over the wblock:
   Drawings which contain xrefs are automatically updated if the original external reference is modified.

6 A worked example will be used to demonstrate external references.

7 The procedure may seem rather involved as it requires the user to save and open several drawings, but the final result is well worth the effort.

8 For the demonstration we will:
   a) create a wblock
   b) use the wblock as an xref to create two drawing layouts
   c) modify the original wblock
   d) view the two drawing layouts
   e) use the existing C:\MYCAD folder.

Getting started

1 Open your A3PAPER standard drawing sheet and refer to Figure 29.4.

2 Make a new current layer with:
name: XREF; colour: red; linetype: continuous.

3 Draw: *a*) A circle of radius 21
         b) An item of text, middled on the circle centre with height 5 and rotation
             angle 0. The text style is to be ST3 and the item of text is AutoCAD and
             blue – Figure 29.4(a).

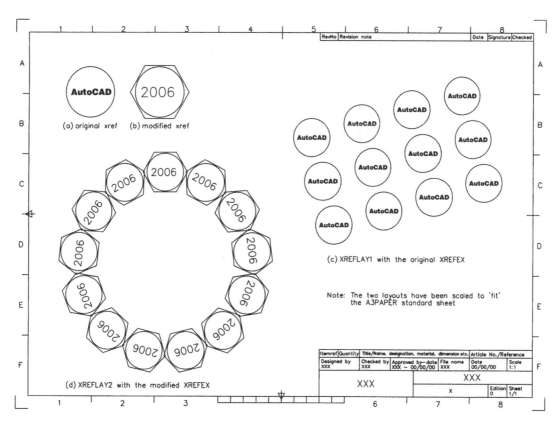

Figure 29.4   The external reference exercise.

Creating the xref (a wblock)

1 At the command line enter **WBLOCK <R>** and:
   prompt   Write Block dialogue box
   respond   *a*) Source: objects
              b) Base point: pick point and pick circle centre point
              c) Objects: select objects, pick circle and text then right-click
              d) Objects: delete from drawing active
              e) File name and path: C:\MYCAD\XREFEX
              f) Insert units: Millimeters
              g) pick OK.

2 A preview of the wblock will be displayed and a blank screen returned, due to the delete
  from drawing option being active.

Inserting the xref (drawing layout 1)

1   *a*) Menu bar with **File-Close** with no to any save changes prompt.
       b) Menu bar with **File-Open** and select your A3PAPER standard sheet with layer OUT current.

2   Menu bar with **Insert-External Reference** and:

prompt	Select Reference File dialogue box *(looks familiar?)*
respond	*a*) Look in: navigate and pick C:\MYCAD if this folder is not current
	b) scroll and pick XREFEX
	c) pick Open
prompt	External Reference dialogue box
with	Name: XREFEX
	Found in: C:\MYCAD\XREFEX.dwg
	Saved path: C:\MYCAD\XREFEX.dwg
respond	*a*) Reference Type: Attachment active (black dot)
	b) Path type: full path active
	c) all on-screen options active, i.e. tick
	d) dialogue box similar to Figure 29.5
	e) pick OK
prompt	**Attach Xref 'XREFEX': C:\MYCAD\XREFEX.dwg**
and	**'XREFEX' loaded**
then	Specify insertion point and enter **50,50 <R>**
prompt	Enter X scale factor and enter **1 <R>**
prompt	Enter Y scale factor and enter **1 <R>**
prompt	Specify rotation angle and enter **0 <R>**
and	the named external reference (XREFEX) will be displayed at the insertion point entered (the complete process seems similar to inserting a wblock?).

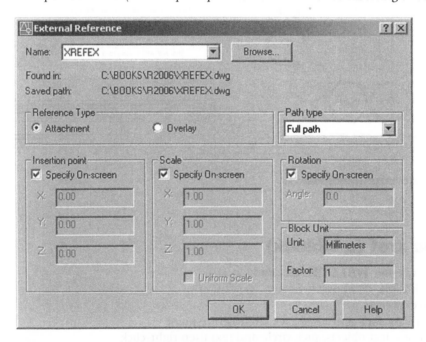

Figure 29.5   The External Reference dialogue box for XREFEX (note that my path is not C:\MYCAD).

3   Now rectangular array the inserted attached xref for:
    a) three rows with row offset: 50
    b) four columns with column offset: 60
    c) Angle of array: 15.

4   Save the layout as **C:\MYCAD\XREFLAY1** – Figure 29.4(c).

Inserting the xref (drawing layout 2)

1 Close the existing drawing then re-open A3PAPER.

2 At the command line enter **XREF <R>** and:
   prompt   Xref Manager dialogue box
   respond   **pick Attach**
   prompt   Select Reference File dialogue box
   respond   *a*) navigate and pick XREFEX from your C:\MYCAD folder
              b) pick Open
   prompt   External Reference dialogue box
   respond   *a*) Reference Type: Attachment
              b) De-activate the three on-screen prompts (no tick)
              c) Insertion point at X = 200, Y = 250, Z = 0
              d) Uniform scale of 1
              e) Rotation angle: 30
              f) pick OK.

3 Now polar array the inserted attached xref with:
   a) centre point: 200,150
   b) number of items: 13
   c) angle to fill: 360
   d) rotate items as copied active.

4 Save the layout as **C:\MYCAD\XREFLAY2**.

Modifying the original xref

1 Close the current drawing.

2 *a*) Open the original XREFEX drawing from C:\MYCAD.
   b) This is the original wblock.

3 *a*) Change the text item to 2006 with style ST2, height 8 and colour green.
   b) Draw a hexagon, centred on the circle and circumscribed in a circle of radius 22.
      Change the colour of this hexagon to blue. These modifications are shown in
      Figure 29.4(b).

4 Menu bar with File-Save to automatically update **C:\MYCAD\XREFEX**.

Viewing the original layouts

1 Close the existing drawing.

2 Menu bar with **File-Open** and:
   a) pick XREFLAY1
   b) note the Preview then pick Open
   c) interesting result?

3 Menu bar with **File-Open** and:
   a) pick XREFLAY2
   b) note the Preview then pick Open
   c) again an interesting result? – Figure 29.4(d).

4 *a*) The layout drawings should display the modified XREFEX without any 'help' from us.
   b) This is the power of external references.
   c) Surely this is a very useful (and dangerous concept)?

5 *a*) Menu bar with **Format-Layer** and note the layer: **XREFEX|XREF**.
   b) This indicates that an external reference (xrefex) has been attached to layer xref.
   c) The (|) is a pipe symbol indicating an attached external reference.

This completes our simple investigation into xrefs.

Assignment

A single external reference activity which requires the creation of several wblocks, attaching these as xrefs to a drawing and then modifying the original wblocks. There is quite a bit of work involved, but the end results is worth the effort.

Activity 45: Weather map

1   Refer to Activity 45(A) and using the grid information given, create:
    a) an outline map of a land mass and save as MAP
    b) the seven ORIGINAL weather symbols and make wblocks of these seven weather symbols using the names given and your own insertion point.

2   Close all drawings then open the saved MAP drawing.

3   *a*) Refer to Activity 45(B) and attach the seven weather symbols as xrefs, creating a weather map of your own design.
    b) When the weather map is complete, save with a suitable name. Remember that you can copy attached xrefs.

4   Open each saved weather symbol and modify as Figure 45(A) then re-save.

5   Open the saved weather map to display the modified weather symbol map.

6   Note that you could also create the seven weather symbols as blocks then re-define these blocks, but that would defeat the purpose of the activity?

Chapter 30

Dynamic blocks

1   Three types of blocks have been investigated, these being:
   a) Block
      1. allow user created objects to be multiple inserted into the current drawing
      2. the Design Center allows these blocks to be inserted into any opened drawing
      3. attributes can be attached to a block definition
      4. when the original block object is modified, the current drawing will display the changes.
   b) Wblock
      1. a drawing file which can be inserted into any opened drawing
      2. all users have access to the drawing file
      3. modifying the original drawing file will not change any drawing using the inserted wblock.
   c) Xref
      1. a drawing file and can be inserted into any opened drawing
      2. all users have access to the drawing file
      3. modifying the original xref will automatically update all drawings using the inserted xref.
   d) these blocks can be considered as 'static', the user having to modify the original objects if an alteration is required.

2   A dynamic block is made and inserted as 'ordinary' blocks but:
   a) it has flexibility and intelligence
   b) it can be modified while the user is working on a drawing
   c) the block geometry can be modified with grips and properties.

Dynamic block terminology

The terminology associated with dynamic blocks are:

1   Action
   a) defines how the geometry will change when the properties are altered
   b) a dynamic block must have at least one associated parameter action.

2   Parameter
   a) defines properties of position, distance and angles
   b) custom grips and properties are used to manipulate the block.

3   Block Editor
   a) used to create a dynamic block
   b) the dynamic block can be a new set of objects or an existing block with dynamic behaviour added
   c) geometry can also be created
   d) parameters and actions added to a block make the block dynamic.

Creating a dynamic block

1 Open your A3PAPER standard sheet and set both GRIPS and PICKFIRST to ON.

2 Refer to Figure 30.1 and:
   a) draw the beam shape as Figure 30.1(a) using your discretion for sizes not given
   b) do not add dimensions
   c) make a block of the beam outline with:
   name: BEAM-1
   insertion point: use as indicated
   units: millimeters
   retain: your decision
   description: add your own.

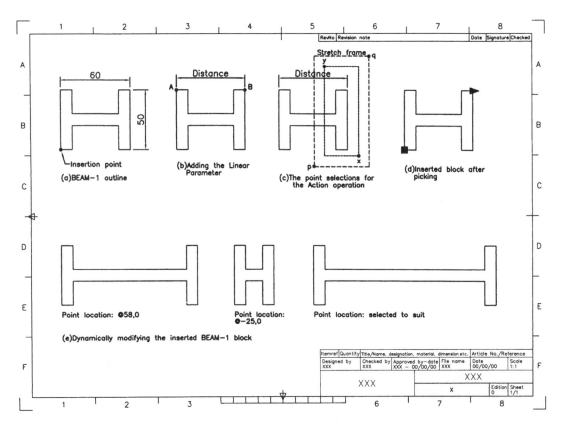

Figure 30.1 Dynamic block exercise 1 with BEAM-1.

3 Menu bar with **Tools-Block Editor** and:
   prompt   Edit Block Definition dialogue box
   respond   **pick BEAM-1** (Figure 30.2) then pick OK
   prompt   Block Editor opened
   with   *a*) AutoCAD message perhaps – select No
               b) a display of the beam outline
               c) the Block Authoring Palettes with three tabs: Parameters, Actions,
                   Parameter Sets.

4 Figure 30.3 identifies the Block Editor icons.

5 Activate the **Parameters tab** then select the Linear Parameter from the palette and:
   prompt   Specify start point or [Name/Label/Chain/Description/
               Base/Palette/Value set]

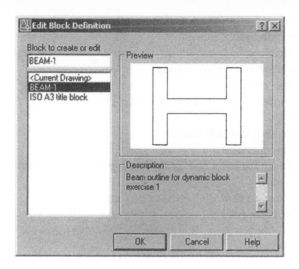

Figure 30.2 The Edit Block Definition dialogue box for BEAM-1.

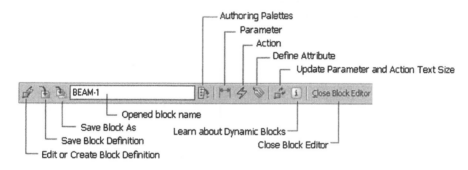

Figure 30.3 The Block Editor icons.

respond	**Endpoint icon and pick point A** indicated in Figure 30.1(b)
prompt	Specify endpoint
respond	**Endpoint icon and pick point B**
prompt	Specify label location
respond	**pick above** as indicated in Figure 30.1(b)
and	*a*) command line returned
	b) Distance 'dimension' added between the two selected points
	c) blue grip triangles added and the linear parameter start and end points
	d) the alert icon (!) displayed, indicating that no action is associated with the parameter.

5 Ensure GRIPS are on and pick the Distance dimension name and:

prompt	grips boxes added to the Distance parameter
respond	**right-click**
prompt	shortcut menu
respond	*a*) pick Grip Display
	b) pick 1
and	linear parameter displayed with one blue grip triangle (at point B).

6 Activate the **Actions tab**, pick Stretch Action and:

prompt	Select parameter
respond	**pick the Distance parameter**
prompt	Specify parameter point to associate with action or [sTart point/Second point]

and	circular marker displayed at points A and B as cursor moved
respond	**pick the blue grip triangle**
prompt	`Specify first corner of stretch frame or [Cpolygon]`
respond	**pick point p** as indicated in Figure 30.1(c)
prompt	`Specify opposite corner`
respond	**pick point q** to include the beam and parameter
prompt	`Select objects`
respond	**pick point x inside the stretch frame**
prompt	`Specify opposite corner`
respond	**pick point y inside to include the beam and parameter then right-click**
prompt	`Specify action location or [Multiplier/Offset]`
respond	**pick a point at the blue grip**
and	*a*) command line returned
	b) Block Editor still active.

7 Figure 30.4 displays the Block Editor at this stage of the dynamic block creation.

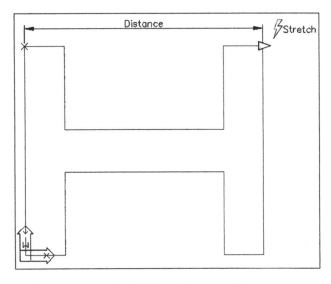

Figure 30.4   The Block Editor for the BEAM-1 block after the parameter and action operations.

8 Menu bar with Insert-Block and using the Insert dialogue box:
  a) select BEAM-1 (it should be the default)
  b) specify Insertion point on-screen active, i.e. tick
  c) specify Scale on-screen not active with uniform scale of 1
  d) specify Rotation on-screen not active with rotation angle of 0
  e) pick OK
  f) specify insertion point: pick a suitable point
  g) the BEAM-1 block will be inserted full-size at the selected point
  h) at present there is no dynamic block effect.

9 Repeat the Insert command another two times and insert BEAM-1 full-size with 0 rotation.

10 As grips are on, pick the first inserted block and:

prompt	`grip box and arrow added to the block` as Figure 30.1(d)
respond	**pick the triangular grip and it becomes hot, i.e. red**
prompt	`Specify point location`
enter	**@58,0 <R> then ESC.**

11 Select the second inserted block and:
    a) make the triangular grip hot
    b) enter **@−25,0** as the point location then ESC.

12 Stretch the third inserted block by selecting a suitable point – Figure 30.1(e).

13 This completes the first exercise, so save as **C:\folder\DYNEX1** for future reference.

Dynamic block exercise 2

1 Either   *a*) continue from the previous exercise
   or   *b*) open the DYNEX1 drawing.

2 Erase all objects from the screen.

3 Refer to Figure 30.5 and:
   a) draw the basic shape as Figure 30.5(a)
   b) do not add dimensions
   c) create two blocks of the complete shape with names BEAM-2A and BEAM-2B
   d) make the lower left vertex the insertion point for each block creation.

4 Menu bar with **Tools-Block Editor** and:
   prompt   Edit Block Definition dialogue box with three created block
              names listed
   respond pick BEAM-2A then OK
   prompt   Block Editor with Block Authoring Palette and beam outline displayed
   respond *a*) activate the Parameters tab
              b) pick Linear Parameter

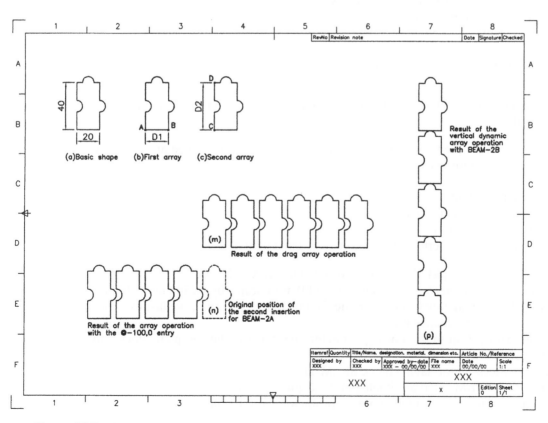

Figure 30.5   Dynamic block exercise 2 – array actions.

    c) start and end points: pick points A and B as Figure 30.5(b)

    d) label location: pick as D1 in Figure 30.5(b)

    e) Leave the two triangular grid markers

    f) activate the Actions tab

    g) pick Array Action

    h) select parameter: pick the Distance parameter added

    i) select objects: window the complete shape and the distance parameter then right-click

    j) enter column distance: enter 25 <R>

    k) specify action location: pick near the added grip

    l) select Close Block Editor from the menu bar

    m) pick Yes to save changes to BEAM-2A

    n) drawing screen returned.

5 Menu bar with Insert-Block and:

    a) block name: scroll and pick BEAM-2A

    b) insertion point: on-screen active

    c) scale: on-screen not active with uniform scale of 1

    d) rotation: on-screen not active with rotation angle of 0

    e) pick OK from the dialogue box

    f) select a suitable insertion point towards the centre of the screen as Figure 30.5(m).

6 Repeat the insertion command and insert BEAM-2A below the first insertion as Figure 30.5(n).

7 *a*) pick the first inserted BEAM-2A block to display the two grid markers

    b) make the right triangular grid hot

    c) with ORTHO on, drag to the right and then <R>

    d) pick the second inserted BEAM-2A block

    e) make the left triangular grid hot

    f) enter @ −100,0 then <R>.

8 The results of these single row arrays are displayed in Figure 30.5.

9 *Questions*

    a) the second inserted BEAM-2A block will not be displayed after the dynamic arrays

    b) why is this?

10 The above steps are the basics for creating a dynamic array (and indeed any dynamic block operation, these steps being:

    a) create the block

    b) open the Block Editor

    c) add parameters and actions

    d) save the dynamic block

    e) insert the block and perform the actions associated with it.

11 With the Block Editor, pick the second created block BEAM-2B and:

    a) add a linear parameter between points C and D as indicated by D2 in Figure 30.5(c)

    b) modify the block to display the top triangular grip only (as step 5 in the first exercise)

    c) the array column distance is to be 45

    d) add an array action to the linear parameter and window the complete block and distance

    e) close the Block Editor, saving block BEAM-2B

    f) insert block BEAM-2B at a suitable point – Figure 30.5(p)

    g) pick the inserted block and drag out vertically for the dynamic array.

12 This second exercise is now complete, so save as **C:\folder\DYNEX2**.

Specifying parameter sizes

1   The first dynamic block exercise used the stretch action with the linear parameter.

2   Block BEAM-1 was stretch with dragging out the grips or by keyboard entry.

3   Specific sizes for the stretch operation can be automatically attached to the block.

4   Close any existing drawing, open DYNEX1 and refer to Figure 30.6.

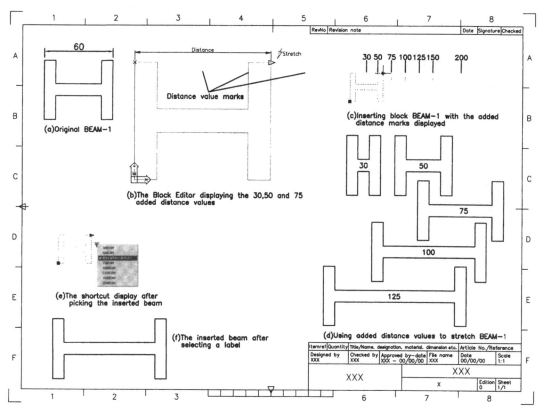

Figure 30.6   Dynamic block exercise, specifying sizes and creating a lookup table.

5   Erase all objects then menu bar with **Tools-Block Editor** and:
    prompt   Edit Block Definition dialogue box
    respond   **pick BEAM-1 then OK**
    prompt   Block Editor
    and   BEAM-1 with parameter and action from Exercise 1 displayed
    respond   a) pick the linear Distance parameter added to the block definition
                b) right-click
    prompt   shortcut menu
    respond   **pick Properties**
    and   Properties palette displayed.

6   Using the Properties palette, pick **Value Set** and:
    prompt   it is highlighted
    respond   a) pick Dist type
                b) pick the scroll arrow
                c) pick List
                d) pick Dist value list then...

prompt	Add Distance Value dialogue box
respond	*a*) distances to add: enter 30,50,75,100,125,150,200
	b) pick Add – Figure 30.7(a)
	c) pick OK
prompt	Properties palette
with	entered distances displayed as Figure 30.7(b)
respond	*a*) close the properties palette
	b) press ESC to cancel the grip effect
and	the Block Editor will be displayed with some small vertical lines, these being the entered distances relative to the stretch action – Figure 30.6(b)
respond	close the Block Editor, with Yes to saving the BEAM-1 definition.

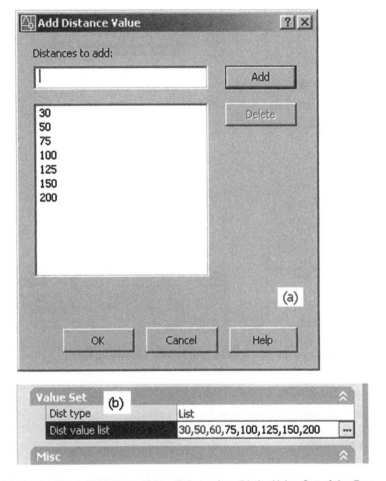

Figure 30.7   (a) The Add Distance Value dialogue box (b) the Value Set of the Properties palette.

7   Insert block BEAM-1 at a suitable point then:
   a) pick the block to display the grip effect
   b) make the triangular grip hot and:
      prompt   the added distance values displayed as Figure 30.6(c)
      respond   **drag the block to the desired distance mark**
   c) repeat the insertion and stretch the beam to other distance marks
   d) Figure 30.6(d) displays five beam insertions using the specified sizes.

8   Save the drawing at this stage as **C:\folder\DYNEX3**.

Adding a lookup table

1   Continue from the previous exercise or open DYNEX3 and refer to Figure 30.6.

2   Refer again to Figure 30.6, menu bar with **Tools-Block Editor** and:
   prompt   Edit Block Definition dialogue box
   respond   **pick BEAM-1 then OK**
   prompt   Block Editor
   and   BEAM-1 with parameter, action and distances from previous exercises
   respond   *a*) activate the Parameters tab
             b) pick Lookup Parameter
   prompt   Specify parameter location or [Name/Label/Description/
             palette]
   respond   *a*) pick any suitable point
             b) activate the Actions tab
             c) pick Lookup action
   prompt   Select parameter
   respond   **pick the Lookup parameter added to the block definition**
   prompt   Specify action location
   respond   pick below the added lookup parameter
   and   Block Editor displayed.

3   When the lookup action location has been selected:
   prompt   Property Lookup Table dialogue box
   respond   **pick Add Properties**
   prompt   Add Parameter Properties dialogue box
   respond   **pick Linear then OK**
   prompt   Property Lookup Table dialogue box
   with   Distance added to the Input Properties pane
   respond   *a*) pick the ▶ symbol
             b) pick the displayed drop arrow
   prompt   list of all specified sizes displayed
   respond   *a*) select the 30 size
             b) pick the corresponding row in the Lookup Properties pane
             c) enter 30BEAM as the label
   then   *a*) pick the next row below the 30 distance in the Input Properties pane
             b) pick the ▶ symbol
             c) pick the displayed drop arrow
             d) select the 50 size
             e) pick the corresponding row in the Lookup Properties pane
             f) enter the label: 50BEAM
   continue with the above sequence of six steps until the dialogue box is displayed
             similar to Figure 30.8
   then   *a*) from the Lookup Properties pane, pick Read only
             b) pick the drop arrow
             c) pick Allow reverse lookup then OK
   and   Block Editor returned
   respond   Close the Block Editor, saving the BEAM-1 definition
   and   drawing screen returned.

4   Menu bar with Insert-Block and insert block BEAM-1 at any suitable point.

5   As grips are active, pick any part of the inserted beam and:
   prompt   beam highlighted with grip points and lookup grip
             displayed
   respond   **pick the lookup grip**
   prompt   shortcut menu with labels as Figure 30.6(e)

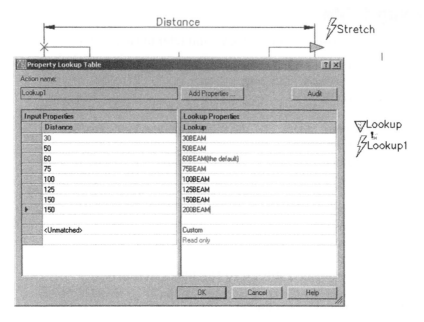

Figure 30.8 The Property Lookup Table dialogue box from the Block Editor, assigning labels to the entered specific sizes.

respond	**pick any label**
and	the inserted beam will be stretched as appropriate
respond	ESC to end the sequence.

Notes

1   This introduction the dynamic blocks is now complete.

2   Not all of the parameters/actions have been discussed, but the user should now investigate these in their own time.

Chapter 31

The Design Center

1 The AutoCAD Design Center is a drawing management system with several powerful advantages to the user including:
   a) the ability to browse and access different drawing sources
   b) access drawing content including Xrefs, blocks, hatches and symbol libraries
   c) view object definitions (e.g. layers) prior to inserting, attaching or copy and paste into the current drawing
   d) search for specific drawing content
   e) redefine a block definition
   f) create shortcuts to drawings and folders
   e) drag drawings, blocks and hatches to a tool palette.

2 In this chapter we will investigate several of these topics.

Getting started

1 Open your A3PAPER standard sheet and cancel any floating toolbars

2 Menu bar with **Tools-Design Center** and:
   prompt   Design Center window
   respond   **study the layout and leave displayed**.

3 The first time that the Design Center is activated, it may be docked at the side of the drawing screen area but it can be moved and resized by:
   a) left-click in the Design Center title bar and hold down the button
   b) drag into the drawing area
   c) resize to suit.

4 The Design Center window consists of the following:
   a) the Design Center title bar at the left or right depending on position
   b) a toolbar with several icon selections
   c) four tabs: Folders, Open Drawings, History, DC Online
   d) the tree side and hierarchy area on the left
   e) the content side with icons on the right
   f) an information bar below the tree view and palette
   e) an Auto-hide and Properties option.

5 *a*) The Design Center window is fully displayed in Figure 31.1(a).
   b) if the Auto-hide is selected, the window 'collapses' and is displayed as Figure 31.1(b)
   c) to fully display the window, click Auto-hide again.

Using the tree view and hierarchy

1 The tree side of the Design Center window consists of a **hierarchy** list of names with a (+) or a (−).

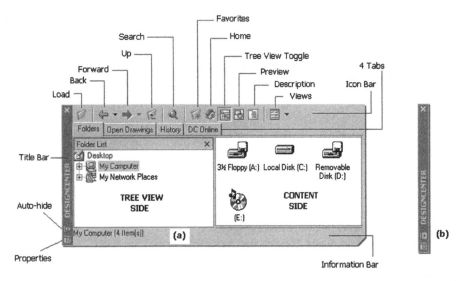

Figure 31.1 The Design Center window (a) fully displayed and (b) with Auto-hide on.

2 If a name in the tree side is selected (left-click) then the content side of the window will reveal the contents of the selected item.

3 *a*) If the (+) beside an item is selected then that item is '**explored**' and the tree side of the window will list the 'contents' of the selected item
  b) The selected item's (+) will be replaced with a (−) indicated that it has been explored.

4 *a*) If the (−) is then selected, the explored effect is removed and the (+) is returned
  b) The selection of the (−) is termed '**collapsing**'
  c) The user is thus **navigating** with the Design Center.

5 *Note*: While 'explore' is the technical term associated with these activities, I prefer 'expand' as it compliments the 'collapse' term.

6 Refer to Figure 31.2 and:
  a) note the Design Center layout as opened – Figure 31.2(a). Any display difference is not important
  b) explore My Computer (if available) by picking the (+) to give a tree expansion similar to Figure 31.2(b)
  c) explore the C: drive by selecting the (+) to give the folders on the C drive of your system – Figure 31.2(c). This will probably differ to what I have displayed
  d) explore C:\MYCAD by selecting the (+) to give a listing of the files you have saved, similar to Figure 31.2(d)
  e) note that in Figure 31.2 I have only displayed the tree view side of the Design Center. This was to ensure all the displays 'fitted' into a single layout
  f) move the pointer arrow onto MYCAD and right-click.
    prompt   Shortcut menu
    respond **pick Explore**
    and   content side of the Design Center window will display icons from the selected folder.

7 *Task*
  a) collapse your named folder by picking the (−) at the name
  b) collapse the C: drive
  c) collapse My computer and the Design Center should be displayed 'as opened' – Figure 31.2(a).

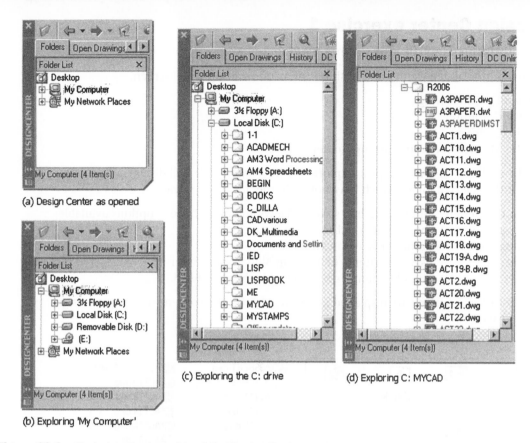

(a) Design Center as opened

(b) Exploring 'My Computer'

(c) Exploring the C: drive

(d) Exploring C: MYCAD

Figure 31.2   Exploring the tree side of the Design Center.

The Design Center menu bar and tabs

1   The menu bar of the Design Center allows the user access to the several icons and tabs.

2   These are displayed and listed in Figure 31.1 and are:
   a) *Icons*

Load	allows the user to navigate to files and load them into the contents area
Back	returns to the most recent location in the history list
Forward	returns to the next later location in the history list
Up	will toggle the Design Center 'up a level'
Search	allows the user to search for named drawings, blocks and non-graphical objects using specified criteria
Favourites	lists the contents of favourites folder
Home	returns the Design Center to the user home folder
Tree View Toggle	hides and displays the tree side of the Design Center
Preview	displays and hides a preview for a selected item in a pane below the content area
Description	display and hides a text description for a selected item in a pane below the content
Views	allows the user to access different displays for the content area.

   b) *Tabs*

Folders	displays the hierarchy of files and folders on the user computer
Open Drawings	displays all drawings currently open in the current drawing session
History	displays a list of files most recently opened with Design Center
DC Online	accesses the Design Center online web page

Design Center exercise 1

1 AutoCAD should still be active with A3PAPER displayed.

2 Activate the Design Center and position the window in centre of screen.

3 Expand the following:
   a) My Computer
   b) the C: drive
   c) the MYCAD (or your working) folder.

4 Right-click on MYCAD and pick Explore and the content side will display information about the selected folder which may not be in the 'form' we want.

5 From the Design Center menu bar, scroll at Views and note the four options available:
   a) Large icon: thumbnail sketches of the files
   b) Small icon: names only with extension
   c) List: lists all contents of folder in alphabetical order
   d) Details: gives file name, size, type and when modified.

6 The four View options are displayed in Figure 31.3 for the content side of the Design Center window.

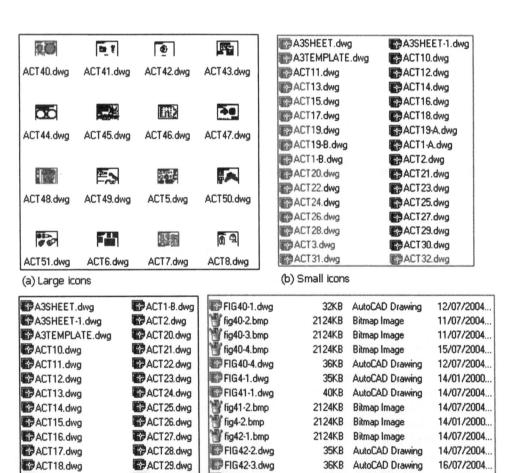

Figure 31.3 The Design Center menu bar VIEW options.

7 *a*) from the Design Center menu bar select Preview and Description
  b) the content side will display two additional areas
  c) these can be resized by dragging the lower 'border' up or down.

8 With large icons from VIEW:
  a) scroll and pick (left-click) any drawing icon from the content side
  b) a preview and description (if applicable) will be displayed similar to Figure 31.4(a).

9 *a*) Scroll and pick another icon item and (if possible) pick another type of file format
  b) Figure 31.4(b) displays a BMP file of Figure 10.2 the Layer Properties Manager dialogue box. This dialogue box was 'screen dumped' from AutoCAD into a word-processing package for inclusion in the book.

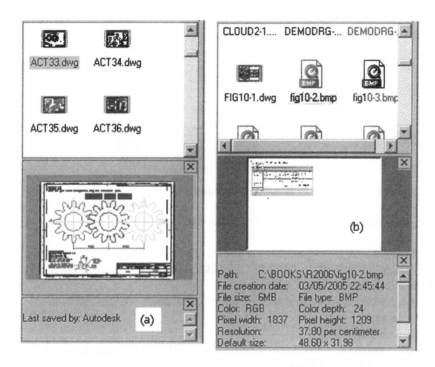

Figure 31.4   The Design Center window with the Preview and Description options active, displaying: (a) ACT33: gear wheel design activity; (b) Figure 10-2: a BMP screen dump of the Layer Properties Manager dialogue box.

Design Center exercise 2

1 Still with the Design Center displayed and A3PAPER opened?

2 Ensure that Preview and Description are 'active'

3 *a*) Scroll in the tree side until the block exercise drawing from Chapter 28 is displayed
  b) this was Figure 28.3 in the book
  c) explore the drawing name with a right-click and also by picking the (+) name
  d) both the tree and contents side of the Design Center will display:
     Blocks, Dimstyles, Layers, Layouts, Linetypes, Tablestyles, Textstyles, Xrefs
  e) the tree side displays a list and the content side displays icons.

4 *a*) From the content side right-click on Blocks then pick explore to display information about the blocks in the named drawing
  b) Five block are listed: SEAT, TABLE1, TABLE2, TABLE3 and ISO A3 title block
  c) 1. The seat and table blocks were made by us, but not the ISO A3 block
     2. ISOA3 title block is an AutoCAD block from the A3PAPER standard sheet
     3. We will discuss it in more detail in another chapter.

5   *a*) Pick (left-click) the block TABLE3 and note the display – Figure 31.5
    b) Note that you may have to resize the preview area for this exercise.

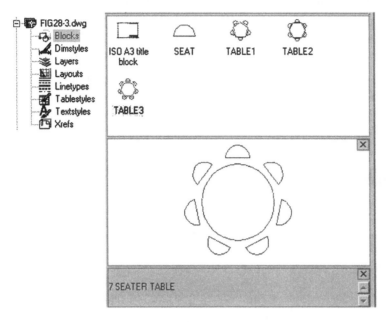

Figure 31.5   The explored figure 28.3 with Block also explored.

6   Explore with a right-click the other items from the expanded tree side and:
    Selection   *Result*
    Dimstyles   A3DIM listed and perhaps others, e.g. ISO-25
    Layers   1. CL, CONS, DIMS, HID, OUT, SECT, TEXT – our seven created layers
                    2. 0 and Defpoints – AutoCAD layers
                    3. FRAMES 025, 050, 070, 200, Title Block, Viewports – AutoCAD template layers
    Layouts   ISO A3 Title Block listed
    Linetypes   ByBlock, ByLayer, CENTER2, Continuous, HIDDEN2 and perhaps others, e.g. CENTER, HIDDEN
    Tablestyles   Standard only listed
    Textstyles   1. ST1, ST2, ST3 – our 3 created text styles
                    2. Standard and perhaps others, e.g. ISO Proportional
    Xrefs   none listed – which should be obvious?

7   From the Design Center menu bar select:
    a) Up: displays contents of explored Figure 28.3 drawing
    b) Up: displays contents of the MYCAD folder
    c) Up: displays the C: drive expansion
    d) Up: displays the My Computer icons
    e) *Note*: if you may have explored other folders/files during this exercise and will need to select the 'Up' menu bar option more than the four times listed.

Design Center exercise 3

1   Still have the Design Center displayed from exercise 2?

2   From the Design Center menu bar pick the **Search icon** and:
    prompt   Search dialogue box
    respond   *a*) scroll at Look for and pick Xrefs
                 b) scroll at In and pick Local disk (C:)
                 c) pick Browse.

prompt	Browse for Folder dialogue box
respond	explore the C: drive, pick MYCAD then pick OK
prompt	Search dialogue box
with	named folder name displayed and Xrefs tab active
respond	a) at Search for the name: enter *
	b) pick **Search Now**
and	the search will 'interrogate' the MYCAD folder for Xref names
then	Search dialogue box
with	a list of the 'found' Xrefs similar to Figure 31.6.

3 a) The Xref names displayed in the Search dialogue box should be XREFLAY1 and XREFLAY2 from the Chapter 29 exercise and the Xrefs from *Activity 45* if this was completed.

b) My Figure 31.6 has the folder name as C:\BOOKS\R2006 (instead of MYCAD).

4 Now cancel the Search dialogue box the Design Center to leave the blank A3PAPER standard sheet as opened.

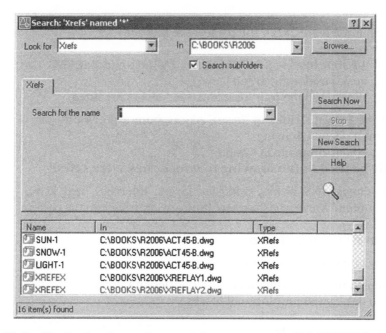

Figure 31.6   The Design Center Search dialogue box The item of the search is 'XREFS' in the C:\BOOKS\R2006 folder.

Design Center exercise 4

1 a) Close any existing drawings with no save to the standard sheet
   b) Menu bar with **File-New** and select **Start from Scratch, Metric then OK**.

2 a) Menu bar with Format-Layers, Format-Text Styles, Format-Dimension Styles
   b) Note the layers, text styles and dimension styles in this current drawing
   c) These features (unless altered by the user) should be the AutoCAD defaults of:
   Layers: 0
   Text Styles: Standard with txt.shx font
   Dimension Style: ISO-25.

3 Activate the Design Center and position to suit, i.e. docked or floating.

4 Explore the following by picking the (+) at the appropriate folder/file:
   a) My Computer
   b) the C: drive

c) the MYCAD folder

d) the A3PAPER drawing file.

5 From the explored A3PAPER tree hierarchy, explore Layers to display the layers from the A3PAPER drawing in the content side of the Design Center.

6 From the content side:
   a) left-click on the OUT layer
   b) hold down the left button
   c) drag into the drawing area and release the left button
   d) repeat 'pick-and-drag' for layers HID and CL.

7 Explore Textstyles and 'pick-and-drag' the three created text styles into the drawing area.

8 From the AutoCAD menu bar with Format-Layers and Format-Text Style, check that the selected layers and text styles are now available for use in the new drawing.

9 Collapse the explored A3PAPER from the tree side, i.e. pick the (−) at the drawing name.

10 Menu bar with **Insert-Block** and there are no block names listed in the current drawing – obviously?

11 *a*) From the tree side, explore Figure 28.3 – the conference room layout
   b) Explore Blocks
   c) Left-click on each of the four blocks (SEAT, TABLE1, TABLE2 and TABLE3) and drag into the drawing area, positioning to suit
   d) Menu bar with Insert-Blocks and the four 'pick and drag' blocks are listed.

12 *a*) We have thus used the Design Center to 'insert' blocks from a previously saved drawing into the current opened drawing
   b) This is contrary to the statement made in Chapter 28
   i.e. '**blocks can only be used in the drawing in which they were saved**'.

13 *a*) The Design Center allows ANY blocks, wblocks, layers, linetypes, text styles, dimension styles, layouts and Xrefs to be 'inserted' from any drawing into the current drawing
   b) this is a very powerful aid to any CAD user.

14 This exercise is complete and does not need to be saved.

15 *a*) collapse all the expanded tree hierarchies and close the Design Center
   b) the screen should still display the four inserted blocks.

Design Center exercise 5

1 *a*) Close any existing drawings with no save
   b) Menu bar with **File-New** and select **Start from Scratch, Metric then OK**.

2 Draw the following:
   a) a 100 square as four lines
   b) a 30-radius circle at the square centre
   c) fillet a corner with radius 30 and chamfer a corner with distances of 20.

3 *a*) with the menu bar sequence Dimension-Style, note the current dimension style
   b) I had ISO-25 as stated in exercise 4.

4 Use this dimension style and add the following dimensions, denoted in Figure 31.7(a):
   a) baseline linear
   b) circular – both diameter and radius
   c) angular.

5 Activate the Design Center and explore:
   a) My Computer
   b) the C: drive

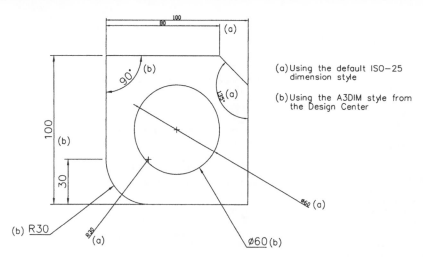

Figure 31.7 Using a dimension style from the Design Center.

    c) your named folder (MYCAD)
    d) the A3PAPER standard sheet.

6 *a*) Right-click, explore Dimstyles and A3DIM should be displayed in content side
    b) Drag-and-drop A3DIM into the current drawing
    c) Close the Design Center.

7 Menu bar with **Dimension–Dimension Style** and
    a) the Dimension Style Manager dialogue box should display the A3DIM style
    b) set A3DIM current
    c) close the dialogue box.

8 Now add similar dimensions as step 4, denoted in Figure 31.7(b).

9 This exercise has shown another use for the Design Center.

10 The exercise is complete. Save if required.

Design Center exercise 6

1 In this example we will use the Design Center to investigate the 'stored' AutoCAD blocks so:
    a) close all existing drawing
    b) open your A3PAPER standard sheet with layer OUT current.

2 Activate the Design Center and position to suit.

3 Explore the following trees:
    a) C: drive
    b) Program Files – read note **
    c) AutoCAD 2006
    d) Sample
    e) Design Center.

4 *Note* **
    a) the folder Program Files is where my AutoCAD 2006 has been installed
    b) your system may use a different folder name.

5 *a*) Explore Analogue Integrated Circuits.dwg
    b) Explore the blocks within this drawing
    c) Figure 31.8 should be the result of the various 'explore' operations.

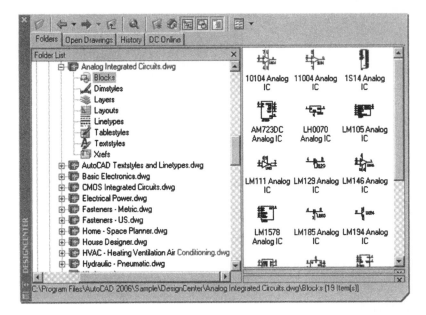

Figure 31.8 The AutoCAD blocks from the folder C:\Program Files\AutoCAD 2006\Sample\ DesignCenter\Analog Intergrated Circuits.dwg.

6   *a*) Refer to Figure 31.9 and drag-and-drop into the opened A3PAPER the first named block from every drawing of the explored Sample-Design Center hierarchy

    b) Remember that they are blocks and you may have to scale during/after insertion

    c) The following table is a list of these first named blocks:

Ref	*File name*	*Block name*
a)	Analogue Integrated Circuits	10104 Analogue IC
b)	Basic Electronics	Battery
c)	CMOS Integrated Circuits	14160 CMOS IC
d)	Electrical Power	Alarm
e)	Fasteners – Metric	Cross Pt Flathead Screw
f)	Fasteners – US	Hex Bolt ½ inch side
g)	Home – Space Planner	Bed – Queen
h)	Home Designer	Bath Tub – 26 × 60
i)	HVAC	Air Injector
j)	Hydraulic–Pneumatic	Air Regulator
k)	Kitchens	Base Cabinet
l)	Landscaping	Car-Sedan side
m)	Pipe Fittings	Cross-flanged
n)	Plant Process	Autoclave
o)	Welding	Double Bevel Weld.

7   *a*) The method of inserting blocks from the Design Center has so far only considered drag-and-drop

    b) There is another method which is more accurate than that used.

8   From the Design Center:

    a) explore the **AutoCAD Landscaping.dwg file**

    b) explore Blocks

    c) from the content side navigate, right-click on **Shrub-Elevation** and:

      prompt   Shortcut menu

      respond   **pick Insert Block**

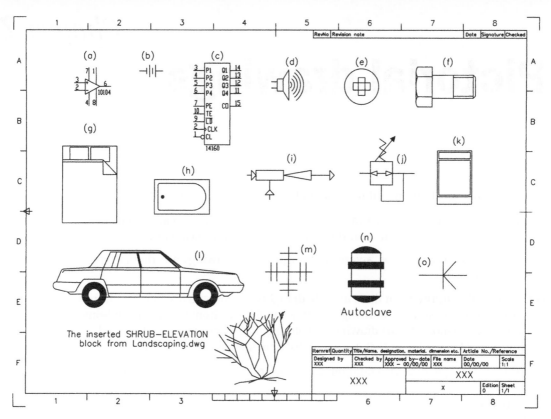

Figure 31.9 Stored AutoCAD blocks – first named.

prompt	Insert dialogue box
respond	1. Insertion at suitable X- and Y-values
	2. Scale at 0.1
	3. Rotation angle: 5
and	the selected block will be inserted at the desired point.

9   This exercise is now complete and can be saved if required.

This chapter has been an introduction to the AutoCAD 2006 Design Center. Hopefully having tried the exercises, you will be able to use it to improve your drawing productivity.

Chapter 32

Pictorial drawings

1   Pictorial drawings allow two-dimensional objects to be displayed in 3D.

2   This is a very useful concept as it can convey additional information about a component which is not always apparent with the traditional orthographical views.

3   The reader should always be aware that all pictorial drawings are themselves two-dimensional.

4   I am constantly surprised that there are still 'draughtspersons' and teachers/lecturers who refer to an isometric as a 3D drawing. Nothing could be far from the truth. An isometric is a 2D representation of a 3D drawing and although as it appears to be displayed in 3D, the user should never forget that it is a 'flat 2D' drawing without any 'depth'.

5   There are several pictorial visualisations available and in this chapter we will investigate three:
   a) isometric
   b) oblique
   c) planometric.

Isometric drawings

1   Isometric drawings are created with the X- and Y-axes at 30 degrees.

2   User entry and accuracy is achieved with polar co-ordinates.

3   Isometric drawings are created 'full-size'.

4   AutoCAD has the facility to display an isometric grid as a drawing aid.

Setting the isometric grid

1   There are two methods for setting the isometric grid:
   a) using a dialogue box
   b) via the keyboard.

2   Open your A3PAPER then menu bar with **Tools-Drafting Settings** and:
   prompt   Drafting Settings dialogue box with four tab selections:
                a) Snap and Grid
                b) Polar Tracking
                c) Object Snap
                d) Dynamic Input
   respond   **pick the Snap and Grid tab and:**
                a) Snap On(F9) active – tick
                b) Grid On(F7) active
                c) Snap type and Style with:
                    1. Grid snap active – black dot
                    2. Isometric snap active
                d) set Grid Y Spacing: 10
                e) set Snap Y Spacing: 5

f) Angle, X-base, Y-base: all 0 – Figure 32.1

g) Ensure Polar Tracking, Object Snap, Dynamic Input aids are off

h) pick OK

and　　the screen will display an isometric grid of 10 spacing, with the on-screen cursor 'aligned' to this grid with a snap of 5.

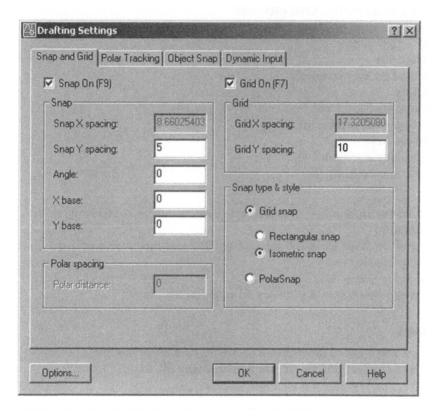

Figure 32.1　The Drafting Settings (Snap and grid tab) dialogue box.

3　*a*) Use the Drafting Settings dialogue box to 'turn the isometric grid off,' i.e. pick the Rectangular snap and set the grid spacing to 10 and the snap spacing to 5.

　　b) The screen will display the standard grid pattern.

4　For the keyboard entry method, at the command line enter **SNAP <R>** and:

prompt	`Specify snap spacing or [ON/OFF/Aspect/Rotate/Style/Type]`
enter	**S <R>** – the style option
prompt	`Enter snap grid style [Standard/Isometric]`
enter	**I <R>** – the isometric option
prompt	`Specify vertical spacing<5.00>`
enter	**10 <R>**.

5　The screen will again display the isometric grid pattern with the cursor 'snapped to the grid points'.

6　Leave this isometric grid on the screen.

Isoplanes

1　AutoCAD uses three 'planes' called isoplanes when creating an isometric drawing, these being named top, right and left. The three planes are designated by two of the X,- Y- and Z-axes as shown in Figure 32.2(a) and are:

　　a) isoplane top: XZ axes

b) isoplane right: XY axes

c) isoplane left: YZ axes.

2   When an isoplane is 'set' or 'current', the on-screen cursor is aligned to the axes of the isoplane axis as displayed in Figure 32.2(b).

3   The isoplane can be set by one of three methods:

   a) at the command line enter **ISOPLANE <R>** and:

   prompt   Enter isometric plane setting [Left/Top/Right]

   enter   **R <R>** – right plane

   b) Using a 'toggle' effect by:

   1. holding down the **Ctrl key** (control)

   2. pressing the **E key**

   3. toggles to isoplane left

   4. Ctrl E again – toggles to isoplane top.

   c) Using the F5 function key:

   1. press F5 – toggles isoplane right

   2. press F5 – toggles isoplane left, etc.

4   *Notes*

   It is user preference as to what method is used to set the isometric grid and isoplane, but my recommendation is:

   a) set the isometric grid ON from the Drafting Settings dialogue box with a grid spacing of 10 and a snap spacing of 5

   b) toggle to the required isoplane with Ctrl E or F5

   c) isoplanes are necessary when creating 'circles' in an isometric 'view'.

Isometric circles

1   Circles in isometric are often called **isocircles** and are created using the ellipse command and the correct isoplane MUST be set. Try the following exercise:

2   Set the isometric grid on with spacing of 10 and toggle to isoplane top.

3   *a*) using the isometric grid as a guide and with the snap on, draw a cuboid shape as Figure 32.2(c)

   b) the size of the shape is not important at this stage – only the basic shape.

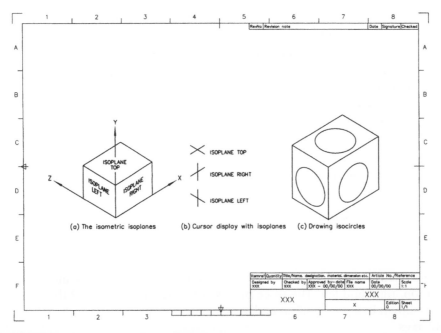

Figure 32.2   Three isometric concepts.

3 Select the **ELLIPSE icon** from the Draw toolbar and:

prompt   `Specify axis endpoint of ellipse or`
        `[Arc/Centre/ Isocircle]`

enter   **I <R>** – the isocircle option

prompt   `Specify centre of isocircle`

respond **pick any point on top 'surface'**

prompt   `Specify radius of isocircle`

respond **drag and pick as required**.

4 Toggle to isoplane right with Ctrl E.

5 At the command line enter **ELLIPSE <R>** and:

prompt   `Specify axis endpoint of ellipse or`
        `[Arc/Centre/ Isocircle]`

enter   **I <R>**

prompt   `Specify centre of isocircle` and **pick a point on 'right side'**

prompt   `Specify radius of isocircle` and **drag/pick to suit**

6 Toggle to isoplane left, and draw an isocircle on the left side of the cuboid.

7 The cuboid now has an isometric circle on the three 'sides'.

8 Now continue with the example which follows.

Isometric example

1 Erase any objects from the screen or re-open your A3PAPER standard sheet and refer to Figure 32.3.

2 Set the isometric grid on, with a grid spacing of 10 and a snap spacing of 5.

3 With the **LINE icon** draw:
First point: pick towards lower centre of the screen
Next point and enter @80<30 <R>
Next point and enter @100<150 <R>

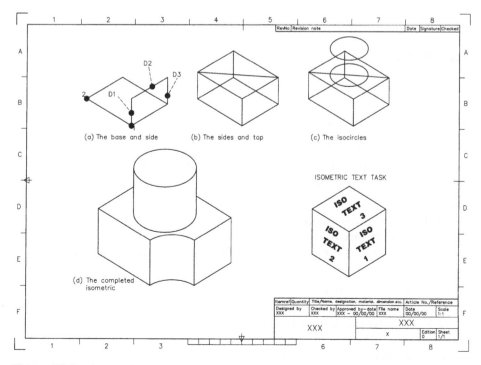

Figure 32.3   Isometric exercise.

Next point and enter @80<−150 <R>
Next point and enter @100<−30 <R>
Next point and enter @50<90 <R>
Next point and enter @80<30 <R>
Next point and enter @50<−90 <R><R> – Figure 32.3(a).

4 With the **COPY icon**:
   a) objects: pick lines D1, D2 and D3 then right-click
   b) base point: endpoint icon and pick point 1
   c) displacement: endpoint icon and pick point 2.

5 Draw the two lines (endpoint–endpoint) to complete the sides and top then draw a top diagonal line – Figure 32.3(b).

6 *a*) toggle to isoplane top
   b) with the **ELLIPSE-Isocircle** command:
      1. midpoint icon and pick diagonal line to position the centre
      2. enter a radius of 30
   c) copy the isocircle:
      1. from the diagonal midpoint
      2. by @50<90 – Figure 32.3(c).

7 *a*) erase the diagonal
   b) draw in the two 'cylinder' sides with quadrant snap, picking the top and bottom isometric circles
   c) trim objects to these lines
   d) erase unwanted objects.

8 Draw two additional 30-radius isometric circles and modify to give the completed model as Figure 32.3(d).

9 *Task*: Text with Isometrics
   a) Adding text to an isometric drawings requires the user to set text styles to suit the appropriate 'face' of the component.
   b) During the text chapter, we created two styles which would suit isometric text.
   c) Refer to the task drawing in Figure 32.3 and:
      1. create three suitable text styles
      2. add text to the three faces on an isometric cube.

Oblique drawings

1 An oblique pictorial is created from the normal x-axis orientation and with the y-axis at 45 degrees to the x-axis.

2 Sizes in the x-axis are full, while the y-axis can be full or half-size.

3 Figure 32.4 displays the basic oblique concepts.

4 AutoCAD does not have the facility to display an oblique grid, but as the y-axis is at 45 degrees, the standard grid can be used as a drawing aid.

5 With oblique drawings:
   a) accuracy is obtained with polar co-ordinates
   b) the oblique component 'front' face is identical to the front orthographic view
   c) as the Y-axis is at 45 degrees to the X-axis, there are always two possible oblique solutions.

6 To demonstrate how an oblique pictorial is obtained:
   a) open you A3PAPER and set a standard 10 grid with 5 snap
   b) refer to Figure 32.5 and using the sizes in Figure 32.5(a), create the front 'face' of the component – Figure 32.5(b)
   c) copy this front face to another part of the screen.

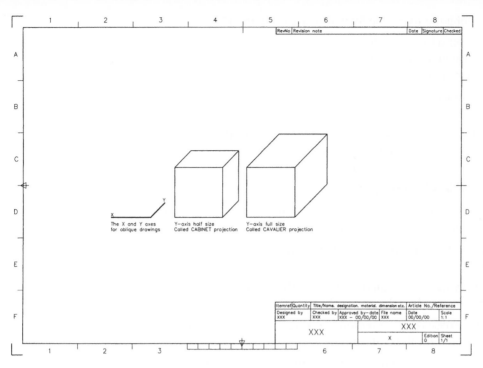

Figure 32.4 Oblique pictorial concepts.

7 The component is 50 'wide', so copy:
   a) one complete front face from an endpoint by @25<45
   b) the other from an endpoint by 25<135 – Figure 32.5(c).

8 Draw in the edge lines then erase and trim unwanted objects to display the complete component – Figure 32.5(d).

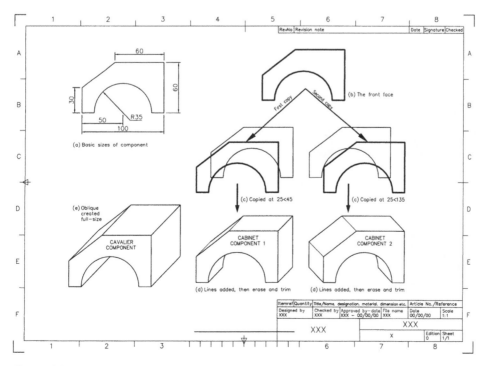

Figure 32.5 Creating an oblique pictorial.

9 Create another oblique using the 50 width size – Figure 32.5(e).

10 Decide for yourself which is better:
   a) the half-size cabinet projection
   b) the full-size cavalier projection.

11 The exercise is now complete.

12 *Notes*:
   a) I have not discussed how to draw an oblique circle in the Y-axis
   b) This is not as easy as creating isometric circles.

Planometric drawings

1 Planometric projections have recently become popular with the introduction of customised CAD systems in kitchen and bedroom design.

2 This type of projection allows a 'birds-eye' view of a complete room, floor layout, etc.
  to the customer.

3 Planometric drawings are created with both the X- and Y-axes at 45 degrees to the
  horizontal and this can be achieved in AutoCAD by modifying the standard grid.

4 The following exercise is rather tedious, so I will let you decide if you want to attempt
  it, although the final result is interesting.

5 Open your A3PAPER and menu bar with **Tools-Drafting Settings** and:

   prompt   Drafting Settings dialogue box
   respond   1. make the Snap and Grid tab active
               2. set a rectangular snap
               3. set the grid X and Y spacing to 5
               4. set the snap X and Y spacing to 2.5
               5. set the angle to 45
               6. pick OK.

6 Refer to Figure 32.6 and zoom in on a small area of the screen.

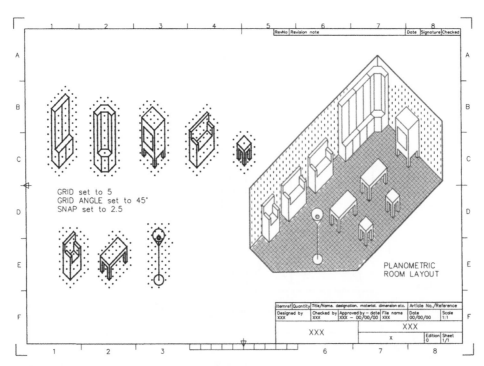

Figure 32.6 The planometric pictorial exercise.

7 Using the grid points displayed as a guide, design some planometric furniture.

8 Design a planometric room layout using your furniture and add hatching to the floor and walls.

9 Save if required.

Assignments

Three pictorial assignments have been included which I have tried to make these interesting.

1 *Activity 46*: *Bicycle spanner*
   a) Draw the two orthographic views, then create both an isometric and an oblique pictorial using the sizes given
   b) The isometric hexagon is interesting to complete.

2 *Activity 47*: *Garden wall*
   a) Design an isometric garden wall block using the basic 100 square × 40 sizes displayed
   b) Scale the wall block by 0.25 then create the wall
   c) I have added some isometric text for effect.

3 *Activity 48*: *Plug*
   a) Using the sizes given, draw the two views of the component
   b) Draw an isometric and oblique pictorial
   c) The larger square is interesting to create, especially as it is not dimensioned. You need to think.

Chapter 33

Model and paper space

1 AutoCAD has two drawing environments:
   a) model space: to draw the component
   b) paper space: to layout the drawing paper for plotting.

2 The two drawing environments are independent of each other and the concept is suitable for both 2D drawings and 3D modelling.

3 Until now:
   a) all work has been completed in model space
   b) we have had the facility to use paper space without being aware of this fact.

4 *Notes*
   a) the user must realise that this chapter is an introduction to the model/paper space concept
   b) paper space is particularly targeted for plotting and as I have no idea what type of plotter the reader has, I have assumed that no plotter is available. This does not affect any of the exercises
   c) when the paper space environment is entered, a new icon will be displayed
   d) the model space and paper space icons are shown in Figure 33.1
   e) the icon display is dependent on whether the UCS icon style has been set to 2D or 3D.

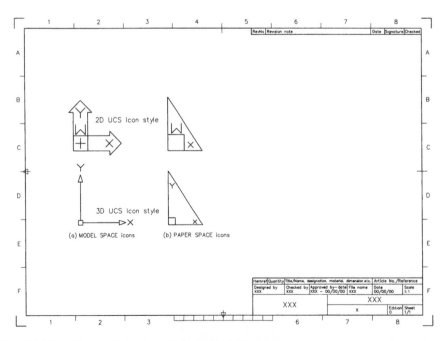

Figure 33.1 The model space and paper space icons.

5 Before paper space can be used, the paper space environment must be activated by:
   a) picking MODEL from the status bar
   b) picking one of the layout tabs from the drawing screen
   c) entering TILEMODE then 0 at the command line.

Investigating the A3PAPER standard sheet

1 The A3PAPER standard sheet was created from an AutoCAD template file and has both the model and paper space environments active but we have only used the model environment for creating all drawing work.

2 To investigate the model and paper concept:
   a) close any existing drawings
   b) open your WORKDRG drawing which was created on A3PAPER
   c) note the 2D model space icon display
   d) note that the drawing area is 'enclosed' within a thick black border
   e) the Layout tab should display two layouts: (a) Model and (b) ISO A3 Title Block.

3 Left-click on MODEL in the Status bar and:
   a) the selected button now displays PAPER
   b) note the 2D paper space icon
   c) the thick black border is not displayed.

4 We have now entered the paper space environment.

5 With layer OUT current:
   a) draw a line from 0,0 to 420,297
   b) this line indicates that the white area displayed is A3 size
   c) try and erase any part of WORKDRG – you cannot.

6 Enter model space by:
   a) entering **MS <R>** at the command line
   b) the PAPER button in the Status bar now displays MODEL
   c) the model space icon is again displayed
   d) try and erase the line drawn in paper space – you cannot
   e) thus any objects created in model space cannot be modified with paper space active and vice-versa.

7 *a*) When working in paper space, viewports are used and a viewport can be defined as:
      'A bounded region within a drawing sheet that displays parts of a model space drawing'
   b) Viewports are generally created on their own layers
   c) The A3PAPER standard sheet has a viewport layer.

8 *a*) The ISO A3 Title Block layout tab should be active
   b) left-click the Model tab and WORKDRG only will be displayed
   c) re-activate the ISO A3 Title Block tab to display the original layout.

Creating a new paper space layout

1 We now want to create a new drawing layout with viewports arranged to our specification, so:
   a) close the existing WORKDRG drawing with no save
   b) open the A3PAPER standard sheet with the ISO A3 Title Block layout tab active (the default).

2   *a*) Make a new layer named SHEET both your own colour selection and with continuous linetype.
    b) make layer viewport active.

3   Menu bar with **Tools-Wizards-Create Layout** and:

    prompt   Create Layout - Begin dialogue box
    respond   **alter Layout1 name to MYNEWA3 then pick Next**
    prompt   Create Layout - Printer dialogue box
    respond   **pick None then Next**
    prompt   Create Layout - Paper Size dialogue box
    respond   *a*) scroll at paper size and pick ISO A3 (420.00 × 297.00MM)
                b) paper space drawing units: Millimetres
                c) pick Next.
    prompt   Create Layout - Orientation dialogue box
    respond   **pick Landscape then Next**
    prompt   Create Layout - Title Block dialogue box
    respond   **pick None then Next**
    prompt   Create Layout - Define Viewports dialogue box
    respond   **pick None then Next**
    prompt   Create Layout - Finish dialogue box
    respond   **pick Finish**.

4   The drawing screen will be returned with:
    a) a white area – the drawing paper
    b) a dotted line area – the plottable area
    c) a new tab – MYNEWA3
    d) the paper space icon displayed (relative to your UCS style icon)
    e) PAPER is displayed in the Status bar.

5   As the paper space icon is displayed, the user is 'in paper space'.

The sheet layout

1   The new MYNEWA3 layout has no viewports so refer to Figure 33.2 and:
    a) paper space active with layer viewport current
    b) menu bar with **View-Viewports-1 Viewport** and:
    prompt   Specify corner of viewport and enter **10,10 <R>**
    prompt   Specify opposite corner and enter **210,145 <R>**
    and   a rectangular viewport is created.

2   Menu bar with **View-Viewports-1 Viewport** selection and create another three viewports using:

	First	*Second*	*Third*
corner:	220,75	190,180	20,165
opposite corner:	370,170	280,250	170,270

3   *a*) make layer SHEET current
    b) with the RECTANGLE command draw a rectangle with the first corner point at 0,0 and the other corner point at 380,282

4   We have now:
    a) created a new layout tab MYNEWA3
    b) customised a paper space drawing sheet with four rectangular viewports
    c) added a sheet border to our drawing paper
    d) have no drawing displayed.

5   Save this layout with a suitable name – you may want to use it again.

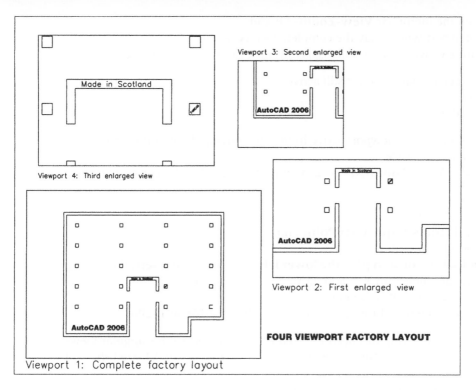

Figure 33.2 The MYNEWA3 factory layout.

The drawing

1   *a)* Rather than draw a new component we will insert an already completed and saved (I hope) drawing into our created paper space MYNEWA3 layout
    b) the drawing selected for this purpose is the large scale factory layout (LARGESC) from Chapter 26
    c) still refer to Figure 33.2.

2   *a)* enter model space with a left-click on PAPER in the Status bar
    b) the traditional model space icon will be displayed in the viewports.

3   Make the lower left viewport active by:
    a) moving the pointer into the largest rectangular area
    b) left-click
    c) cursor cross-hairs displayed and the viewport border is highlighted.

4   Make layer OUT current then menu bar with **Insert-Block** and:
    prompt   Insert dialogue box
    respond   *a)* pick Browse
                  b) navigate to your C:\MYCAD folder if it is not current
                  c) navigate and pick LARGESC (or your entered name)
                  d) pick Open.
    prompt   Insert dialogue box with Name: LARGESC
    respond   *a)* ensure Specify On-screen active, i.e. ticks
                  b) ensure Explode active, i.e. tick
                  c) pick OK.
    prompt   Specify insertion point for block and enter **0,0 <R>**
    prompt   Specify scale factor for XYZ axes and enter **1 <R>**
    prompt   Specify rotation angle and enter **0 <R>**.

5   Now select from the menu bar **View-Zoom-All** and
    a) the active viewport will display the complete factory layout
    b) the other three viewports may still display nothing or perhaps some lines.

6   With the large rectangular viewport still active:
    a) erase all text
    b) freeze layer DIMS.

7   *a*) Make each of the other viewport active in turn by moving the pointing arrow into
        the viewport and left-click
    b) menu bar with **View-Zoom-All** in each viewport to display the complete factory
        layout.

8   With the lower left viewport active:
    a) menu bar with **Tools-Inquiry-ID Point** and:
    prompt   Specify point
    respond   **Endpoint icon and pick the lower left vertex on the inserted drawing**
    and   command line with X = 800.00 Y = 3200.00 Z = 0.00.
    b) these were the co-ordinates used to start the large-scale drawing
    c) identify the co-ordinates of the top right vertex. They are 15800,14200.

9   *a*) Thus a very large drawing has been inserted into a paper space rectangular view-
        port of size 200 × 135 and this viewport has been created on an A3 sized sheet of
        paper
    b) this is the power of using model-paper space.

Using the viewports

1   *a*) With the lower left viewport active, menu bar with **View-Zoom-Extents**
    b) the factory layout will 'fill the viewport'
    c) menu bar with **View-Zoom-Scale** and:
    prompt   Enter a scale factor and enter **0.02 <R>**
    and   the factory layout is displayed in the centre of the viewport.

2   With the lower right viewport active select **View-Zoom-Window** from the menu
    bar and:
    prompt   Specify first corner and enter **4000,1600 <R>**
    prompt   Specify opposite corner and enter **13000,9000 <R>**.

3   With the top right viewport active, zoom a window from 500,2500 to 9000,9000.

4   Make layer TEXT current and the lower left viewport active then select from menu
    bar **Draw-Text-Single Line Text** and create an item of text using:
    style: ST3
    start point: right-justified on the point 6500,3750
    height: 450
    rotation: 0
    item of text: AutoCAD 2006
    and: the entered item of text will be displayed in all viewports.

5   Still with layer text current, add another two items of text using:

	First text	*Second text*
style	ST1	ST2
start point	middled on 10450,7650	7250,8250
height	50	150
rotation	45	0
item	FARCAD	Made in Scotland.

6 With the top left viewport active, zoom a window from 6000,7250 to 11000,8500.

7 Make layer OUT current and the lower right viewport active.

8 Draw two lines:
a) first point at 5750,6250, next point at 12000,6250
b) first point at 5750,7250, next point @0,8000.

9 a) use the TRIM command to trim these lines for an opening into the factory – displayed in all viewports
b) you may have to zoom in on the wall area to complete the trim, but remember to zoom previous.

Using the paper space environment

1 Enter paper space and with layer TEXT current, add the following text items to the layout:

Style	Start	Ht	Rot	Text item
a) ST1	10,2	6	0	Viewport 1: Complete factory layout
b) ST1	220,65	5	0	Viewport 2: First enlarged view
c) ST1	190,255	4	0	Viewport 3: Second enlarged view
d) ST1	20,155	4	0	Viewport 4: Third enlarged view
e) ST3	215,25	5	0	FOUR VIEWPORT FACTORY LAYOUT.

2 Save the drawing as the exercise is now complete.

Model-Paper space example 2

1 The previous exercise created the viewports prior to 'inserting' a drawing into the layout.

2 In this exercise we will:
a) open a previously created drawing
b) modify the paper space layout
c) investigate model and paper space dimensioning.

3 Open WORKDRG (again) and erase any text and dimensions to display:
a) the original red outline and two red circles
b) four green centre lines.

4 Enter paper space and:
a) make layer viewports current
b) freeze the four FRAME layers and the Title Block layer
c) erase the displayed viewport and WORKDRG will 'disappear'
d) thaw the five frozen layers.

5 We now want to create three rectangular viewports within the displayed drawing area, so:
a) menu bar with **View-Viewports-1 Viewport** and:
prompt   Specify corner of viewport and enter **15,15 <R>**
prompt   Specify opposite corner and enter **235,285 <R>**
and   a viewport is created between the entered points with WORKDRG displayed.
b) menu bar with **View-Viewports-Named Viewports** and:
prompt   Viewports dialogue box
respond   1. select the New Viewports tab
            2. at Standard viewports, pick Two: Horizontal

> 3. Viewport Spacing: 0.00
> 4. Setup: 2D
> 5. Change view to: 'Current' – Figure 33.3
> 6. pick OK

prompt   `Specify first corner or [Fit]`

enter   **235,50 <R>**

prompt   `Specify opposite corner` and enter **405,275 <R>**

and   two new viewports are created at the right of the drawing area with WORKDRG displayed.

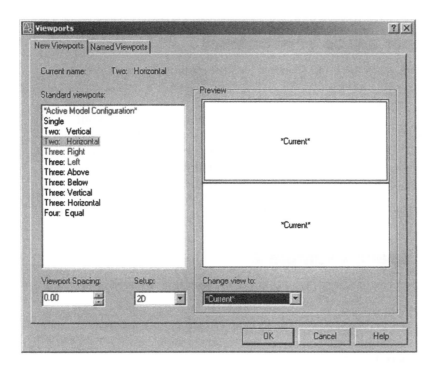

Figure 33.3 The Viewports dialogue box.

6   Enter model space with **MS <R>** and:
    a) make layer OUT current and the large left viewport active
    b) draw a hexagon, centred on 130,202.5 and inscribed in a 5.5 radius circle
    c) add an item of text, middle justified on 130,202.5, style ST3 with height 3 and 0 rotation, the text item being CAD (remember to use the correct layer)
    d) full circle polar array the hexagon and text item about the point 130,185 for 5 items with rotation.

7   Still with the left viewport active, make layer text current and add the following text:
    a) style ST2 and centred on the point 130,225
    b) height 8 and rotation 0
    c) text item: **AutoCAD** then **2006**.

8   At this stage the three viewports display the same 'full size' view of WORKDRG.

9   *a*) With the left viewport active, zoom-extents then zoom to a scale of 1.75
    b) activate the top right viewport and zoom-window the circle with the hexagonals
    c) in the lower right viewport, zoom-all then zoom-window the 'top part' of WORKDRG

d) your drawing layout should resemble Figure 33.4, i.e. we have displayed a full size WORKDRG and zoomed in to two different areas of the drawing

e) the term of the drawing created is a multi-viewport layout.

Dimensioning in model and paper space

1. When a multi-viewport layout has been created in paper space, many users are unsure whether dimensions should be added in model space or paper space.

2. We will investigate both concept with our existing screen layout.

3. Model space with the left viewport active and layer DIMS current.

4. Refer to the Figure 33.4 and diameter dimension the circle and linear dimension as indicated.

5. *a)* The dimensions added are model space dimensions and are GLOBAL, i.e. displayed in all viewports
   b) these dimensions are represented by in Figure 33.4 by (a)
   c) this can be one of the major 'problems' with multi-viewport layouts.

6. Enter paper space with **PS <R>** or select MODEL from the Status bar

7. At the command line enter **DIMASSOC <R>** and:
   prompt   Enter new value for DIMASSOC
   enter   **1 <R>**.

8. *a)* Dimension the same two objects as before, i.e. a linear and a diameter
   b) being in paper space, the dimensions are only displayed 'where picked'
   c) the two dimensions are represented in Figure 33.4 by (b)
   d) they are both obviously wrong. The circle diameter should be 50 and not 45.83.

9. At the command line enter **DIMASSOC <R>** and:
   prompt   Enter new value for DIMASSOC
   enter   **2 <R>**.

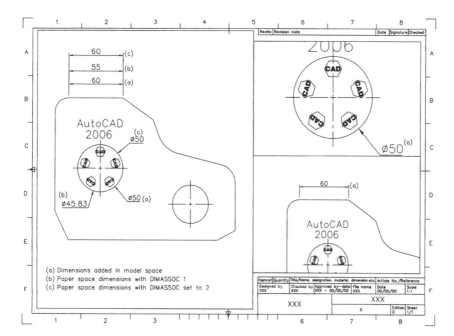

Figure 33.4   Creating a multi-viewport layout and adding dimensions.

10 *a*) Now dimension the same two objects as before, represented in Figure 33.4 by (c)
    b) these dimensions are correct.

11 *a*) Correct paper space dimensioning is thus obtained with the system variable DIMASSOC set to 2
    b) this should be the actual default value.

12 Right-click ISO A3 Title Block from the layout tabs and:
prompt	shortcut menu
respond	pick Rename
prompt	Rename Layout dialogue box
respond	*a*) alter name to: WORKDRG
	b) pick OK
and	the tab displays the entered name.

13 The exercise is now complete and can be saved, but not as WORKDRG.
    a) question: with multi-viewport layouts, will you dimension in model space or paper space?

Final thoughts

1 Paper space layouts should be used at all times?

2 Dimensions should always be added to a paper space layout?

3 Will you dimension in model space or paper space?

   I will leave these thoughts for the user to think about!

Chapter 34

Templates, standards and sheets

These three topics will be separately discussed in this chapter.

Templates

1 We have been using templates for our exercises and activities without discussing the concept in detail.

2 My reasons for this were to allow the user to:
   a) become proficient at draughting with AutoCAD
   b) understand attributes
   c) understand the concepts of paper space layouts.

What is a template?

1 A template is a prototype drawing, i.e. it is similar to our A3PAPER standard sheet which has been used when every new exercise/activity has been started.

2 The term prototype/standard drawing refers to a drawing which has various default settings, e.g. layers, text styles, dimension styles, etc. The drawing has thus been customised to company/user requirements.

3 With AutoCAD, all drawings are saved with the file extension **.dwg** while template files have the extension **.dwt**.

4 Any drawing can be saved with the .DWG or .DWT extensions.

5 Template drawings (files) are used to 'safeguard' the prototype drawing being mistakenly overwritten.

6 *a*) AutoCAD has templates which conform to several drawing conventions.
   b) the drawing conventions available are: ANSI, DIN, Gb, ISO and JIS
   c) all templates are available in several paper sizes
   d) other template drawings are also available
   e) template files can be opened from the Startup dialogue box.

7 We will investigate:
   a) completing and saving a drawing using an opened template file
   b) creating our own template file.

Investigating the ISO A3 AutoCAD template file

1 The ISO A3 template file has been saved as our A3PAPER standard/prototype drawing file.

2 *a*) We have modified this file, e.g. layers, text styles, dimension style have been customised to our own requirements.
   b) We have not yet investigated the title block of this file.

3 Open your A3PAPER drawing and make paper space active.

4 The ISO A3 template consists of:
   a) a white A3 sized drawing sheet bounded by four corner edges
   b) a black viewport which is highlighted when model space is active
   c) a dotted rectangle for guidance with plotting
   d) a title block with various attributes
   e) drawing paper grid reference markers.

5 *The title block*:
   a) when opened, the title block is displayed with the attribute default values as Figure 34.1(a)
   b) in paper space, menu bar with **Modify-Explode** and:

prompt	Select objects
respond	**pick any object of the title block then <R>**
and	title block displayed with attribute tags as Figure 34.1(b).

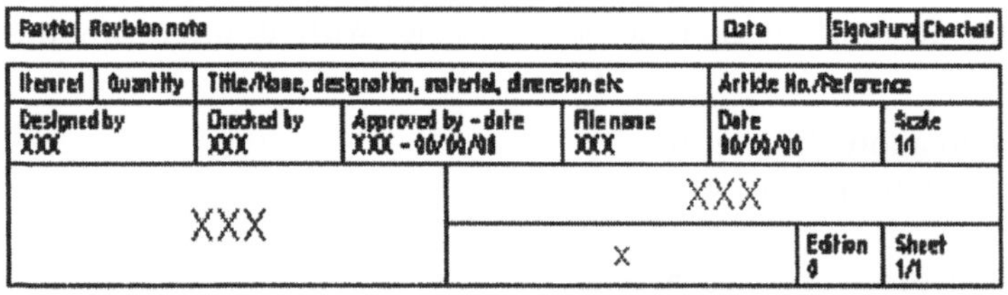

(a)The title block when the ISO A3 file is opened

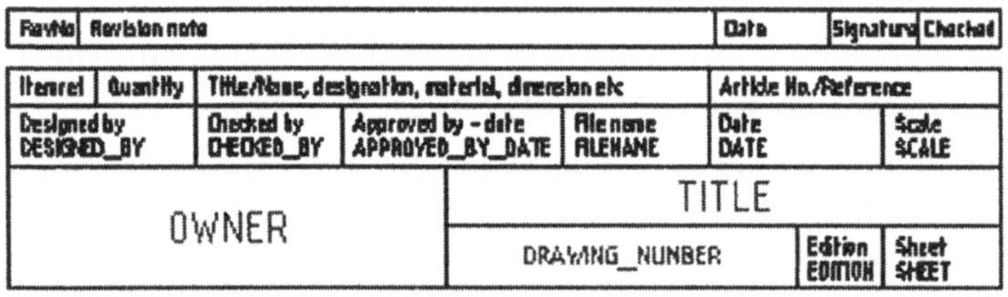

(b)The title block after the explode command

Figure 34.1 The ISO A3 file title block.

4 *The title block attributes*:
   a) with paper space still active, enter **ATTEDIT <R>** at the command line and:

prompt	Select block reference
respond	**pick any word in the title block**
prompt	That object is not a block
respond	**press ESC**.

   b) the reason for the 'not a block' prompt is that we have exploded the title block.
   c) as there is no way to 'unexplode a block' we would need to close the A3PAPER drawing (no save) then re-open it again.
   d) with the re-opened A3PAPER file displayed and paper space active, enter **ATTEDIT <R>** and:

prompt	Select block reference
respond	**pick any X in the title block**
prompt	Edit Attributes dialogue box
with	11 attributes which can be edited by the user (need to pick Next) e.g.:

              1. attribute 1: File name, default XXX
              2. attribute 2: Drawing number, default X

       3. attribute 10: Edition, default 0

       4. attribute 11: Sheet, default 1/1

respond   study the dialogue box (including Next)

then   **pick Cancel at present**.

e) Figure 34.2 displays the:

  1. edit Attribute dialogue boxes combined into one display

  2. 11 attributes defaults replaced by numbers 1–11 for reference – Figure 34.2(a)

  3. actual title block with the corresponding eleven numbered attributes – Figure 34.2(b).

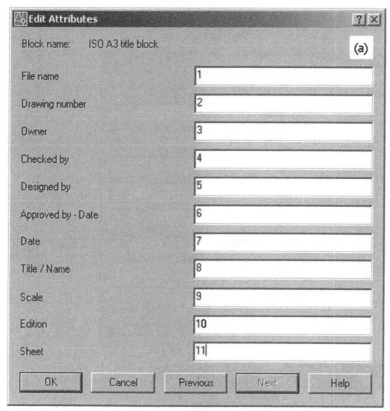

Figure 34.2   The Edit Attribute dialogue box and ISO A3 Title Block with numbered attributes.

5   The ISO A3 layers

   a) the original ISO A3 template file had seven layers, these being:

layer name	*colour*	*use*
layer 0	white/black	all drawings have this layer
FRAME 025	white/black	attributes: 1,4,5,6,7,9,10,11 as Figure 34.2
FRAME 050	magenta	attributes: 2,3,8 and paper grid markers
FRAME 070	blue	title block lines and sheet border
FRAME 200	colour 8	A3 sheet corner markers
Title block	white/black	for the inserted block
Viewport	white/black	for drawing

   b) we added six additional layers (OUT, CL, etc.) and saved as A3PAPER.

6 Editing the attributes: two methods available:
   a) with paper space still active, enter **ATTEDIT <R>** at the command line and:

prompt	Select block reference
respond	**pick any part of the title block**
prompt	Edit Attributes dialogue box as described in step 4(d)
respond	1. Owner tag: alter the XXX value to **FARCAD**
	2. Drawing number tag: alter the X value to **Standard sheet**
	3. Title/Name tag: alter the XXX value to **A3PAPER** – Figure 34.3(a)
	4. pick OK
and	title block displayed with the three entered attribute values as Figure 34.3(b).

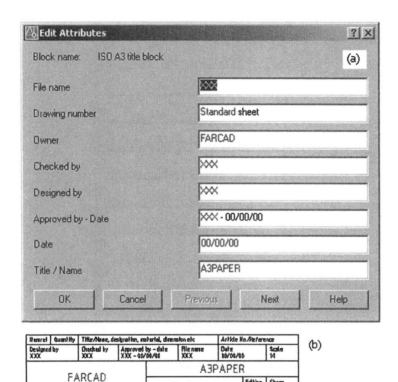

Figure 34.3 Editing three of the title block attributes using ATTEDIT.

   b) from the menu bar select **Modify-Object-Attribute-Single** and:

prompt	Select a block
respond	**pick any XXX text item in the title block**
prompt	Enhanced Attribute Editor dialogue box
with	1. Tab options: Attribute, Text Options and Properties
	2. Attribute tab active with Tag, Prompt and Value details
and	the three previously entered attribute values displayed
respond	1. resize the dialogue box to display the 11 attributes, FILENAME-SHEET
	2. drag the tag and prompt division lines to display the complete details
	3. pick the FILENAME line (highlighted) and alter the XXX value to **R2006/AB/01**
	4. alter the other tag values using:

tag	value
CHECKED_BY	RMF
DESIGNED_BY	HTC

APPROVED_BY_DATE	1/2/3
DATE	02-03-04
TITLE	STEAM EXPANSION BOX
SCALE	1:1
EDITION	REV 5
SHEET	1 of 5 – Figure 34.4(a)

> 5. pick Apply then OK
>
> *and*   the title block will be displayed with the entered values.

7   *Modifying the entered attributes*:

     a) the entered attribute values can be modified to user requirements

     b) repeat the menu bar selection of **Modify-Object-Attribute-Single**, select a text item from the title block and:

> *prompt*   Enhanced Attribute Editor dialogue box
>
> *respond*   1. pick the OWNER line and it highlights
>
>   2. pick the Tab Options tab and alter:
>
>   *a*) Text Style: scroll and pick ST3
>
>   *b*) Height: alter to 10
>
>   *c*) Oblique Angle: alter to 10
>
>   3. pick the Properties tab and alter Colour to Red
>
>   4. pick Apply then OK
>
> *and*   the title block will be displayed with the entered modifications – Figure 34.4(b).

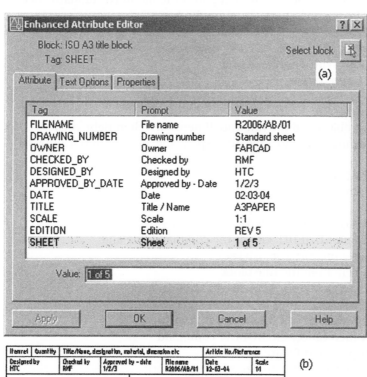

Figure 34.4   The Enhanced Editor dialogue box and the title block with all the edited attributes.

8 *Altering the layout tab*:
   a) the layout tabs should display Model and ISO A3 Title Block
   b) right-click on ISO A3 Title Block from the Layout tabs and:

prompt	Shortcut menu
respond	**pick Rename**
prompt	Rename Layout dialogue box
respond	1. alter name to A3CUSTOMISED
	2. pick OK.

9 Saving the modified standard sheet
   a) save your title block and layout tab medications as both a drawing file and a template file using the name **C:\named folder\A3PAPERMOD**
   b) we now have:
   1. an A3PAPER standard sheet template and drawing file with the original ISO A3 title block
   2. a customised A3PAPERMOD standard sheet template and drawing file.

Using the modified standard sheet

1 All drawing work has been completed with the A3PAPER standard sheet.

2 The title block has not been modified during the exercises and activities.

3 The modified standard sheet can now be used for all future drawing work and the title box altered to suit customer needs as required.

4 As an exercise, we will 'insert' an activity drawing into the modified standard sheet. The method used differs from previous insertion methods so:
   a) open your modified standard sheet A3PAPERMOD with model space active
   b) open any activity drawing (I selected ACT 23) with model space active
   c) menu bar with Edit-Copy with Base Point and:

prompt	Specify base point
enter	0,0 <R>
prompt	Select objects
respond	window the object and the title (no dimensions) then right-click

   d) close the activity drawing with no save to display the opened A3PAPERMOD drawing
   e) menu bar with Edit-Paste as Block and:

prompt	Specify insertion point
enter	0,0 <R>
and	the copied objects are 'inserted' into the drawing

   f) enter paper space and enter DDEDIT <R> at the command line and:

prompt	Select an annotation object
respond	pick a text item from the title block
prompt	Enhanced Attribute Editor dialogue box
with	entered values
respond	1. alter the values to suit the 'inserted' drawing
	2. pick OK then cancel the DDEDIT command
and	drawing displayed as Figure 34.5.

   g) Activity 23 was 'copied-and-pasted' into AutoCAD as a block
   1. is it a drawing and can it be exploded?
   2. does it have all the properties of a 'real' drawing?

This completes the template exercises.

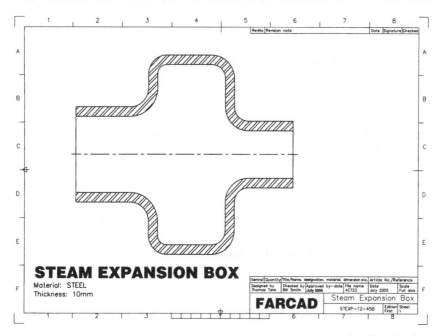

Figure 34.5 Using A3PAPERMOD with an activity drawing.

CAD standards

1 CAD standards allow the user to create a file that defines certain properties, these being layers, dimension styles, text styles and linetypes.

2 This file is saved as a drawing standards file with the extension **.dws**.

3 When a standards drawing has been saved it can be compared with other saved standards drawings and any differences in the four properties will be highlighted.

4 As all of our exercises have been saved with the .dwg extension, it will be necessary to create a standards drawing.

5 We will use our A3PAPER standard sheet for this purpose and to demonstrate how CAD standards drawings are used, we will:
   a) save our original A3PAPER drawing as a .dws file
   b) modify some of the properties in the A3PAPER drawing
   c) save this modified drawing as another .dws file
   d) compare the original standards file with the modified file.

Getting started

1 Close any existing drawings then open your A3PAPER (original) standard sheet.

2 Menu bar with **File-Save As** and from the Save Drawing As dialogue box:
   a) file type: scroll-pick **AutoCAD Drawing Standards (*.dws)**
   b) file name: enter A3PAPER (if not already the default)
   c) save in: MYCAD folder
   d) pick Save.

3 The original A3PAPER drawing will still be displayed and will now be modified.

4 Menu bar with **Format-Layers** and from the Layers Properties Manager dialogue box, alter the following layer properties:

Layer name	Property to alter	New property
CONS	linetype	PHANTOM2 (may need loaded)
TEXT	colour	green

DIMS   lineweight   0.3
TEST   new layer with defaults

5   Menu bar with **Format-Text Style** and using the Text Style dialogue box, alter the following:

Style name   *Font name*
Standard   gothice.shx

6   Menu bar with **Dimension-Style** and using the Dimension Style Manager dialogue box:
   a) highlight the A3DIM style
   b) right click and pick rename
   c) alter name to DIMNEW
   d) pick Set Current then close the dialogue box.

7   With these modifications to the A3PAPER standard sheet, menu bar with **File-Save As** and:
   a) file type: scroll-pick **AutoCAD Drawing Standards (*.dws)**
   b) file name: enter TRIAL
   c) save in: scroll and pick MYCAD folder
   d) pick Save.

Comparing standards

1   Close all existing drawings then menu bar with **File-Open** and:
   prompt   Select File dialogue box
   respond   *a*) file type: Standards (*.dws)
             b) look in: named folder (MYCAD)
             c) pick TRIAL then Open.

2   The TRIAL standards file will be displayed (check that this file has the modified layers).

3   To compare 'standards' files, menu bar with **Tools-CAD Standards-Configure** and:
   prompt   Configure Standards dialogue box
   with   Two tabs: Standards and Plug-ins
   respond   *a*) ensure Standards tab active
             b) pick +;
   prompt   Select Standards File dialogue box
   respond   *a*) ensure your named folder (MYCAD) is current
             b) file type: Standard (*.dws)
             c) pick A3PAPER then Open

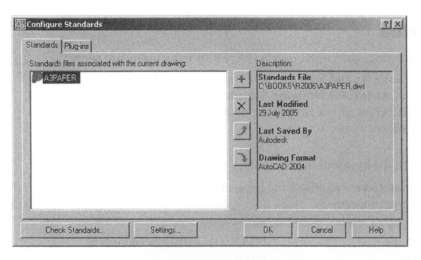

Figure 34.6   The Configure Standards dialogue box for A3PAPER.dws.

prompt	Configure Standards dialogue box.
with	*a*) Standards tab active
	b) A3PAPER listed
	c) Description of File, Last Modified, Format
	d) dialogue box as Figure 34.6 and realise that my display will differ from your display
respond	**pick Check Standards**
prompt	Check Standards dialogue box
with	*a*) Problem: Dimstyle 'DIMNEW'
	Name is non-standard
	b) Replace with:
	1. A3DIM Dimstyle from A3PAPER Standards File
	2. perhaps other named dimension styles
respond	**pick A3DIM then Fix**, i.e. we are replacing the DIMNEW dimension style with the A3DIM style from the A3PAPER standards file
then	Check Standards dialogue box
with	*a*) Problem:
	Layer 'DIMS'
	Properties are non-standard
	b) Replace with:
	1. DIMS layer from A3PAPER Standards file (highlighted and ticked)
	2. perhaps other named dimension styles
	c) Preview of changes information – Figure 34.7

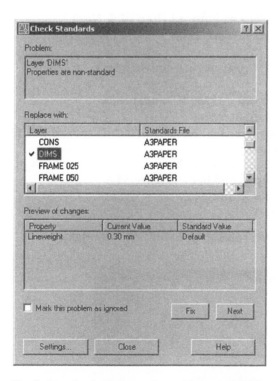

Figure 34.7   The Check Standards dialogue box with layer 'DIMS' problem displayed.

respond	**pick DIMS then Fix,** i.e. we are replacing the 0.30 mm lineweight value on layer DIMS with the default value from the A3PAPER standards file
then	Check Standards dialogue box
with	*a*) Problem:
	Layer 'TEXT': Properties are non-standard

b) Replace with:

   TEXT layer from A3PAPER standards file (highlighted and active)

c) Preview of changes information

respond **pick TEXT then Fix,** i.e. we are replacing the green colour on layer TEXT with the Blue default from A3PAPER

then `Check Standards dialogue box`

with *a*) Problem:

   Layer 'CONS': Properties are non-standard

b) Replace with:

   CONS layer from A3PAPER standards file

c) Preview of changes information

respond **pick CONS then Fix,** i.e. we are replacing the PHANTOM2 linetype on layer CONS with the Continuous default from the A3PAPER standards file

then `Check Standards dialogue box`

with *a*) Problem:

   Layer 'TEST': Properties are non-standard

b) Replace with:

   layer names from A3PAPER

c) Preview is blank as TEST is a layer in the TRAIL file only

respond **pick Mark this problem as ignored then pick Next**

then `Check Standards dialogue box`

with *a*) Problem: Linetype 'PHANTOM2': Properties are non-standard

b) Replace with: linetype names from A3PAPER

c) Preview is blank

respond **pick Mark this problem as ignored then pick Next**

then `Check Standards dialogue box`

with *a*) Problem:

   Textstyle 'Standard': Properties are non-standard

b) Replace with:

   Standard text style from A3PAPER standards file

c) Preview of changes information

respond **pick Standard then Fix,** i.e. we are replacing gothice.shx font with A3PAPER Standards font

then `Checking Complete dialogue box`

with details about the problems and fixes – Figure 34.8

respond *a*) read the dialogue box information

   b) pick OK then close.

This CAD standards exercise is now complete. Do not save any changes.

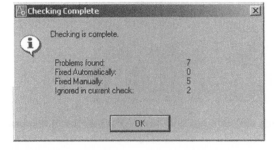

Figure 34.8   The Checking Complete message box.

Sheet Sets

1   Any major drawing project consists of a series of drawings/layouts and organising these into a 'usable order' can be very time-consuming.

2 AutoCAD 2006 allows multiple drawing files to be organised into a sheet set, this being achieved with the Sheet Set Manager.

3 New sheets can be created from 'scratch' and added to the existing sheet set and existing drawing files can be imported into the sheet set.

4 In simple terms:
   a) **a sheet set is a named collection of drawing sheets**
   b) the Sheet Set Manager allows the user to organise, display and manage sheet sets
   c) **each sheet in a sheet set is a layout in a drawing (DWG) file**.

The Sheet Set Manager

1 To create a sheet set from our existing drawings open your A3PAPER drawing file.

2 Menu bar with **Tools-Sheet Set Manager** and:
   prompt   Sheet Set Manager palette
   with   *a*) Sheet List control at top
               b) Several buttons which vary depending on tab selected
               c) Three tabs: Sheet List, View List and Resource Drawing
   and   Figure 34.9 displays the basic Sheet Set Manager palette details

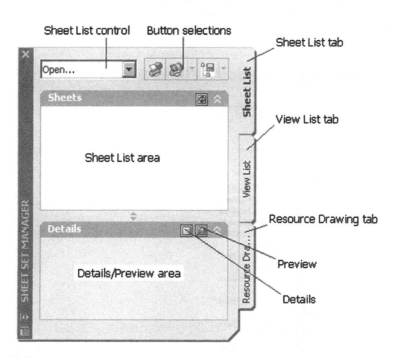

Figure 34.9 The basic Sheet Set Manager palette.

   respond   *a*) scroll at the sheet list control
               b) pick New Sheet Set
   prompt   Create Sheet Set - Begin dialogue box
   respond   *a*) Existing drawings active
               b) pick Next
   prompt   Create Sheet Set - Sheet Set Details dialogue box
   respond   *a*) Name of new sheet set: enter **MINE-1**
               b) Description: enter to suit, e.g. **MY FIRST ATTEMPT USING DRAW-
                  INGS FROM MY FOLDER**
               c) Store sheet set data file (.dst) here:
                  1. ensure named folder listed

      2. it should be if you have opened your A3PAPER from your named folder

      3. if it is not listed then navigate and select it.

   d) Read the displayed note, it may be useful

   e) pick Next

prompt	`Create Sheet Set - Choose Layouts dialogue box`
respond	**pick Import Options**
prompt	`Import Options dialogue box`
respond	**all options active then pick OK**
prompt	`Create Sheet Set - Choose Layouts dialogue box`
respond	**pick Browse**
prompt	`Browse for Folder dialogue box`
respond	*a*) navigate to your named folder
	b) pick OK
prompt	`Create Sheet Set - Choose Layouts dialogue box`
with	*a*) hierarchy of all drawings with layouts displayed
	b) active boxes (tick) at the folder, all drawing files and layouts – Figure 34.10
respond	pick Next
prompt	`Create Sheet Set - Confirm dialogue box`
with	hierarchy structure of the MINE-1 sheet set
respond	**pick Finish**

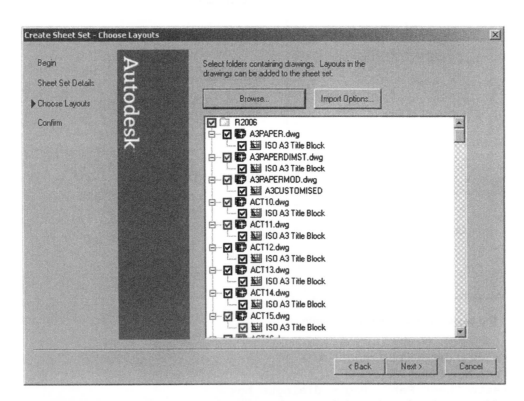

Figure 34.10 The Create Sheet Set (Choose Layouts) dialogue box after the named folder selection.

3   The sheet set palette will be displayed with the listed layout data.

4   Figure 34.11 displays the sheet set palette with a selected file and:
    a) details active
    b) preview active.

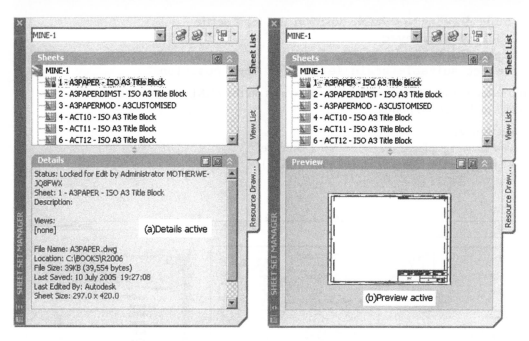

Figure 34.11 The sheet set palette with the drawing list.

5 Now cancel the Sheet Set Manager palette and close A3PAPER with no changes but with AutoCAD still active.

Adding a View List to the palette

1 AutoCAD still active, so menu bar with **File-Open Sheet Set** and:
  prompt   `Open Sheet Set dialogue box`
  respond   *a*) navigate to your named folder (if not current)
             b) pick MINE-1
             c) pick Open
  prompt   `Sheet Set Manager palette with MINE-1 active`
  respond   *a*) ensure Sheet List tab active
             b) pick A3PAPER – ISO A3 Title Block, Details
  and   Figure 34.11 displays the Sheet Set palette for Details and Preview
  then   *a*) right-click on A3PAPER – ISO A3 Title Block
             b) pick Open from the shortcut menu
  and   the file is opened with appropriate layout tab active.

2 Ensure that paper space is active.

3 Pick the **Resource Drawings tab** from the palette and:
  prompt   `Sheet Set palette`
  respond   **pick Add New Location** (two left-clicks or right-click and select)
  prompt   `Browse for Folder dialogue box`
  respond   *a*) named folder active
             b) pick Open
  prompt   `Sheet Set Manager palette displayed with list of draw-`
            `ings from named folder`
  respond   *a*) scroll and select any activity drawing
             b) right-click
             c) from shortcut menu select Place on Sheet
  and   selected activity drawing displayed attached to pointer
  prompt   `Specify insertion point at command line`

> *respond* **move drawing to suitable position and let it settle**
> *prompt* `tooltip displayed`
> *respond* **right-click to display scales**
> *prompt* `select suitable scale (e.g. 1:4) then position.`

4 Repeat step 3 and place another four activity drawings on the MINE-1 sheet selecting suitable scales and insertion points.

5. Figure 34.12 displays:
   a) five activity drawings placed on the MINE-1 sheet set
   b) the View List tab of the Sheet Set palette for MINE-1.

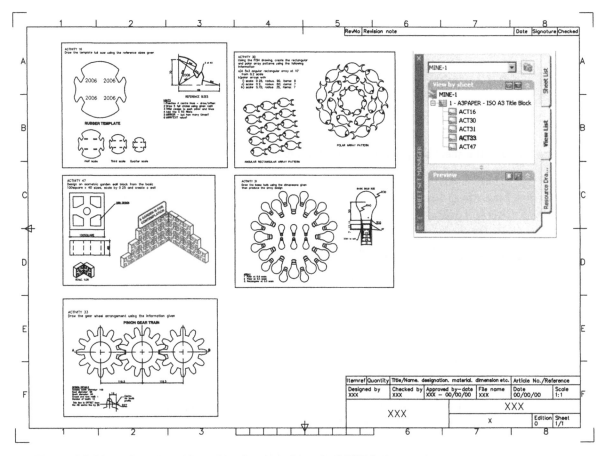

Figure 34.12   A3PAPER with activity drawings from the MINE-1 sheet set.

6 From the View List tab with a right-click the user has access to several options including:
   a) display any drawing
   b) rename and renumber the named drawing list.

Notes

1 The sheet set demonstration is now complete.

2 While this has been a brief introduction to sheet sets, the user should realise that a powerful drawing management tool is available.

3 Large design projects have many drawing sheets and these can be integrated into a few sheets sets.

4 The View List option allows the user to display views from several different drawings.

Chapter 35

Toolbars and tool palettes

1 Toolbars have been used since the first line and circle were drawn and most users will probably be more than satisfied with the existing toolbar icon buttons.

2 It is possible to customise toolbars to user requirements thereby improving drawing productivity.

3 Tool palettes can also be customised for user blocks and hatch patterns.

Creating a new toolbar

1 Open your A3PAPER file (drawing or template) with model space active.

2 Select from the menu bar **Tools-Customize-Interface** and:

   prompt   Customize User Interface dialogue box (may take a few seconds to open)

   with   *a*) two tab selections: Customise and Transfer

            b) three 'sections':

               1. Customizations in All CUI Files: with hierarchical tree structure

               2. Command List: with categories and navigation ability

               3. Properties: with a general comment area

   respond *a*) right-click on Toolbars

            b) **pick New** then pick Toolbar

   and   Toolbars list is expanded with new Toolbar1 added – Figure 35.1(a)

   respond right-click the new Toolbar1 name

   prompt   shortcut menu

   respond *a*) pick Rename

            b) enter MYTBAR then <R>

   and   new toolbar displayed with entered name.

3 To add command icons to the new toolbar, scroll at Command List Categories, select Draw and all AutoCAD draw commands listed in alphabetical order.

   respond *a*) scroll and click Line

            b) hold down the left button and drag the command to below the MYT-BAR name

            c) when a black separator line is displayed, release the left button

   and   Line command added to the new MYTBAR toolbar

   respond *a*) at command list, scroll and pick Center,Radius

            b) hold down the left button and drag the command to below the MYT-BAR name

            c) when a black separator line is displayed, release the left button

   and   the circle command (Center,Radius) added to the new MYTBAR toolbar.

4 Repeat the select and drag operations for the Polygon and Polyline draw commands.

   and   when complete, the CUI tree hierarchy should be displayed as Figure 35.1(b).

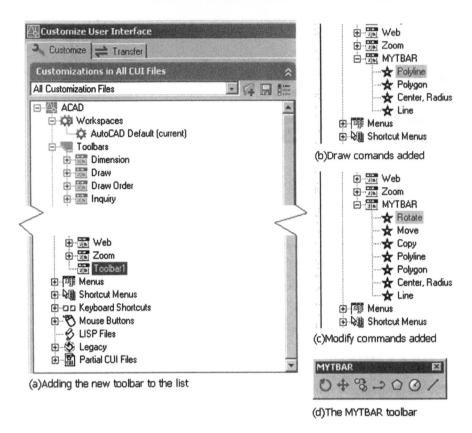

(a)Adding the new toolbar to the list

(b)Draw comands added

(c)Modify commands added

(d)The MYTBAR toolbar

Figure 35.1   Creating the new toolbar MYTBAR using the CUI dialogue box.

5   At Command List Categories, scroll and select Modify
respond   *a*) pick and drag Copy, Move and Rotate to your new toolbar – Figure 35.1(c)
             b) pick OK from the dialogue box
and   your new toolbar will be displayed as Figure 35.1(d).

Adding a fly-out

1   Right-click in any displayed toolbar and:
prompt   shortcut menu
respond   **pick Customize**
and   Customize User Interface dialogue box displayed.

2   Right-click the MYTBAR name and:
prompt   shortcut menu
respond   **pick New then Flyout**
and   tree hierarchy displayed as Figure 35.2(a).

3   At Command List Categories, scroll and select Dimension and:
prompt   list of dimension commands displayed
respond   select Linear, hold down left button and drag below Toolbar2 name as
             Figure 35.2(b).

4   *a*) Repeat select and drag for the radius, angular and leader dimension commands
    b) Pick OK
    c) The new MYTBAR toolbar will be displayed as Figure 35.2(c) with the dimension
         fly-out.

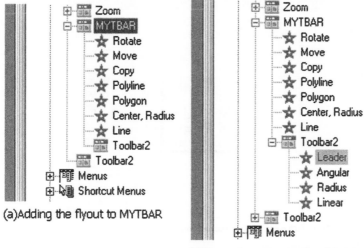

(a)Adding the flyout to MYTBAR

(b)Dimension commands added

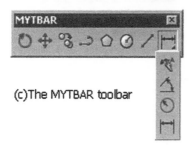

(c)The MYTBAR toolbar

Figure 35.2 The MYTBAR toolbar with the complete fly-out.

Customising an icon button

1 Activate the Customize User Interface dialogue box.

2 Scroll at Command List Categories, pick Help and:
   prompt   all help commands displayed
   respond   scroll, pick Zoom Window and drag below the MYTBAR name – Figure 35.3(a).

3 In the Button Image pane of the CUI dialogue box, pick Edit and:
   prompt   Button Editor dialogue box
   respond   *a*) pick Grid
            b) pick Clear
            c) select the red colour
            d) 1. using your 'artistic' ability, design a logo for the zoom-window command
               2. note that Figure 35.3(b) is my simple attempt
   then   when the button design is complete, pick Save As and:
   prompt   Create File dialogue box
   respond   *a*) navigate to your named folder
            b) enter file name: MYTBARZOOM
            c) file type: BMP
            d) pick Save
            e) close the Button Editor
   prompt   CUI dialogue box
   with   *a*) the Button Image displaying the new design
            b) the new design icon is added to the icon display as Figure 35.3(c)

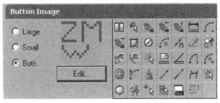

(a)Zoom Window added

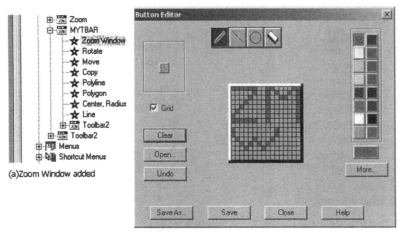

(b)The Button Editor dialogue box with my design

(c)The Button Image pane of the CUI with the new designed icon

(d)MYTBAR toolbar

Figure 35.3   Customising an icon button.

respond   pick OK from the CUI dialogue box
and   the MYTBAR toolbar will be displayed as Figure 35.3(d).

4   Now use your new toolbar to draw lines and circles and add dimensions.
    This simple exercise into creating new toolbars is now complete but there may be a problem:
    a) the new icon for the zoom window command will display your design at all times
    b) do you leave it as designed or change back to the original magnifying glass image
    c) your decision now.

Tool palettes

1   A tool palette is a draughting aid which can be customised by the user with the addition of drawings, blocks and hatch patterns.

2   These AutoCAD objects can be dragged from the Design Center or cut and pasted from other tool palettes.

3   To view the default tool palette, A3PAPER standard sheet should still be opened.

4   Menu bar with **Tools-Tool Palettes Window** and:
    prompt   Tool Palettes display for all palettes – Figure 35.4
    with   seven tab options:
               a) Annotation
               b) Architectural
               c) Mechanical
               d) Electrical
               e) Civil/Structural
               f) Hatches
               g) Command Tools
    respond use the tool palette tabs to drag some objects into the current drawing
    then   cancel the tool palette window.

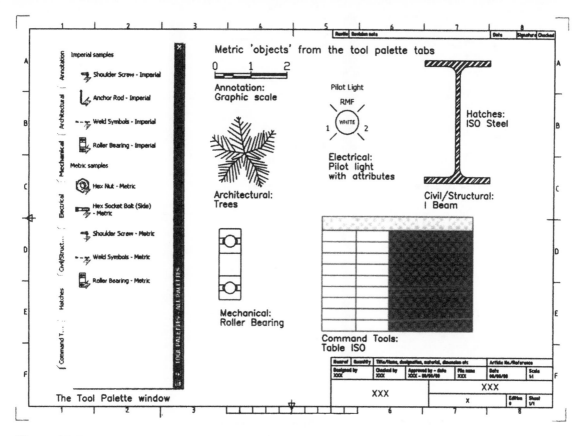

Figure 35.4 The Tool Palette window with an object from each tab placed in an AutoCAD drawing.

5   Figure 35.4 displays the tool palette window as a screen dump pasted into an AutoCAD drawing and a drop-and-drag object from each of the tabs (scaled as appropriate).

Creating a new tool palette

1   *a*) Ensure a blank A3PAPER sheet is displayed
      b) Activate the Design Center.

2   From the Design Center, explore:
      a) your MYCAD folder on the tree side to display the folder contents with large icons
      b) your equivalent to Figure 28.3 with the seat and table blocks
      c) explore Blocks from the tree side to display the named blocks in the content side.

3   **Right-click in the content side** and:
prompt	shortcut menu
respond	**pick Create Tool Palette**
prompt	Tool Palette window displayed
with	*a*) the seven default tabs
	b) a new tab (Figure 28.3 in my case) – Figure 35.5(a)
	c) four blocks added to this palette
respond	**right-click the new tab name**
prompt	shortcut menu
respond	*a*) pick Rename Tool Palette
	b) alter name to **MYPAL** then <RETURN>
and	the new tool palette will be renamed as Figure 35.5(b).

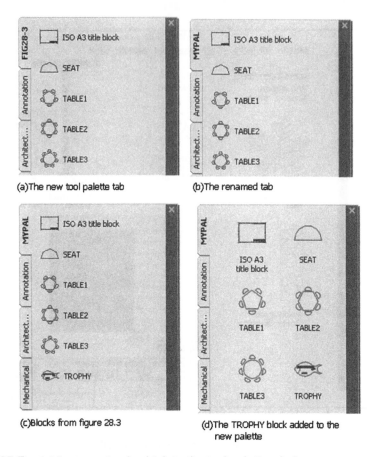

(a)The new tool palette tab

(b)The renamed tab

(c)Blocks from figure 28.3

(d)The TROPHY block added to the new palette

Figure 35.5 Adding a customised tab to the tool palette window.

4   Using the Design Center:
    a) explore the fisherman's trophy drawing from Chapter 28, i.e. Figure 28.11
    b) explore the blocks from this drawing and TROPHY displayed in the content side
    c) left-click on TROPHY and drag into the MYPAL tab – Figure 35.5(c).

5   Right-click the MYPAL tab and from the shortcut menu:
    a) pick View Options
    b) select Icon with Text
    c) increase the image size
    d) pick OK.

6   The new tab will display the icons at the increased scale – Figure 35.5(d).

7   The blocks in the MYPAL tab can now be inserted into any AutoCAD drawing.

8   Cancel the Tool Palette and Design Center windows to complete this exercise.

Finally

1   This chapter has been a brief introduction to using and customising toolbars and tool palettes.

2   They are very powerful draughting aids if used correctly.

3   You should be able to use these draughting aids to increase your drawing productivity.

Chapter 36

File formats

1 All drawings created with AutoCAD:
   a) are saved in named folders as a file
   b) have a drive, folder name, file name and extension, e.g. **C:\MYCAD\ACT1.dwg.**

2 The file types used so far have had extensions:
   a) .dwg
      1. the standard file format for saving vector graphics
      2. all AutoCAD drawings are saved with this extension.
   b) .dwt
      1. a drawing template file
      2. has settings customised to user requirements
      3. any drawing can be saved as a template file.
   c) .dws
      1. drawings standards file format
      2. is used to define common properties and to compare inconsistencies between drawings.

3 Other file formats are available with AutoCAD and several of these will now be investigated.

Backup files

1 Backup files have the extension **.bak**.

2 They are:
   a) automatically created by AutoCAD when an opened drawing is saved with the same file name
   b) are stored in the current folder with the drawing files
   c) can be renamed (and then opened) as drawing (.dwg) files.

3 To rename a backup file:
   a) open Windows Explorer (or similar) and navigate to your named folder
   b) view the displayed icons as a detailed list
   c) arrange the icons by type
   d) scroll until the .bak files are displayed similar to Figure 36.1(a)
   e) right-click any backup file name and:

prompt	Shortcut menu
respond	**pick Rename**
prompt	text box placed around the selected file name
respond	enter the required name, e.g. **MYNEW1.dwg then <R>**
and	*a*) file is renamed as Figure 36.1(b)
	b) preview of renamed file is displayed
respond	a double left-click on the renamed file will open it in AutoCAD.

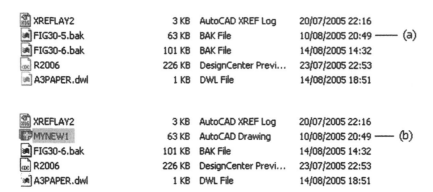

Figure 36.1 Renaming a backup file as a drawing file.

Temporary files

1 When automatic save is active, drawings are saved at user specified intervals with the extension .sv$.

2 Temporary files are:
   a) automatically deleted when an opened drawing is closed
   b) not deleted in event of a power failure.

3 Temporary files can be renamed with the extension .dwg before AutoCAD is exited.

DXF files

1 DXF are Drawing Interchange Format files:
   a) are use for exporting drawing data to other CAD systems and/or CADCAM/CNC systems
   b) consist of ASCII or binary code
   c) have the extension **.dxf**.

2 To export a drawing file in DXF format, open your saved WORKDRG and:
   a) model space active
   b) ensure that several dimensions are displayed
   c) add a title, e.g. WORKING DRAWING using the ST3 text style format
   d) identify the large circle centre – 130,185.

3 Menu bar with **File-Save As** and:
   prompt   File Save As dialogue box
   respond   *a*) ensure your named folder is current
                    b) File type: scroll and pick AutoCAD 2004 DXF (*.dxf)
                    c) File name: WORKDRG.dxf
                    d) pick Save
   and   command line returned.

4 Menu bar with **File-Close** and:
   prompt   AutoCAD message about saved format
   respond   read the message and decide on the response.

5 Menu bar with **File-Open** and:
   prompt   Select File dialogue box
   respond   *a*) named folder current
                    b) File type: scroll and pick DXF (*.dxf)
                    c) pick WORKDRG then open

and   saved dxf file displayed

respond   *a*) is the drawing correct? i.e. text, dimensions, centre lines, etc.

          b) identify the large circle centre – correct co-ordinates?

6   The above steps is the sequence for saving and opening a DXF file.

DWF files

1   DWF are Design Web files.

2   They are compressed 2D vector files and:

     a) are created from drawing (.dwg) files

     b) can be published and viewed on the web

     c) can be viewed with AutoCAD Viewer.

3   Close all drawing files, open any saved activity, (e.g. ACT35) and remove all text and dimensions from the activity drawing.

4   Menu bar with **File-Plot** and:

prompt   `Plot dialogue box`

respond   *a*) Printer/plotter: scroll and select **DWF6ePlot.pc3**

          b) Paper size: scroll and select **ISO A3 (420.00 × 297.00 MM)**

          c) Plot area: What to plot: Display

          d) Plot offset: Centre the plot active

          e) Plot style table: None

          f) Plot scale: Fit to paper active (tick)

          g) Plot options: set to own requirements

          h) Drawing orientation: Landscape active

          i) pick Preview

and   centred preview of activity drawing displayed

respond   **right-click**

prompt   `shortcut menu`

respond   **pick Plot**

prompt   `Browse for Plot File dialogue box with file type *.dwf`

respond   *a*) scroll to your named folder

          b) file name: ACT35.dwf

          c) pick Save

prompt   `Plot and Publish Job Complete with No errors or warnings`
             `message displayed`

respond   **cancel the message**.

5   Drawing Web Format files can be:

     a) opened and plotted with AutoCAD DWF Viewer (not discussed in this book)

     b) viewed in MS Internet Explorer

     c) published on the World Wide Web or any Intranet network.

Publishing to the Web

1   Web pages of existing drawings can be created.

2   Start a new metric drawing from scratch to display the typical AutoCAD blank screen.

3   Menu bar with **File-Publish to Web** and:

prompt   `Publish to Web dialogue box` (similar to the Layout dialogue box)

respond   *a*) pick Create New Web Page

          b) pick Next

prompt   `Publish to Web - Create Web Page dialogue box`

respond   *a*) Web page name: enter MYPAGE

b) Ensure parent directory is your named folder
c) Add any suitable description – Figure 36.2
d) pick Next

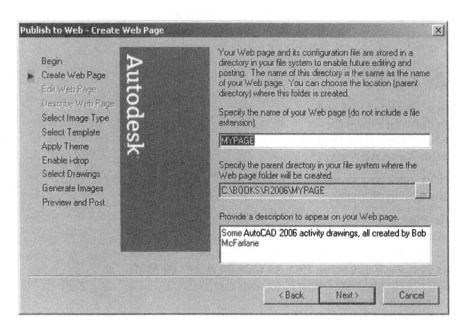

Figure 36.2   Publish to Web – Create Web Page dialogue box.

prompt	Publish to Web – Select Image Type dialogue box
respond	*a*) select type from list: DWF
	b) pick Next
prompt	Publish to Web – Select Template dialogue box
respond	*a*) select template type: Array plus Summary
	b) pick Next
prompt	Publish to Web – Apply Theme dialogue box
respond	*a*) scroll and select Ocean Waves – Figure 36.3
	b) pick Next

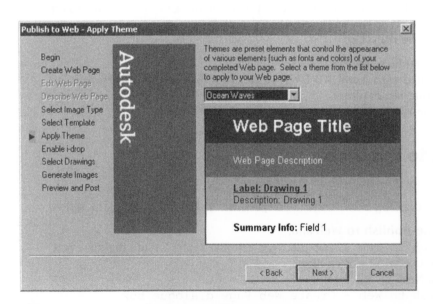

Figure 36.3   The Publish to Web – Apply Theme dialogue box.

prompt	`Publish to Web - Enable i-drop dialogue box`
respond	**activate i-drop then pick Next**
prompt	`Publish to Web - Select Drawings dialogue box`
respond	**at Drawing, select the (...) navigate button**
prompt	`Publish to Web dialogue box`
respond	*a*) scroll to your named folder
	b) select an activity, e.g. ACT33 then Open
prompt	`Publish to Web - Select Drawings dialogue box`
with	ACT33 listed
respond	*a*) Layout: select Model
	b) Label: alter to GEARS (or similar)
	c) Description: enter to your own requirements
	d) pick Add->
and	GEARS added to Image list
now	select several drawings from your list and:
	a) Layout: Model
	b) add a suitable label
	c) add a suitable description
	d) pick Add->
then	pick Next
prompt	`Publish to Web - Generate Images dialogue box`
respond	**Regenerate all images then Next**
prompt	`Plot progress information displayed`
then	Publish to Web – Preview and Post dialogue box
respond	**pick Preview**
and	Internet Explorer with Images of drawings – Figure 36.4
respond	*a*) view your images
	b) close the Internet Explorer to return to AutoCAD
and	select Finish.

4 You have now create a web page which can be:
   a) edited to your requirements
   b) posted as required.

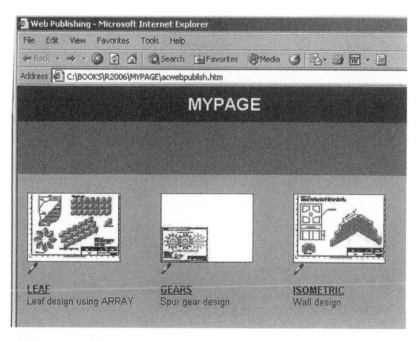

Figure 36.4   The web page preview.

Notes

1 I hope that in this chapter the user has realised that AutoCAD has uses other than drawing.

2 The web page creation is very useful and relatively simple to create.

3 This chapter is now complete.

ACTIVITY 1
a) Draw the simple shapes using the LINE and CIRCLE commands
b) Set the GRID and SNAP to your own requirements
c) Decide if you want to use Polar Tracking or Dynamic Input
d) When complete, save as C:\MYCAD\ACT1

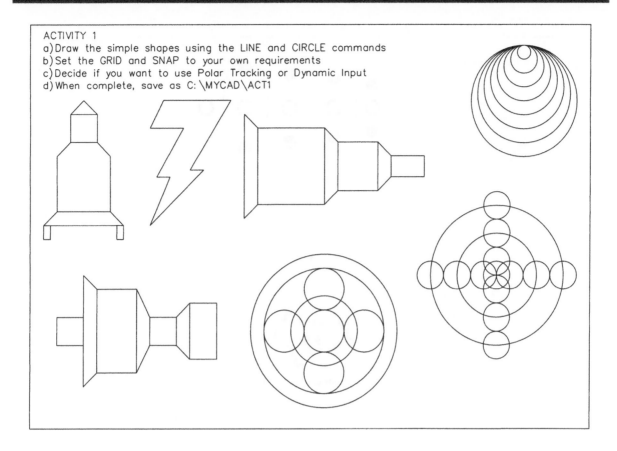

ACTIVITY 2
Draw the three shapes using the sizes given.
The suggested start points are:
A(55,80), B(200,95), C(290,210)

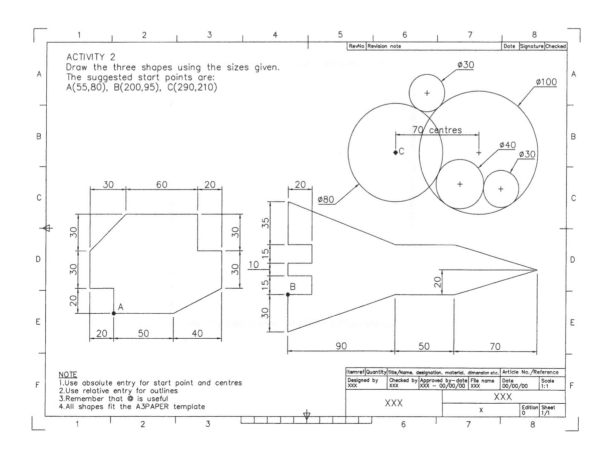

NOTE
1. Use absolute entry for start point and centres
2. Use relative entry for outlines
3. Remember that @ is useful
4. All shapes fit the A3PAPER template

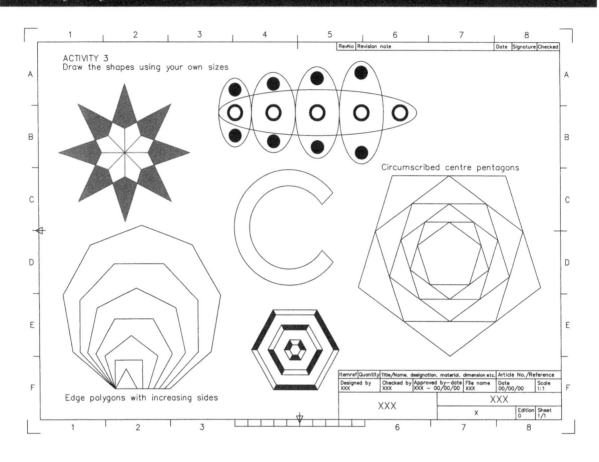

ACTIVITY 3
Draw the shapes using your own sizes

Circumscribed centre pentagons

Edge polygons with increasing sides

Itemref	Quantity	Title/Name. designation, material, dimension etc.			Article No./Reference	
Designed by XXX		Checked by XXX	Approved by—date XXX – 00/00/00	File name XXX	Date 00/00/00	Scale 1:1
		XXX			XXX	
				x	Edition 0	Sheet 1/1

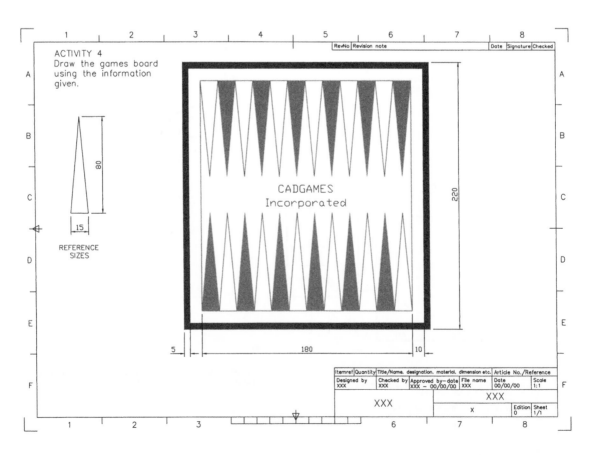

ACTIVITY 4
Draw the games board
using the information
given.

80

15

REFERENCE
SIZES

CADGAMES
Incorporated

220

5   180   10

Itemref	Quantity	Title/Name. designation, material, dimension etc.			Article No./Reference	
Designed by XXX		Checked by XXX	Approved by—date XXX – 00/00/00	File name XXX	Date 00/00/00	Scale 1:1
		XXX			XXX	
				x	Edition 0	Sheet 1/1

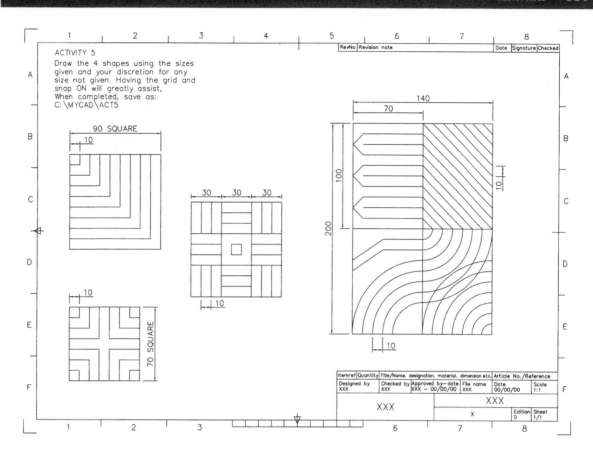

ACTIVITY 5
Draw the 4 shapes using the sizes given and your discretion for any size not given. Having the grid and snap ON will greatly assist, When completed, save as:
C:\MYCAD\ACT5

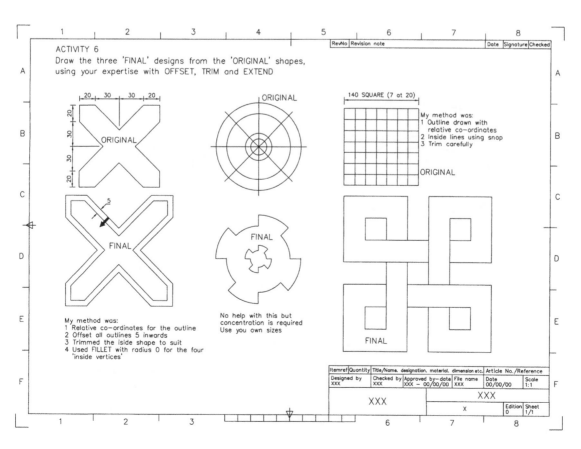

ACTIVITY 6
Draw the three 'FINAL' designs from the 'ORIGINAL' shapes, using your expertise with OFFSET, TRIM and EXTEND

My method was:
1 Relative co-ordinates for the outline
2 Offset all outlines 5 inwards
3 Trimmed the iside shape to suit
4 Used FILLET with radius 0 for the four 'inside vertices'

No help with this but concentration is required
Use you own sizes

140 SQUARE (7 at 20)
My method was:
1 Outline drawn with relative co-ordinates
2 Inside lines using snap
3 Trim carefully

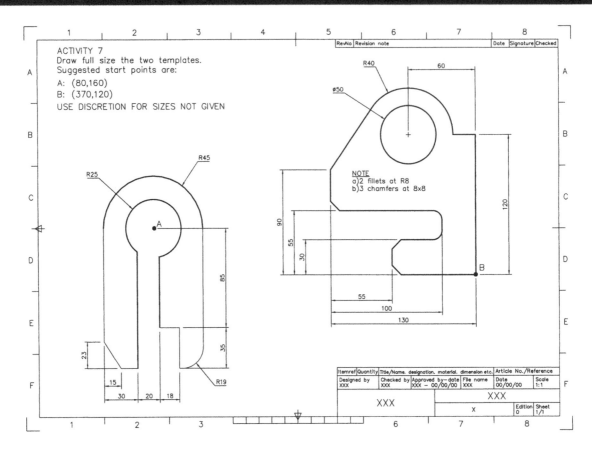

ACTIVITY 7
Draw full size the two templates.
Suggested start points are:
A: (80,160)
B: (370,120)
USE DISCRETION FOR SIZES NOT GIVEN

NOTE
a) 2 fillets at R8
b) 3 chamfers at 8x8

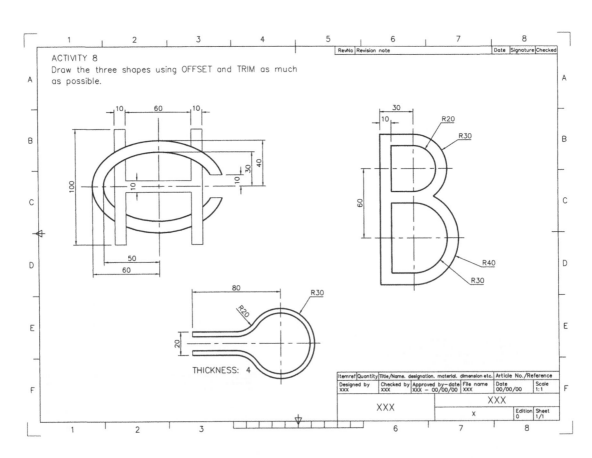

ACTIVITY 8
Draw the three shapes using OFFSET and TRIM as much
as possible.

THICKNESS: 4

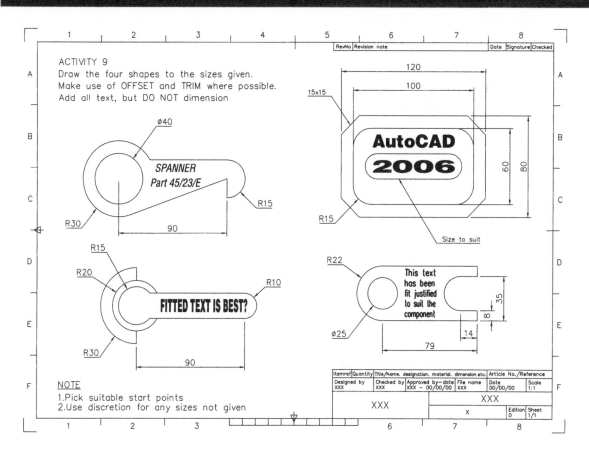

ACTIVITY 9
Draw the four shapes to the sizes given.
Make use of OFFSET and TRIM where possible.
Add all text, but DO NOT dimension

SPANNER
Part 45/23/E

AutoCAD
2006

FITTED TEXT IS BEST?

This text
has been
fit justified
to suit the
component

NOTE
1. Pick suitable start points
2. Use discretion for any sizes not given

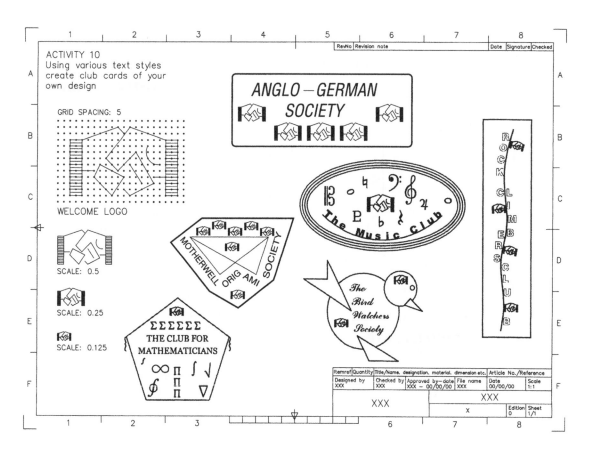

ACTIVITY 10
Using various text styles
create club cards of your
own design

GRID SPACING: 5

WELCOME LOGO

SCALE: 0.5

SCALE: 0.25

SCALE: 0.125

ANGLO—GERMAN
SOCIETY

The Music Club

MOTHERWELL ORIG AMI SOCIETY

ΣΣΣΣΣ
THE CLUB FOR
MATHEMATICIANS

The
Bird
Watchers
Society

ROCK CLIMBERS CLUB

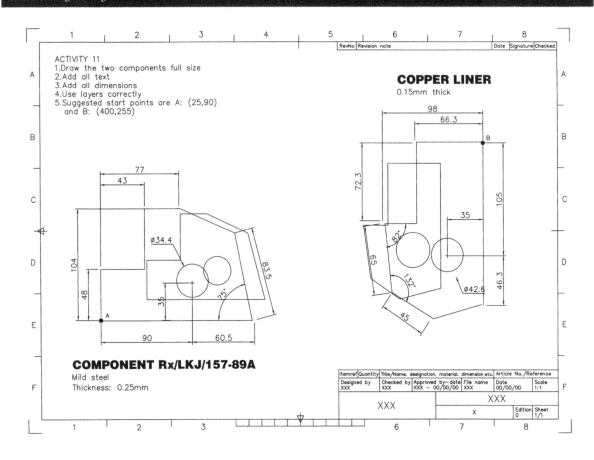

ACTIVITY 11
1. Draw the two components full size
2. Add all text
3. Add all dimensions
4. Use layers correctly
5. Suggested start points are A: (25,90) and B: (400,255)

COPPER LINER
0.15mm thick

COMPONENT Rx/LKJ/157-89A
Mild steel
Thickness: 0.25mm

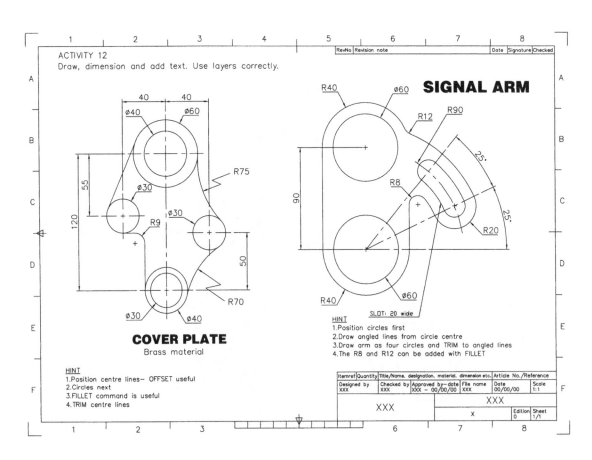

ACTIVITY 12
Draw, dimension and add text. Use layers correctly.

SIGNAL ARM

COVER PLATE
Brass material

HINT
1. Position centre lines— OFFSET useful
2. Circles next
3. FILLET command is useful
4. TRIM centre lines

SLOT: 20 wide

HINT
1. Position circles first
2. Draw angled lines from circle centre
3. Draw arm as four circles and TRIM to angled lines
4. The R8 and R12 can be added with FILLET

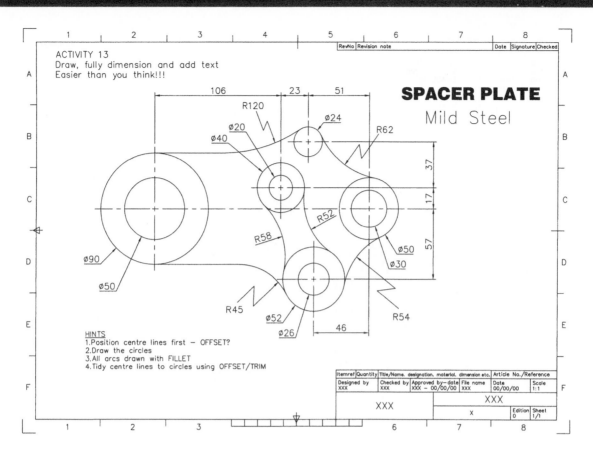

ACTIVITY 13
Draw, fully dimension and add text
Easier than you think!!!

SPACER PLATE
Mild Steel

HINTS
1.Position centre lines first — OFFSET?
2.Draw the circles
3.All arcs drawn with FILLET
4.Tidy centre lines to circles using OFFSET/TRIM

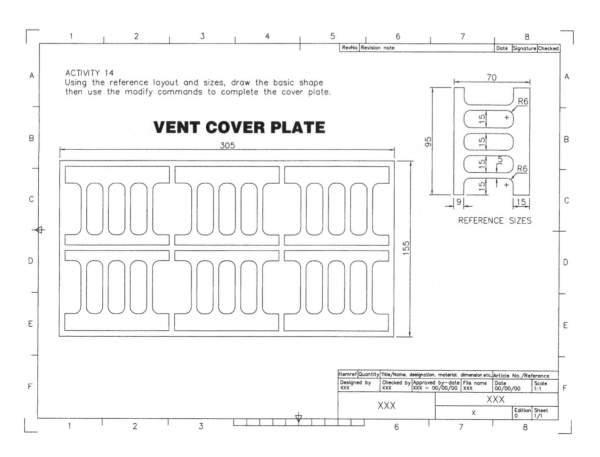

ACTIVITY 14
Using the reference layout and sizes, draw the basic shape
then use the modify commands to complete the cover plate.

VENT COVER PLATE

REFERENCE SIZES

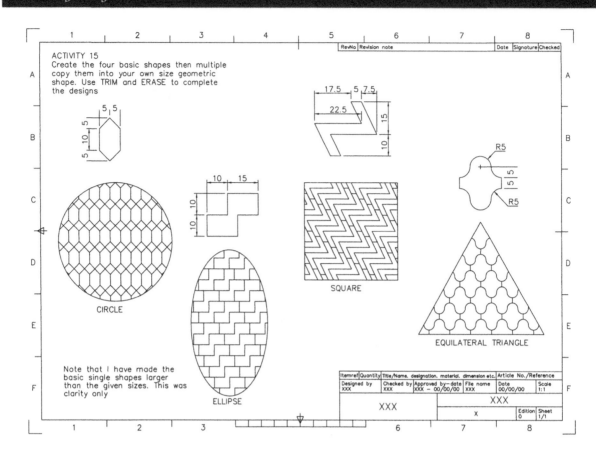

ACTIVITY 15
Create the four basic shapes then multiple copy them into your own size geometric shape. Use TRIM and ERASE to complete the designs

CIRCLE

ELLIPSE

SQUARE

EQUILATERAL TRIANGLE

Note that I have made the basic single shapes larger than the given sizes. This was clarity only

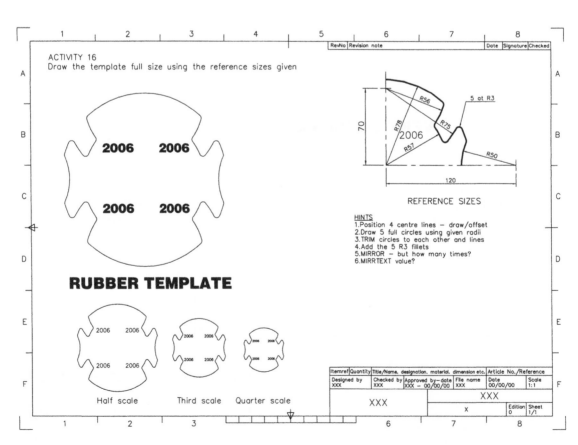

ACTIVITY 16
Draw the template full size using the reference sizes given

2006 2006

2006 2006

RUBBER TEMPLATE

REFERENCE SIZES

HINTS
1. Position 4 centre lines – draw/offset
2. Draw 5 full circles using given radii
3. TRIM circles to each other and lines
4. Add the 5 R3 fillets
5. MIRROR – but how many times?
6. MIRRTEXT value?

Half scale   Third scale   Quarter scale

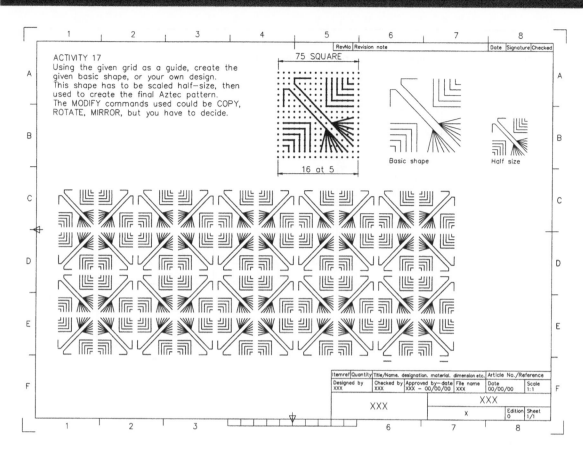

ACTIVITY 17
Using the given grid as a guide, create the given basic shape, or your own design. This shape has to be scaled half-size, then used to create the final Aztec pattern. The MODIFY commands used could be COPY, ROTATE, MIRROR, but you have to decide.

75 SQUARE

16 at 5

Basic shape

Half size

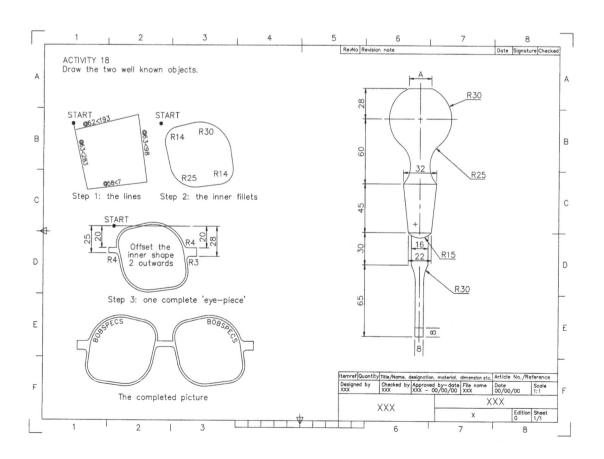

ACTIVITY 18
Draw the two well known objects.

START
@62<193
@63<283
@63<98
@58<7

Step 1: the lines

START
R14   R30
R25   R14

Step 2: the inner fillets

START
25 20
R4
R4

Offset the inner shape 2 outwards

20 28
R4
R3

Step 3: one complete 'eye-piece'

BOBSPECS   BOBSPECS

The completed picture

A
28
R30
60
32
R25
45
+
30
16
22
R15
65
R30
8
8

ACTIVITY 19(a)
The basic sizes and the layout for the robot arm assignment

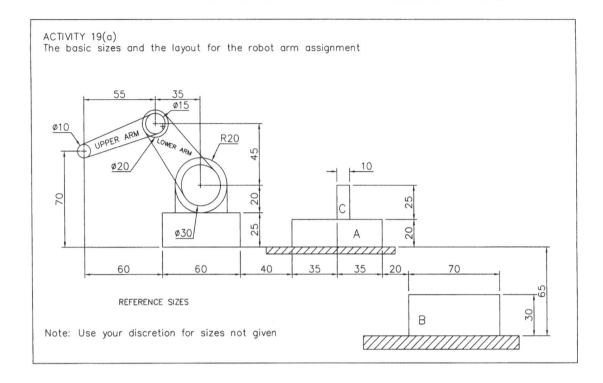

REFERENCE SIZES

Note: Use your discretion for sizes not given

ACTIVITY 19(b)
Using GRIPS, re-align the robot arm to move component (C) from pallet A to pallet B

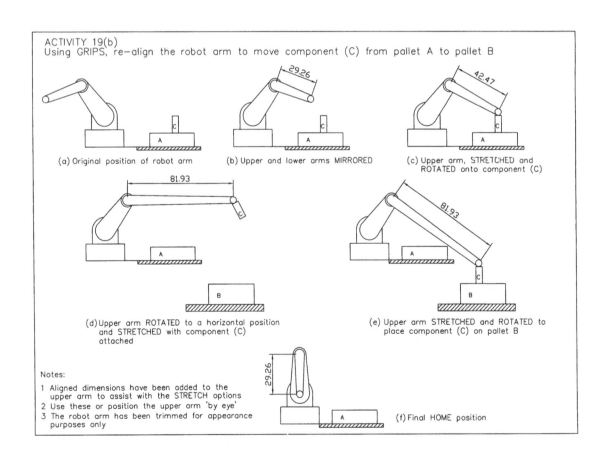

(a) Original position of robot arm

(b) Upper and lower arms MIRRORED

(c) Upper arm, STRETCHED and ROTATED onto component (C)

(d) Upper arm ROTATED to a horizontal position and STRETCHED with component (C) attached

(e) Upper arm STRETCHED and ROTATED to place component (C) on pallet B

(f) Final HOME position

Notes:
1 Aligned dimensions have been added to the upper arm to assist with the STRETCH options
2 Use these or position the upper arm 'by eye'
3 The robot arm has been trimmed for appearance purposes only

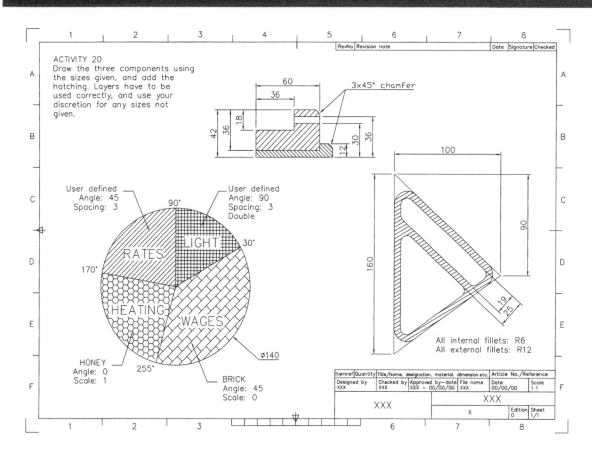

ACTIVITY 20
Draw the three components using the sizes given, and add the hatching. Layers have to be used correctly, and use your discretion for any sizes not given.

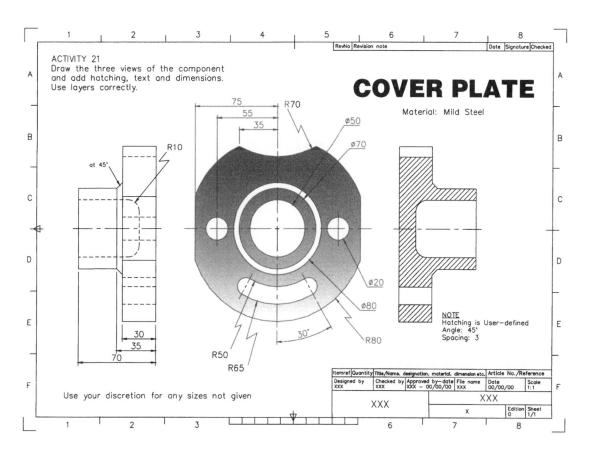

ACTIVITY 21
Draw the three views of the component and add hatching, text and dimensions. Use layers correctly.

COVER PLATE

Material: Mild Steel

NOTE
Hatching is User-defined
Angle: 45°
Spacing: 3

Use your discretion for any sizes not given

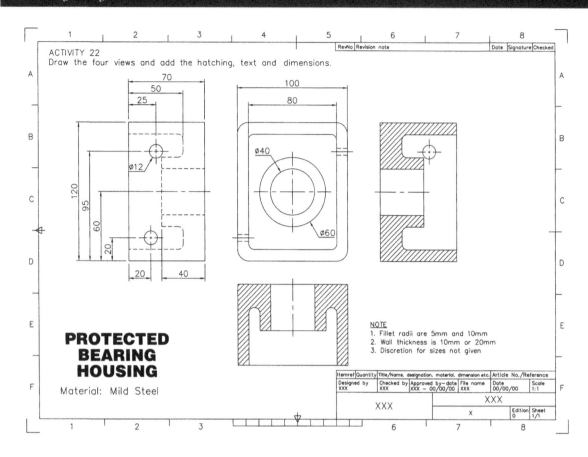

ACTIVITY 22
Draw the four views and add the hatching, text and dimensions.

PROTECTED BEARING HOUSING

Material: Mild Steel

NOTE
1. Fillet radii are 5mm and 10mm
2. Wall thickness is 10mm or 20mm
3. Discretion for sizes not given

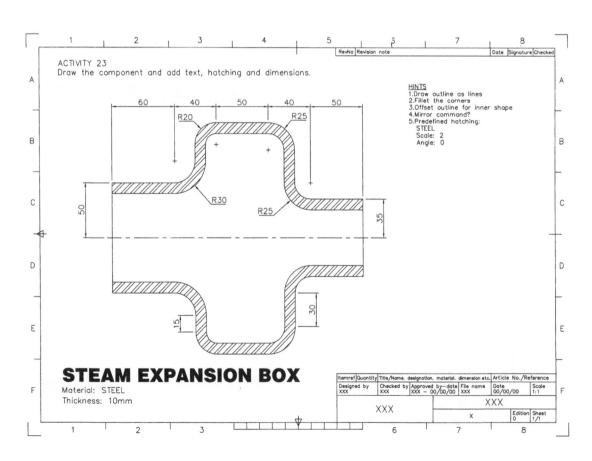

ACTIVITY 23
Draw the component and add text, hatching and dimensions.

HINTS
1. Draw outline as lines
2. Fillet the corners
3. Offset outline for inner shape
4. Mirror command?
5. Predefined hatching:
   STEEL
   Scale: 2
   Angle: 0

STEAM EXPANSION BOX

Material: STEEL
Thickness: 10mm

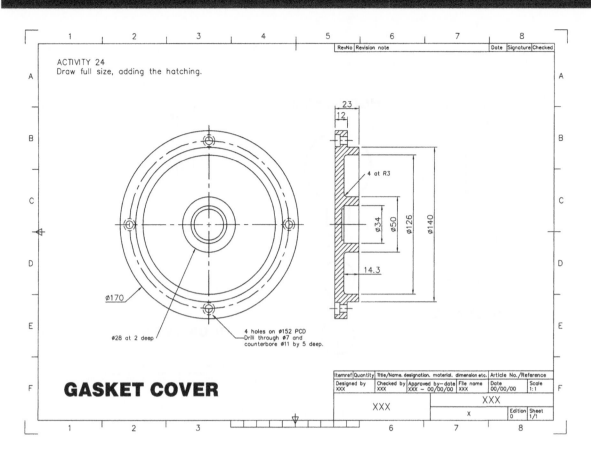

ACTIVITY 24
Draw full size, adding the hatching.

Ø170

Ø28 at 2 deep

4 holes on Ø152 PCD
Drill through Ø7 and
counterbore Ø11 by 5 deep.

23
12
4 at R3
Ø34
Ø50
Ø126
Ø140
14.3

GASKET COVER

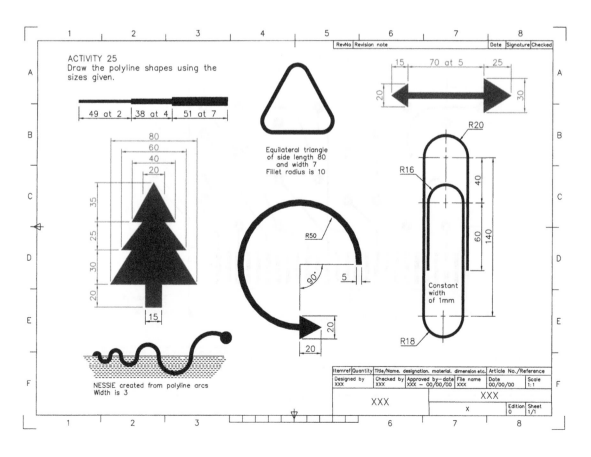

ACTIVITY 25
Draw the polyline shapes using the
sizes given.

49 at 2   38 at 4   51 at 7

80
60
40
20
35
25
30
20
15

NESSIE created from polyline arcs
Width is 3

Equilateral triangle
of side length 80
and width 7
Fillet radius is 10

R50
90°
5
20
20

15   70 at 5   25
20
30

R20
R16
40
60
140
Constant
width
of 1mm
R18

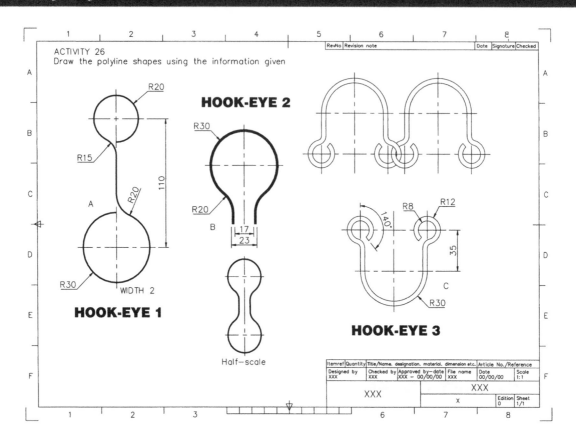

ACTIVITY 26
Draw the polyline shapes using the information given

HOOK-EYE 2

HOOK-EYE 1

HOOK-EYE 3

Half—scale

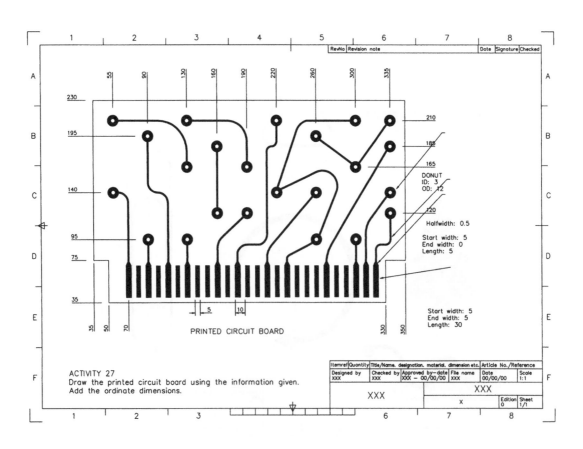

ACTIVITY 27
Draw the printed circuit board using the information given.
Add the ordinate dimensions.

PRINTED CIRCUIT BOARD

DONUT
ID: 3
OD: 12

Halfwidth: 0.5

Start width: 5
End width: 0
Length: 5

Start width: 5
End width: 5
Length: 30

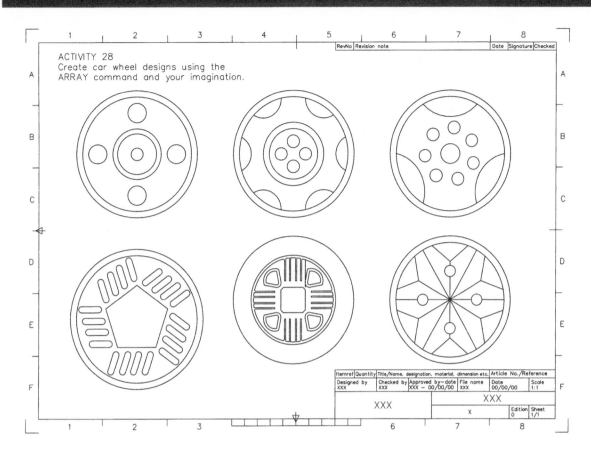

ACTIVITY 28
Create car wheel designs using the
ARRAY command and your imagination.

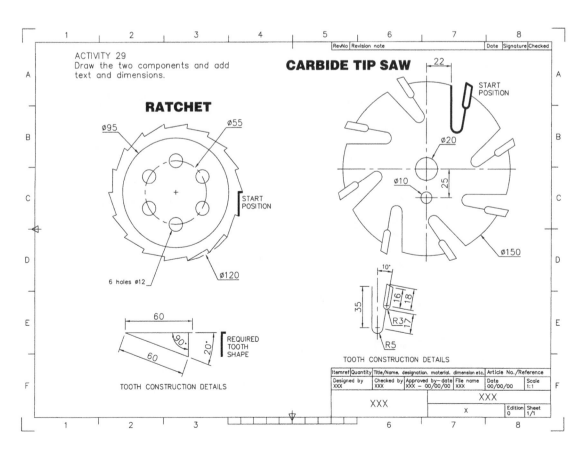

ACTIVITY 29
Draw the two components and add
text and dimensions.

RATCHET

CARBIDE TIP SAW

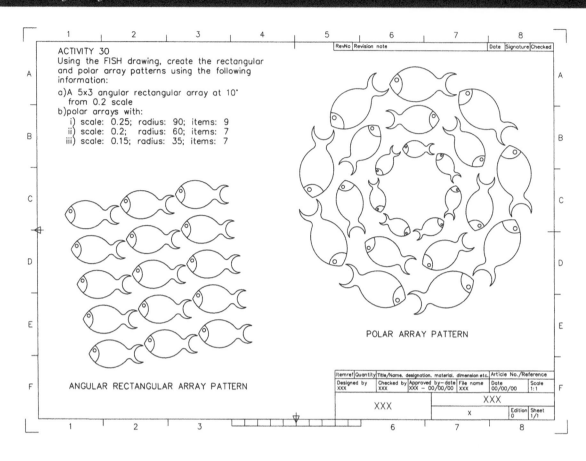

ACTIVITY 30
Using the FISH drawing, create the rectangular and polar array patterns using the following information:

a) A 5x3 angular rectangular array at 10° from 0.2 scale
b) polar arrays with:
  i) scale: 0.25; radius: 90; items: 9
  ii) scale: 0.2;   radius: 60; items: 7
  iii) scale: 0.15; radius: 35; items: 7

POLAR ARRAY PATTERN

ANGULAR RECTANGULAR ARRAY PATTERN

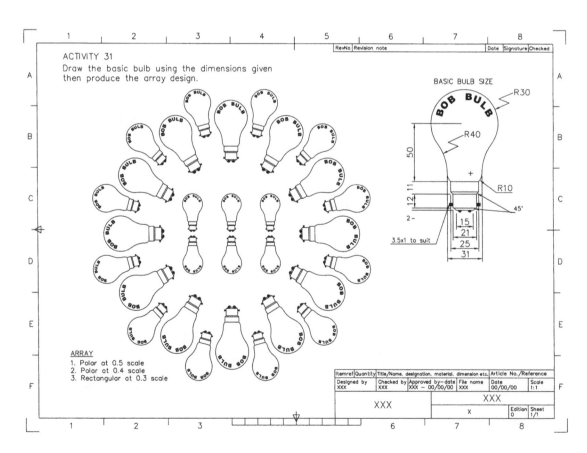

ACTIVITY 31
Draw the basic bulb using the dimensions given then produce the array design.

BASIC BULB SIZE

ARRAY
1. Polar at 0.5 scale
2. Polar at 0.4 scale
3. Rectangular at 0.3 scale

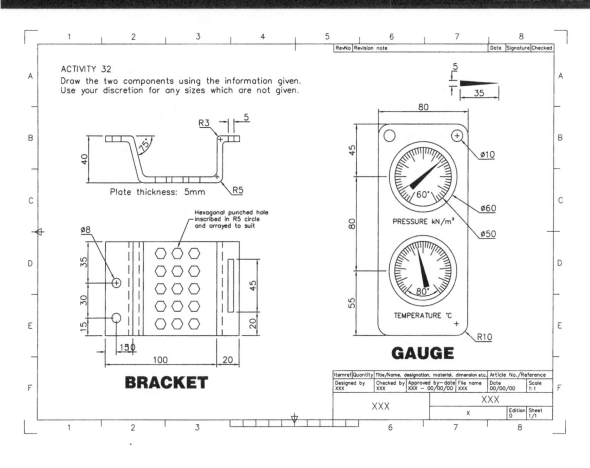

ACTIVITY 32
Draw the two components using the information given.
Use your discretion for any sizes which are not given.

Plate thickness: 5mm

Hexagonal punched hole inscribed in R5 circle and arrayed to suit

BRACKET

PRESSURE kN/m²

TEMPERATURE °C

GAUGE

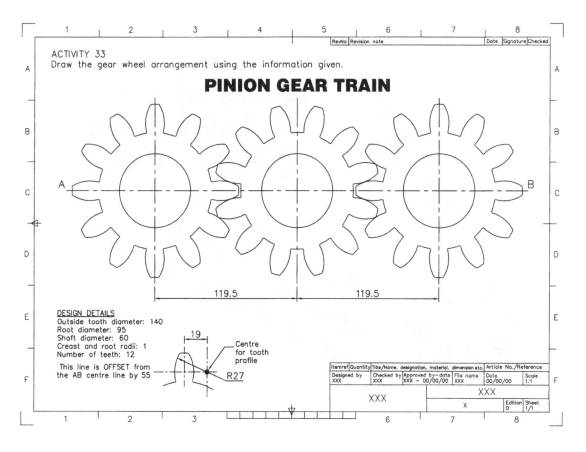

ACTIVITY 33
Draw the gear wheel arrangement using the information given.

PINION GEAR TRAIN

A

B

119.5   119.5

DESIGN DETAILS
Outside tooth diameter: 140
Root diameter: 95
Shaft diameter: 60
Creast and root radii: 1
Number of teeth: 12

This line is OFFSET from the AB centre line by 55

Centre for tooth profile

R27

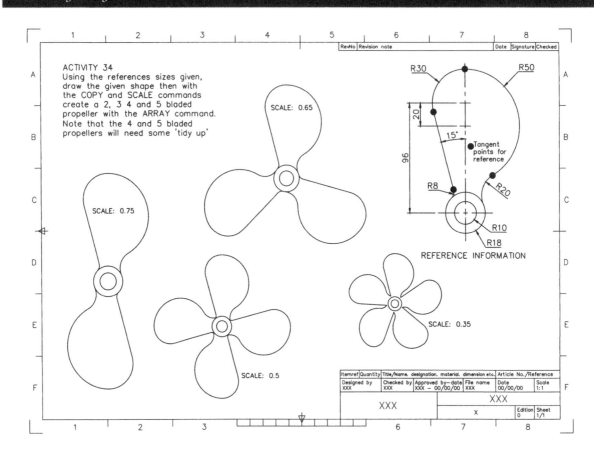

ACTIVITY 34
Using the references sizes given, draw the given shape then with the COPY and SCALE commands create a 2, 3 4 and 5 bladed propeller with the ARRAY command. Note that the 4 and 5 bladed propellers will need some 'tidy up'

SCALE: 0.65

SCALE: 0.75

SCALE: 0.35

SCALE: 0.5

REFERENCE INFORMATION

R30 R50 R8 R20 R10 R18 20 96 15° Tangent points for reference

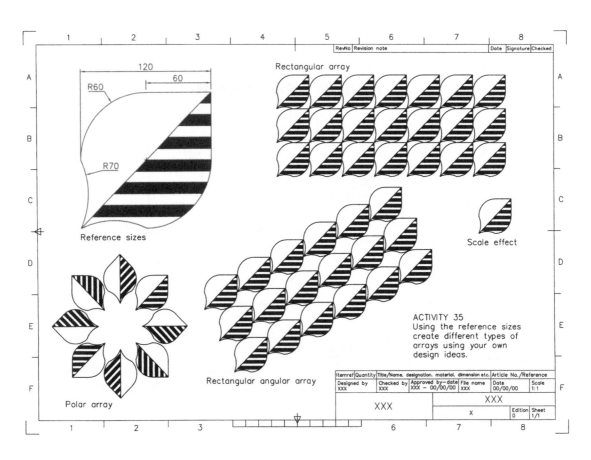

Rectangular array

R60 120 60 R70

Reference sizes

Scale effect

Polar array

Rectangular angular array

ACTIVITY 35
Using the reference sizes create different types of arrays using your own design ideas.

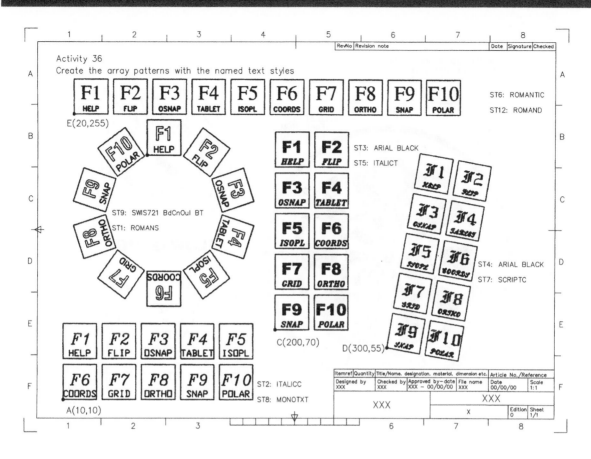

Activity 36
Create the array patterns with the named text styles

E(20,255)

ST6: ROMANTIC
ST12: ROMAND

ST3: ARIAL BLACK
ST5: ITALICT

ST9: SWIS721 BdCnOul BT
ST1: ROMANS

ST4: ARIAL BLACK
ST7: SCRIPTC

C(200,70)

D(300,55)

ST2: ITALICC
ST8: MONOTXT

A(10,10)

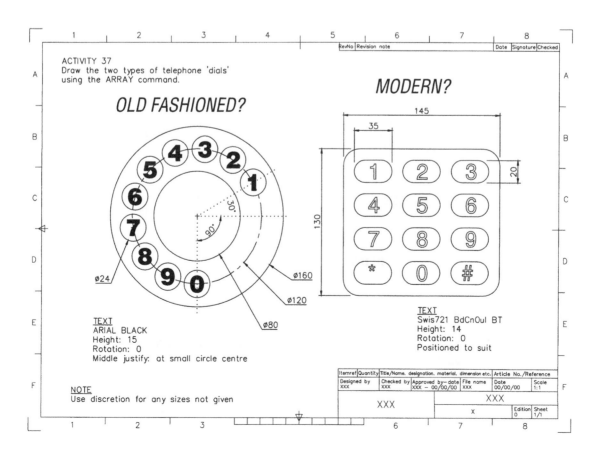

ACTIVITY 37
Draw the two types of telephone 'dials'
using the ARRAY command.

OLD FASHIONED?

MODERN?

145
35
130
20

Ø24
Ø160
Ø120
Ø80

TEXT
ARIAL BLACK
Height: 15
Rotation: 0
Middle justify: at small circle centre

TEXT
Swis721 BdCnOul BT
Height: 14
Rotation: 0
Positioned to suit

NOTE
Use discretion for any sizes not given

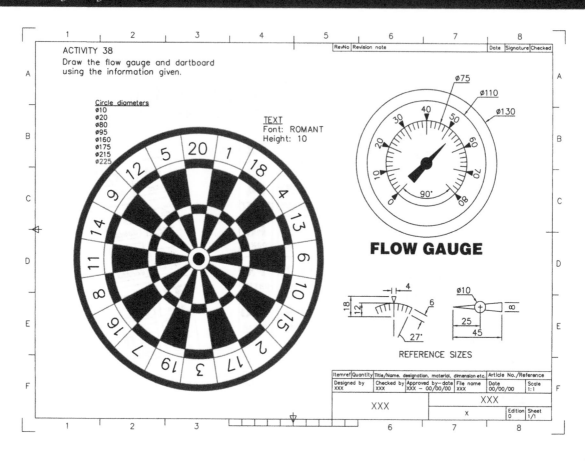

ACTIVITY 38
Draw the flow gauge and dartboard
using the information given.

Circle diameters
ø10
ø20
ø80
ø95
ø160
ø175
ø215
ø225

TEXT
Font: ROMANT
Height: 10

FLOW GAUGE

REFERENCE SIZES

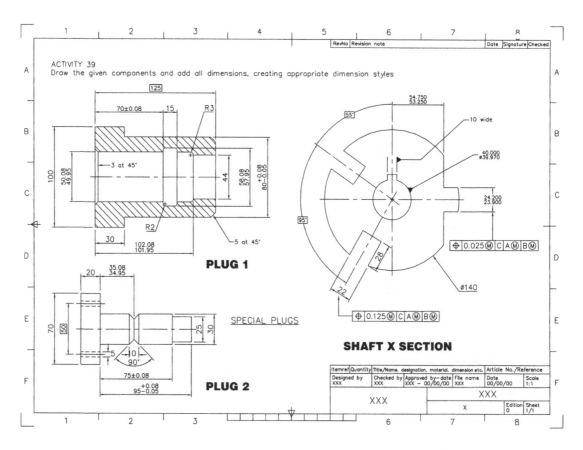

ACTIVITY 39
Draw the given components and add all dimensions, creating appropriate dimension styles

PLUG 1

PLUG 2

SPECIAL PLUGS

SHAFT X SECTION

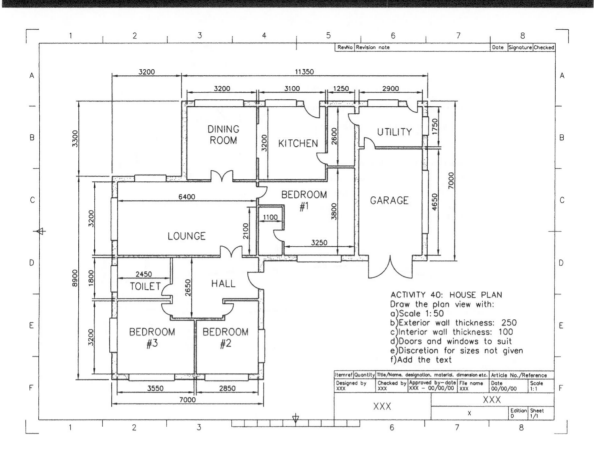

ACTIVITY 40: HOUSE PLAN
Draw the plan view with:
a)Scale 1:50
b)Exterior wall thickness: 250
c)Interior wall thickness: 100
d)Doors and windows to suit
e)Discretion for sizes not given
f)Add the text

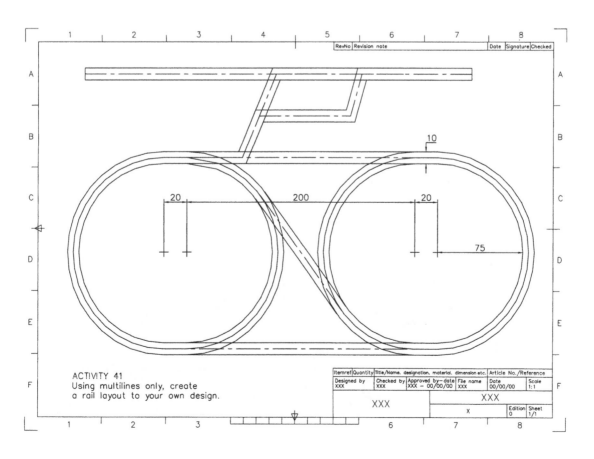

ACTIVITY 41
Using multilines only, create
a rail layout to your own design.

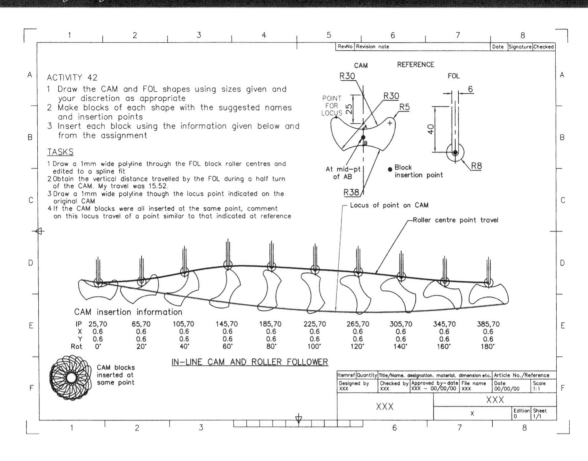

ACTIVITY 42

1 Draw the CAM and FOL shapes using sizes given and your discretion as appropriate
2 Make blocks of each shape with the suggested names and insertion points
3 Insert each block using the information given below and from the assignment

TASKS

1 Draw a 1mm wide polyline through the FOL block roller centres and edited to a spline fit
2 Obtain the vertical distance travelled by the FOL during a half turn of the CAM. My travel was 15.52.
3 Draw a 1mm wide polyline though the locus point indicated on the original CAM
4 If the CAM blocks were all inserted at the same point, comment on this locus travel of a point similar to that indicated at reference

CAM insertion information

IP	25,70	65,70	105,70	145,70	185,70	225,70	265,70	305,70	345,70	385,70
X	0.6	0.6	0.6	0.6	0.6	0.6	0.6	0.6	0.6	0.6
Y	0.6	0.6	0.6	0.6	0.6	0.6	0.6	0.6	0.6	0.6
Rot	0°	20°	40°	60°	80°	100°	120°	140°	160°	180°

IN–LINE CAM AND ROLLER FOLLOWER

CAM blocks inserted at same point

ACTIVITY 43
Modify the LARGSC drawing to include the BEAM block

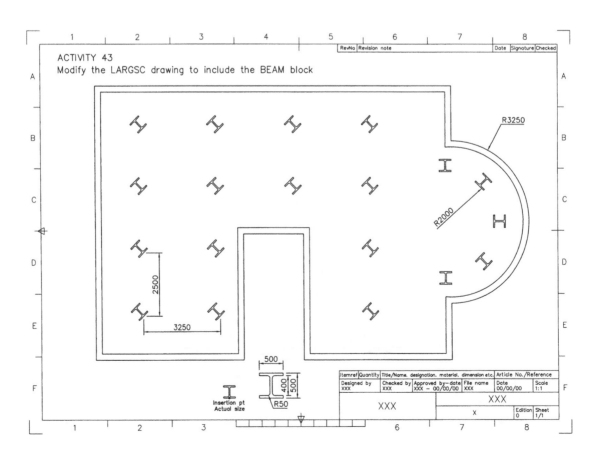

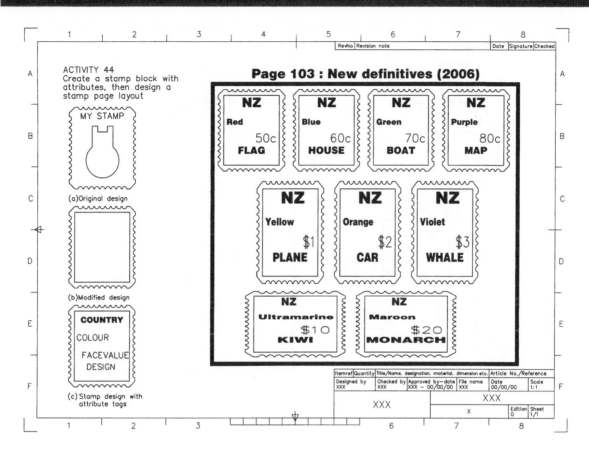

ACTIVITY 44
Create a stamp block with attributes, then design a stamp page layout

MY STAMP

(a) Original design

(b) Modified design

COUNTRY
COLOUR
FACEVALUE
DESIGN

(c) Stamp design with attribute tags

Page 103 : New definitives (2006)

NZ Red	NZ Blue	NZ Green	NZ Purple
50c FLAG	60c HOUSE	70c BOAT	80c MAP

NZ Yellow $1 PLANE

NZ Orange $2 CAR

NZ Violet $3 WHALE

NZ Ultramarine $10 KIWI

NZ Maroon $20 MONARCH

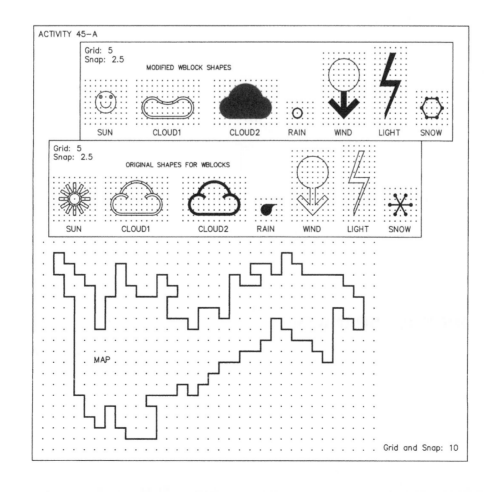

ACTIVITY 45-A

Grid: 5
Snap: 2.5
MODIFIED WBLOCK SHAPES

SUN CLOUD1 CLOUD2 RAIN WIND LIGHT SNOW

Grid: 5
Snap: 2.5
ORIGINAL SHAPES FOR WBLOCKS

SUN CLOUD1 CLOUD2 RAIN WIND LIGHT SNOW

MAP

Grid and Snap: 10

ACTIVITY 45-B

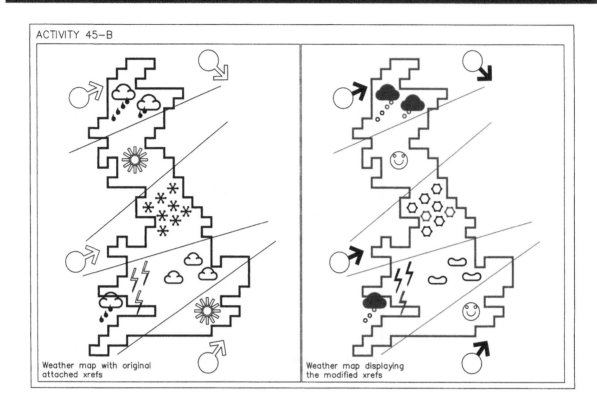

Weather map with original
attached xrefs

Weather map displaying
the modified xrefs

ACTIVITY 46
Draw the two isometric views using the sizes given then
construct both an isometric and oblique pictorial.

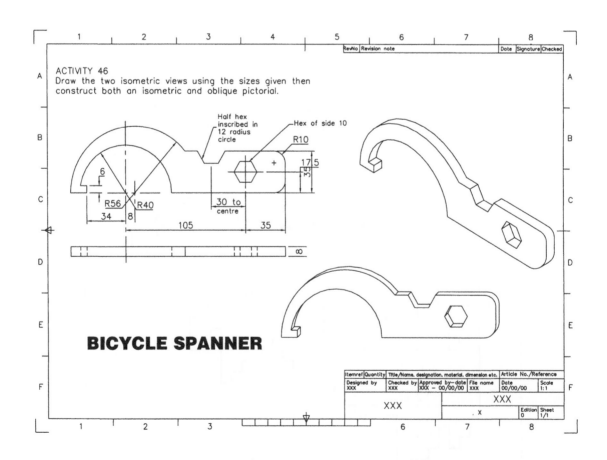

Half hex
inscribed in
12 radius
circle

Hex of side 10

R10

17.5

35

6

R56   R40

34   8

105   35

30 to
centre

8

BICYCLE SPANNER

RevNo	Revision note				Date	Signature	Checked

Itemref	Quantity	Title/Name, designation, material, dimension etc.		Article No./Reference		
Designed by XXX	Checked by XXX	Approved by-date XXX - 00/00/00	File name XXX	Date 00/00/00	Scale 1:1	
	XXX			XXX		
				. X	Edition 0	Sheet 1/1

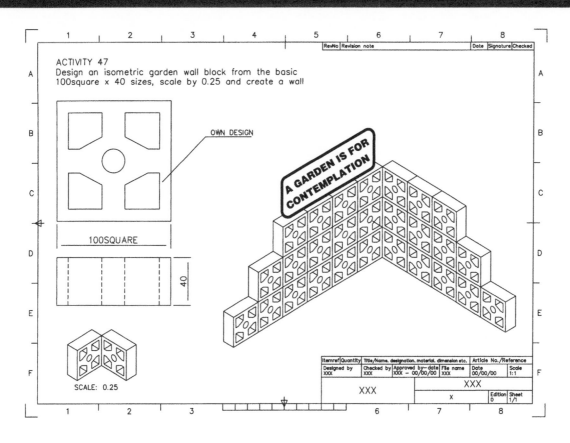

ACTIVITY 47
Design an isometric garden wall block from the basic
100square x 40 sizes, scale by 0.25 and create a wall

OWN DESIGN

A GARDEN IS FOR CONTEMPLATION

100SQUARE

40

SCALE: 0.25

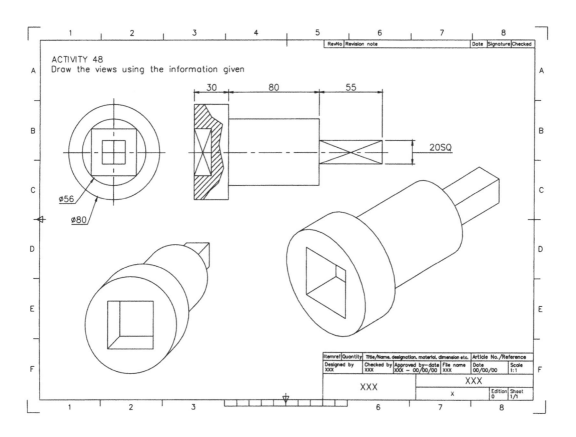

ACTIVITY 48
Draw the views using the information given

30   80   55

20SQ

ø56
ø80

Index